CHEMICAL PROCESS PRINCIPLES

Part Two

THERMODYNAMICS

CHEMICAL PROCESS PRINCIPLES

Part Two

THERMODYNAMICS

BY

OLAF A. HOUGEN

AND

KENNETH M. WATSON

PROFESSORS OF CHEMICAL ENGINEERING

UNIVERSITY OF WISCONSIN

NEW YORK

JOHN WILEY & SONS, INC.

CHAPMAN AND HALL, LIMITED

LONDON

FIFTH PRINTING, SEPTEMBER, 1949

PRINTED IN THE UNITED STATES OF AMERICA

PREFACE

"In the following pages certain industrially important principles of chemistry and physics have been selected for detailed study. The significance of each principle is intensively developed and its applicability and limitations scrutinized." Thus reads the preface to the first edition of *Industrial Chemical Calculations*, the precursor of this book. The present book continues to give intensive quantitative training in the practical applications of the principles of physical chemistry to the solution of complicated industrial problems and in methods of predicting missing physicochemical data from generalized principles. In addition, through recent developments in thermodynamics and kinetics, these principles have been integrated into procedures for process design and analysis with the objective of arriving at optimum economic results from a minimum of pilot-plant or test data. The title *Chemical Process Principles* has been selected to emphasize the importance of this approach to process design and operation.

The design of a chemical process involves three types of problems, which although closely interrelated depend on quite different technical principles. The first group of problems is encountered in the preparation of the material and energy balances of the process and the establishment of the duties to be performed by the various items of equipment. The second type of problem is the determination of the process specifications of the equipment necessary to perform these duties. Under the third classification are the problems of equipment and materials selection, mechanical design, and the integration of the various units into a coordinated plot plan.

These three types may be designated as process, unit-operation, and plant-design problems, respectively. In the design of a plant these problems cannot be segregated and each treated individually without consideration of the others. However, in spite of this interdependence in application the three types may advantageously be segregated for study and development because of the different principles involved. Process problems are primarily chemical and physicochemical in nature; unit-operation problems are for the most part physical; the plant-design problems are to a large extent mechanical.

In this book only process problems of a chemical and physicochemical nature are treated, and it has been attempted to avoid overlapping into the fields of unit operations and plant design. The first part deals primarily with the applications of general physical chemistry, thermophysics, thermochemistry, and the first law of thermodynamics. Generalized procedures for estimating vapor pressures, critical constants, and heats of vaporization have been elaborated. New methods are presented for dealing with equilibrium problems in extraction, adsorption, dissolution, and crystallization. The construction and use of enthalpy-concentration charts have been extended to complex systems. The treatment of material balances has been elaborated to include the effects of recycling, by-passing, changes of inventory, and accumulation of inerts.

In the second part the fundamental principles of thermodynamics are presented with particular attention to generalized methods. The applications of these principles to problems in the compression and expansion of fluids, power generation, and refrigeration are discussed. However, it is not attempted to treat the mechanical or equipment problems of such operations.

Considerable attention is devoted to the thermodynamics of solutions with particular emphasis on generalized methods for dealing with deviations from ideal behavior. These principles are applied to the calculation of equilibrium compositions in both physical and chemical processes.

Because of the general absence of complete data for the solution of process problems a chapter is devoted to the new methods of estimating thermodynamic properties by statistical calculations. This treatment is restricted to simple methods of practical value.

All these principles are combined in the solution of the ultimate problem of the kinetics of industrial reactions. Quantitative treatment of these problems is difficult, and designs generally have been based on extensive pilot-plant operations carried out by a trial-and-error procedure on successively larger scales. However, recent developments of the theory of absolute reaction rates have led to a thermodynamic approach to kinetic problems which is of considerable value in clarifying the subject and reducing it to the point of practical applicability. These principles are developed and their application discussed for homogeneous, heterogeneous, and catalytic systems. Particular attention is given to the interpretation of pilot-plant data. Economic considerations are emphasized and problems are included in establishing optimum conditions of operation.

In covering so broad a range of subjects, widely varying comprehensibility is encountered. It has been attempted to arrange the material in the order of progressive difficulty. Where the book is used for college instruction in chemical engineering the material of the first part is suitable for second- and third-year undergraduate work. A portion of the second part is suitable for third- or fourth-year undergraduate work; the remainder is of graduate level.

The authors wish to acknowledge gratefully the assistance of Professor R. A. Ragatz in the revision of Chapters I and VI, and the suggestions of Professors Joseph Hirschfelder, R. J. Altpeter, K. A. Kobe, and Dr. Paul Bender.

Olaf A. Hougen
Kenneth M. Watson

Madison, Wisconsin
August, 1943

CONTENTS

PART II
THERMODYNAMICS

TABLE OF SYMBOLS

A	area
A	atomic weight
A	component A
A	total work function
a	activity
a_m	external surface per unit mass
a_p	external surface per particle
a_v	external surface per unit volume
B	component B
B	constant of Calingaert–Davis equation
B	thickness of effective film
C	component C
C	concentration per unit volume
C	degrees centigrade
C	number of components
C	over-all rate constant
C_p	heat capacity at constant pressure
C_v	heat capacity at constant volume
C_s	Sutherland constant
c	concentration of adsorbed molecules per unit mass of catalyst
c	specific heat
c	velocity of light
c_p	molal heat capacity at constant pressure
c_v	molal heat capacity at constant volume
c'	surface concentration of adsorbed molecules per unit catalyst area
D	diameter
D_{AB}	diffusivity of A and B
D_p	effective particle diameter equal to diameter of sphere having the same external surface area as particle
D'_p	effective diameter equal to diameter of sphere having same area per unit volume as particle
d	differential operator
E	energy in general
E	energy of activation, Arrhenius equation
E_A	effectiveness factor of catalysis

e	base of natural logarithms
F	degrees of freedom
F	feed rate
F	force
F_e	external void fraction
F_i	internal void fraction
f	friction factor
f	fugacity
f	weight fraction
G	free energy
G	mass velocity per unit area
G	specific gravity
G	free energy per mole
$\bar{\mathrm{G}}$	partial molal free energy
ΔG	change in free energy
(g)	gaseous state
g	degeneracy
g_0	standard gravitational constant, 32.174 (ft/sec)/sec
H	enthalpy
H	Henry's constant
H	humidity
H_c	height of catalytic unit
H_d	height of mass-transfer unit
H_h	height of heat-transfer unit
H_p	height equivalent to a theoretical plate
H_p	percentage humidity
H_R	height of reactor unit
H_r	relative humidity
H_t	height of transfer unit
ΔH	change in enthalpy
ΔH_c	heat of combustion
ΔH_f	heat of formation
ΔH_r	heat of reaction
$\Delta H^\ddagger$	standard enthalpy of activation
H	enthalpy per mole
$\bar{\mathrm{H}}$	partial molal enthalpy
$\Delta\bar{\mathrm{H}}$	partial molal enthalpy change
h	Planck's constant
h	heat-transmission coefficient
I	inert component
I	integration constant
I	moment of inertia

J	Jacobian function
J	mechanical equivalent of heat
j_d	mass-transfer factor in fluid film
j_h	heat-transfer factor in fluid film
K	characterization factor
K	degrees Kelvin
K	distribution coefficient
K	equilibrium constant
K	vaporization equilibrium constant
K_a	equilibrium constant for adsorption
K_c	equilibrium constant, concentration units
K_g	over-all mass-transfer coefficient, pressure units
K_L	over-all mass-transfer coefficient, liquid concentration units
K_p	equilibrium constant, pressure units
K'	surface equilibrium constant
k	forward-reaction velocity constant
k	thermal conductivity
$\mathbf{k}$	Boltzmann constant
k_A	adsorption velocity constant
k'_A	desorption velocity constant
k_G	mass-transfer coefficient, gas film
k_L	mass-transfer coefficient, liquid film
k'	reverse-reaction velocity constant
L	mass velocity of liquid per unit area
L	total molal adsorption sites per unit mass
L_M	molal mass velocity of liquid per unit area
L'	active centers per unit area of catalyst
l	length
l_p	heat of pressure change at constant pressure
l_v	heat of expansion at constant temperature
(l)	liquid state
ln	natural logarithm
log	logarithm to base 10
M	molecular weight
M_m	mean molecular weight
m	mass
m	slope of equilibrium curve dy^*/dx
m	Thiele modulus
N	Avogadro number $6.023(10^{23})$
N	mole fraction
N_t	number of transfer units

N_G	number of transfer units, gas film
N_L	number of transfer units, liquid film
n	number of moles
P	pressure (used only in exceptional cases to distinguish pressure of pure components from partial pressures of some component in solution)
p_f	factor for unequal molal diffusion in gas film
Q	heating value of fuel
Q	partition function
q	heat *added to* a system
q_r	rate of heat flow
R	component R
R	gas constant
r	radius
r_{aA}	rate of reaction or transfer of A per unit area
r_{mA}	rate of reaction or transfer of A per unit mass
r_{vA}	rate of reaction or transfer of A per unit volume
S	component S
S	cross section
S	entropy
S	humid heat
S_p	percentage saturation
S_r	relative saturation
S_v	space velocity
$\Delta S^{\ddagger}$	entropy of activation
s	molal entropy
s	number of equidistant active sites adjacent to each other
(s)	solid state
T	absolute temperature, degrees Rankine or Kelvin
t	temperature, °F or °C
U	internal energy
U	over-all heat-transfer coefficient
U	internal energy per mole
u	velocity
V	molecular volume in Gilliland equation
V	volume
V_r	volume of reactor
v	volume per mole
w	weight
w	work *done by* system
w_e	work of expansion *done by* system

w_f	electrical work *done by* system
w_s	shaft work
$X^{\ddagger}$	activated complex
x	mole fraction in liquid phase
x	mole fraction of reactant converted in feed
x	quality
y	mole fraction in vapor
y^*	mole fraction in vapor, equilibrium value
Z	elevation above datum plane
Z	height or thickness of reactor
z	compressibility factor
z	mole fraction in total system

Dimensionless Numbers

N_{Re}	Reynold's number	$\frac{DG}{\mu}$
N_{Pr}	Prandtl number	$\frac{C_p\mu}{k}$
N_{St}	Stanton number	$\frac{h}{C_pG}$
N_{Sc}	Schmidt number	$\frac{\mu}{\rho D_v}$

Subscripts

A	component A
a	air
B	component B
b	normal boiling point
C	component C
c	critical state
D	component D
D	dense arrangement
e	expansion
f	electrical and radiant
f	formation
f	fusion
G	gas or vapor
H	isenthalpic
L	liquid
L	loose arrangement

p	constant pressure
R	component R
r	reduced conditions
r	relative
S	component S
S	isentropic
s	normal boiling point
s	saturation
T	isothermal
t	temperature
t	transition
V	constant volume
v	vapor
w	water vapor

Greek Symbols

(α)	crystal form
α	coefficient of compressibility
α	proportionality factor for diffusion
α	relative volatility
α	thermal diffusivity
(β)	crystal form
β	coefficient of volumetric expansion
γ	activity coefficient
(γ)	crystal form
Δ	finite change of a property; positive value indicates an increase
δ	change in moles per mole of reactant
δ	deformation vibration
∂	partial differential operator
ϵ	energy per molecule
η	efficiency
θ	fraction of total sites covered
κ	ratio of heat capacities
Λ	heat of vaporization
λ	heat of vaporization per mole
λ	wave length
λ_f	heat of fusion per mole
μ	chemical potential
μ	viscosity
ν	frequency
ν	fugacity coefficient of gas

ν	number of ions
ν	number of molecules
ν	valence or stretching vibration
ω	expansion factor of liquid
ω	wave number
π	total pressure of mixture, used where necessary to distinguish from p
ρ	density
ρ_B	bulk density
ρ_P	particle density
ρ_C	true solid density
Σ	summation
σ	surface tension
σ	symmetry number
τ	time
ϕ	activity coefficient
ϕ	number of phases

Superscripts

$*$	ideal behavior
$*$	equilibrium state
$\circ$	standard state
$'$	pseudo state
$'$	reverse rate
$\ddagger$	standard state of activation

CHAPTER XI

THERMODYNAMIC PRINCIPLES

In its most practical use thermodynamics deals with the limitations imposed on the transformation of energy from one form to another and with the study of the availability of energy for useful work. Thermodynamic principles also provide useful means of predicting with a minimum of experimental data those properties of a substance which are related to energy and the equilibrium conditions which are approached in any physical or chemical process.

Thermodynamic potentials constitute the driving forces which cause all natural processes to progress toward a state of equilibrium. However, the rate at which the state of equilibrium is approached depends not only upon the driving force but also upon a resistance or rate factor. Thermodynamic principles alone do not permit evaluation of these resistances or rate factors. The combined knowledge of the thermodynamic driving forces and of the corresponding resistances or rate factors constitutes the science of *kinetics*.

System and Process. In thermodynamic terminology a *system* refers to a substance or group of substances set apart for study and a *process* to the changes taking place in the system. Thus hydrogen, oxygen, and water constitute a system, and the combustion of hydrogen to form water constitutes the process. A system is *closed* when there can be no exchange of matter with its surroundings and *open* when such exchange is possible. In a multiphase system each separate phase is an open system, since material is free to enter and leave individual phases although the system as a whole may be closed.

A closed system may be free to exchange heat and work with its surroundings. Heat can be transferred through the walls of the vessel enclosing the system, and mechanical work can be performed upon or by the system by means of a piston connecting the vessel with some external mechanism. Devices can also be arranged for the removal or addition of electric energy. A closed system is *thermally isolated* when the enclosing walls are impervious to the flow of heat, *mechanically isolated* when it is enclosed by rigid walls, and *completely isolated* when neither material nor energy in any form can be added or removed. A closed system in contact with a *heat reservoir* is free to receive or lose energy in the flow of heat from or to the reservoir. A closed system in contact with a piston is free to receive or deliver work energy in the movement of the piston. The atmosphere surrounding a closed system may also act as a reservoir for the flow of heat or as a piston for transmitting work from and to the system. The atmosphere also may be made part of the system as in the combustion of a fuel in air. In a consideration of the energy relations of any nonisolated system, the energy of the surroundings must be taken into account as well as that of the system itself.

Availability and Degradation of Energy. It is generally recognized that the most useful forms of energy are those which are capable of doing mechanical work. Thus, the rotation of a shaft or the motion of a piston represent energies of the highest order, which in the absence of friction can be completely converted into useful work. Similarly, electric energy in the absence of electrical resistance and mechanical friction can be converted entirely into mechanical work. Actually, however, because of friction part of the mechanical or electric energy of a machine is always converted into heat. It is also common experience that, whereas heat can be converted into mechanical or electric energy by means of an engine, only part of the heat which flows into the engine can be recovered as work. Thus, although there is an exact quantitative equivalence among the different forms of energy, there is a marked difference in the availability of these forms for useful work. Heat represents the least available form of energy, and the transformation of other forms into heat represents a degradation of energy.

There are other means of degrading energy and rendering it less available for useful work. Thus, the internal energy of a gas at high pressure is more available for performing useful work than that of the same gas at low pressure and at the same temperature. Even though the internal-energy content of the gas is the same at both pressures, as is the case for ideal gases, the availability of the internal energy for further work of expansion is less in the expanded gas. Thus, an isothermal reduction in the pressure of a gas represents a degradation of its internal energy. Similarly, the isothermal mixing of two unlike gases represents a degradation in the internal energy of the system since work is required to separate them.

Reversibility. A *reversible* process is defined as one which proceeds under conditions of balanced forces such that the direction of the process can be reversed by an infinitesimal change in external conditions. An example of a reversible process is the vaporization of a liquid under its own vapor pressure in a cylinder fitted with a frictionless piston and in contact with an isothermal heat reservoir. At any stage an infinitesimal increase in pressure upon the piston will produce condensation, and an infinitesimal decrease will cause vaporization. It follows that *in a* *reversible process an infinitesimal change will tend to restore the original conditions, so that no degradation of energy takes place, and the availability of the energy of the combined system and its surroundings remains constant.*

Actual processes are generally to some extent irreversible and result in a decrease in the availability of energy. The flow of heat from one body to another is always an irreversible process. It is impossible to restore heat to the first body except by transfer from a still hotter body or by the degradation of highly available mechanical energy. Similarly, the dropping of an object without restraint to a lower level is an irreversible process.

Another example of an irreversible process is the free expansion of a gas. If a gas is confined in a closed container of volume V_1 and is allowed to expand into a vacuum chamber such that the new total volume is V_2, and if this system is entirely isolated from its surroundings, by rigid insulating walls, the gas is said to *expand freely.* No work is performed upon the surroundings, no heat flows in or out of the system, and for an ideal gas there is no change in temperature. Hence, the energy content of the expanded gas is the same as that of the original gas. However, in order to return the expanded gas to its original condition the expenditure of external energy as work of compression is required. For an ideal gas all this work manifests itself as heat in compressing the gas, and in isothermal compression this heat is transferred to the surroundings. Thus, in isothermal compression of an ideal gas the energy content remains constant but assumes a higher level of availability.

Other examples of irreversible processes are the mixing of hot and cold fluids, the inelastic deformation of a solid, the magnetic hysteresis of an iron core, the flow of electricity through a resistance, the dissolution of a solid, and spontaneous chemical reactions.

The Second Law of Thermodynamics. The concept of reversibility forms the basis for the second law of thermodynamics. One statement of this law is that *all spontaneous processes are to some extent irreversible and are accompanied by a degradation of energy.* A corollary to this statement is that it is impossible for any self-acting machine to transfer energy from a given state to a higher state of availability.

The validity of the second law of thermodynamics is confirmed by extensive experimental evidence when applied to gross masses of matter in which large numbers of molecules are present. Under these circumstances the laws of probability are followed in establishing the distribution of molecular and atomic energies.

An important deduction from the second law is that any machine which performs work by transforming molecular energy received at fixed intake conditions and rejecting molecular energy in a further degraded state at fixed discharge conditions will convert a maximum percentage of the energy received into work when it operates reversibly and with no net degradation of energy. Any actual self-acting machine must operate with a lower efficiency of transformation than that of a reversible machine. If it were possible to construct a more efficient engine, it could be used to drive the reversible machine in reverse to transfer molecular energy from a given state to the original less degraded state in contradiction of the second law.

The familiar steam engine is an example of such a machine which performs work by receiving steam at a high temperature and rejecting it at a low temperature. The maximum efficiency possible with given intake and exhaust conditions is obtained when such an engine operates reversibly. This theoretical maximum efficiency is always less than 100 per cent because of the energy content of the rejected steam.

ENTROPY

Before undertaking development of the quantitative relationships involved in transformations of energy it is desirable to define an additional thermodynamic function to serve as a measure of the *unavailablity* or *degradation* of the energy of a system. This function is termed *entropy.*

The arbitrary mathematical definition of various thermodynamic functions is an entirely legitimate procedure which may be extended to create as many additional functions as may prove useful. Enthalpy is an example of such a function, and numerous others are introduced in the following pages as opportunities for their uses arise. For mathematical accuracy it is only necessary that each new function be completely defined and used in exact accord with the terms of the definition. However, the intelligent application of such a function is greatly assisted by a thorough understanding of its physical significance. For this reason the physical significance of entropy is first developed as a basis for its mathematical definition.

The physical concept of entropy may be expressed as follows: *Entropy is an intrinsic property of matter so defined that an increase in the unavailability of the total energy of a system is quantitatively expressed by a corresponding increase in its entropy.* Since entropy is an intrinsic property, its magnitude is dependent only on the nature of the matter under consideration and the state in which it exists and not upon its external position or motion relative to other bodies of matter. Thus, the entropy of an elevated weight is no different from that of the same weight in a lower position, and the entropy of a rotating flywheel is equal to that of the same flywheel at rest under the same conditions of temperature and pressure. The same is true for other intrinsic properties such as internal energy and enthalpy.

Since entropy is a measure of unavailability of total energy, it follows that the entropy of a system is increased by the degradation into heat of any higher form of energy which it possesses. This is illustrated by consideration of an isolated system comprising an inelastic weight suspended above a rigid plate. If the weight is allowed to fall, the total energy content of the system is unchanged, but the potential energy of the weight is converted into heat as a result of the inelastic impact on the plate. The temperature of both the weight and the plate are correspondingly increased by the absorption of this heat in the form of internal energy. The process of dropping the weight has not changed the energy content of the system but has degraded it to a less available form and is therefore accompanied by an increase in entropy. Similarly, if a rotating flywheel is stopped by a brake, kinetic energy is degraded into heat, and the entropy of the system is increased. However, if the flywheel were stopped in a reversible manner by means of a frictionless transmission which sets another flywheel in motion, there would be no degradation of energy and no change in entropy.

In both of these illustrative systems an increase in entropy results from the addition of heat to the system through the degradation of a higher form of energy. However, since entropy is an intrinsic property of the matter of the system not influenced by the elevation of the weight or the movement of the flywheel, it follows that *the entropy of a system is increased by the addition of heat through any mechanism or from any source.* Thus, heat might be added to the system containing the elevated weight and the rigid plate increasing the total energy content of the system until a temperature is reached which is equal to that attained after the weight has been permitted to fall in the system when isolated. Since entropy is independent of the position of the weight and the plate, the final entropies of the system are equal, and the increases in entropy accompanying these two processes are also equal, although heat is added to the system by the degradation of its own energy in one case and by increase in its total energy content in the other.

It is evident that the amount of heat added to a system is a partial measure of the magnitude of its increase in entropy. However, the quantity of heat added is not the sole measure of increase in entropy. The unavailability of energy and the entropy of an isolated system are increased by the transfer of heat within the system to a region of lower temperature. Similarly, the addition of heat to a system at a low temperature results in a greater degradation of energy than at a high temperature. It follows that *the increase in entropy accompanying the addition of a given amount of heat to a system is increased by lowering the temperature* at which the heat is added.

In order to complete the definition of entropy, one more factor must be considered in addition to the amount of heat added and the temperature level at which it is added. For example, a gas may be expanded freely to a lower pressure within a closed system which is completely isolated both thermally and mechanically. This is an irreversible process resulting in degradation of the energy of the system and an increase in its entropy. However, no heat is added, and no work is done, and for an ideal gas the process is isothermal. In this irreversible process the entropy increase is not measured by the addition of heat. The same final state might be reached by expanding the gas through an engine within the system and continuously converting the work done into heat by means of friction. In this case heat is added to the system by the degradation of the mechanical work. The amount of heat added increases as the efficiency of the engine is increased and reaches a maximum when the engine operates reversibly. However, entropy is an intrinsic property of matter, the change of which in any process is dependent only upon initial and final states and not upon the path. It is evident that in the process under consideration the amount of heat added to the system can be taken as a measure of the increase in entropy only if the nature of the process is specified. This specification is logically taken to correspond to the maximum possible degradation of higher forms of energy into heat. Thus *the increase in entropy of a system is measured by the amount of heat added only when all changes in the intrinsic states of the matter of the system occur reversibly.*

The three requirements for the quantitative definition of entropy are satisfied by the following equation:

$$dS = \frac{d'q}{T} \quad \text{(reversible change of state)} \tag{1}$$

where S = entropy.

The primed differential symbol d' is used for indicating incremental quantities of heat and work. In contrast to the other terms in the energy equation, neither heat nor work are properties of the system, nor can they in general be expressed as a function of the state of the system. An infinitesimal change in a property such as volume can be expressed as dV which is an *exact* differential with respect to the state of the system. However, since neither q nor w are properties an increment of either is referred to as an *inexact* differential.

An important corollary of Equation (1) which follows from the second law of thermodynamics and the definition of entropy as a measure of degradation is that the entropy change accompanying an irreversible change of state is greater than $d'q/T$, or

$$T\,dS > d'q \quad \text{(irreversible change of state)} \tag{2}$$

Thus, if an isothermal change of state such as the expansion of a gas occurs reversibly the gain in entropy of the gas is compensated by a reduction in entropy of its surroundings resulting from withdrawal of heat, and the combined entropy of the material and its surroundings is unchanged. If the same change in state is accomplished irreversibly, the combined entropy of the gas and its surroundings increases although the increase in entropy of the gas itself is the same as for the reversible change. Consequently less heat is withdrawn from the surroundings.

It is important that the significance of the restriction to reversibility in Equation (1) be appreciated. This restriction does not apply to the manner by which heat is added to the system or the process employed in the degradation of higher forms of energy. Thus, if a block of steel is heated by a flame, its increase in entropy is measured by the amount of heat added even though the addition of the heat from the flame to the solid is highly irreversible. Similarly, the restriction does not apply to changes involving only the relative positions or movement of bodies of matter. The process involving the falling weight which was discussed previously is irreversible, but the accompanying entropy change is measured by the heat added. It is only processes involving changes in the intrinsic state of the matter itself which must be reversible for the application of Equation (1).

Much has been written on the physical significance of entropy. The concept has already been advanced that an increase in the unavailability of the energy of a system is accompanied by an increase in entropy. On this basis entropy may be looked upon as a measure of the unavailability of the energy of a system *of given energy content.* However, the entropy changes which accompany a change in total energy content do not necessarily indicate increased unavailability of energy. For example, if the temperature of a substance is increased by heating, both the total energy and the entropy are increased. In this case the increase in entropy of the substance is accompanied by a decrease rather than an increase in the unavailability of its energy and the increase in entropy of the substance is not a measure of the change in unavailability of its energy. However, if consideration is broadened to include the substance plus its surroundings from which the heat is withdrawn, the entropy of this entire system of constant energy content is a measure of the unavailability of its energy.

A further physical concept of entropy results from its relationship to probability. The existence of such a relationship is evident from the fact that all spontaneous changes are in the direction of maximum probability and are also accompanied by increases in entropy. It follows from the theory of probability that the most probable state in which a system can exist is that having the least order of arrangement or greatest randomness. This concept may be illustrated by consideration of a large number of black and white balls shaken together in a box. The most probable average arrangement is a uniform distribution of the black among the white balls, and the chance of a compartment of the box containing balls of only one color is remote. Thus, a state of randomness, free from orderly arrangement, is the most probable state of any system, and entropy may be looked upon as a measure of randomness which is a minimum for systems in orderly arrangement.

It is evident that in any given change in conditions accompanied by an increase in entropy more heat is added in a system of large mass than in a smaller one. Thus, like internal energy or volume, entropy is an *extensive property* the magnitude of which is dependent upon the mass involved. It follows that the entropy of a system is equal to the sum of the entropies of its separate phases or mechanical parts.

In the foregoing discussion no consideration has been given to the scale or function employed for the quantitative expression of temperature in the definition of entropy. Although any arbitrary scale may be used to indicate relative hotness or coldness, it is evident that only one type of scale or function can be used to express the temperature of Equation (1) if entropy as defined by it is to represent a true measure of the unavailability of energy at different temperature levels. This correct function is termed the *thermodynamic temperature scale.*

At this point it is difficult to arrive at the nature of the thermodynamic temperature by direct derivation, and so it will be tentatively assumed that it is the same as the ideal-gas temperature which is defined in Chapter II, page 30, as $T = pv/R$. Rigorous thermodynamic functions and relations are developed from Equation (1) without consideration of the nature of the temperature scale. From these relations it is proved (page 473) that the thermodynamic and ideal-gas temperature scales are identical. Any size of temperature unit may be used with no change in the relationships involved except a change in the numerical values of temperature and entropy.

Further confirmation of both the assumed nature of the thermodynamic temperature scale and the fact that entropy as defined by Equation (1) is a true measure of the degradation of energy and consistent with the second law may be developed through consideration of ideal heat-engine cycles, such as are discussed in Chapter XIII. These proofs are well summarized by Keenan.[1]

[1] J. H. Keenan, "Thermodynamics," John Wiley & Sons, New York (1941).

Change of Entropy with Change of Phase or Temperature. Since entropy is an extensive property of matter, its magnitude is dependent on the physical state and mass of the material under consideration as well as on the temperature and pressure at which it exists. Any change which occurs in any of these properties is accompanied by a corresponding change in entropy which may be calculated from Equation (1).

The changes in entropy which take place when a substance undergoes a reversible change of phase at constant temperature and pressure as in fusion, evaporation, and transition are calculated by dividing the change in enthalpy by the absolute temperature.

Illustration 1. Calculate the molal entropy changes in the fusion and in the vaporization of ethyl alcohol at atmospheric pressure.

Ethyl alcohol melts at 159°K with a heat of fusion of 1150 cal per g-mole. Since fusion and vaporization are reversible isothermal processes, the molal entropy increase of fusion is

$$\Delta s = \frac{q}{T} = \frac{1150}{159} = 7.23 \text{ cal/(g-mole)(°K)}$$

The heat of vaporization of ethyl alcohol at atmospheric pressure is 9400 cal per g-mole at a normal boiling point of 351°K. Hence, the increase in entropy in vaporization is

$$\Delta s = \frac{q}{T} = \frac{9400}{351} = 26.8 \text{ cal/(g-mole)(°K)}$$

It may be observed that the units of entropy are the same as for heat capacity.

When heat is added with a resultant temperature rise, the increase in entropy must be calculated by integration of Equation (1) over the required temperature range, thus

$$\Delta S = \int_{T_1}^{T_2} \frac{d'q}{T} = \int_{T_1}^{T_2} \frac{C_p\,dT}{T} \qquad (3)$$

Illustration 2. Calculate the increase in entropy when 1 g-mole of CO_2 gas is heated from 0°C to 1000°C at atmospheric pressure. The molal heat capacity of CO_2 expressed in terms of absolute temperature degrees Kelvin is given on page 214 as

$$c_p = 6.85 + 0.008533T - 2.475(10^{-6})T^2$$

$$\Delta s = \int_{T_1}^{T_2} \frac{c_p\,dT}{T} = \int_{273}^{1273} \left(\frac{6.85}{T} + 0.008533 - 2.475(10^{-6})T\right) dT$$

$$= 6.85 \ln \frac{1273}{273} + 0.008533(1273 - 273) - \frac{2.475}{2}(10^{-6})[(1273)^2 - (273)^2]$$

$$= 17.16 \text{ cal/(g-mole) (°K)}$$

In calculating the entropy of a substance with reference to any selected standard conditions, it is necessary to add together the entropy changes for each transition, for each temperature rise, and for changes in pressure. The variation of entropy with pressure is developed in a following section.

Third Law of Thermodynamics. It was proposed by Nernst and subsequently confirmed by extensive experimentation that at the absolute zero of temperature the entropy of any pure crystalline substance free of all random arrangement is zero. This principle is known as the *third law of thermodynamics.* Accordingly, by extending measurements of specific and latent heats down to 0°K absolute values of entropy can be calculated from Equation (1). It is thus evident that the four reference properties of matter, temperature, pressure, volume, and entropy, can all be presented as absolute values, whereas energy contents can be reported only relative to some reference state. It has already been shown that absolute values of temperature, pressure, and volume are required in formulating the properties of matter. It is shown later that the same must be true of entropy.

The utility of the third law in establishing values of absolute entropy brings out the importance of calorimetric measurements at temperatures down to the absolute zero. At these low temperatures the common empirical heat-capacity equations of Chapter VII do not apply. It is a corollary of the third law that the heat capacity of a crystalline substance is zero at the absolute zero temperature. However, the manner in which the heat capacity diminishes with decreasing temperature in the low-temperature range is a unique property, varying widely for different materials. It is impossible to estimate low temperature thermal data by the extrapolation to zero of high temperature measurements.

FREE ENERGY AND TOTAL WORK FUNCTIONS

A thermodynamic function of great importance is defined by the following relation and termed the *free energy G*,

$$G = H - TS \tag{4}$$

or, for changes taking place at constant temperature,

$$\Delta G = \Delta H - T\Delta S \tag{5}$$

Where work is available from a process in addition to the mechanical work of expansion, it follows from the first law discussed in Chapter VII that for a nonflow process

$$q = \Delta U + w_e + w_f \tag{5a}$$

where

w_e = work of expansion done by the system
w_f = *useful work* other than work of expansion, such as electric energy

For a reversible process at constant temperature and pressure $q = T\,\Delta S$ and $w_e = p\,\Delta V$. Hence, under these conditions the first law may be written as

$$\Delta U + p\,\Delta V = T\,\Delta S - w_f = \Delta H \tag{6}$$

By combination of (5) and (6), it follows that

$$\Delta G = -w_f \tag{7}$$

Thus, in a process which occurs reversibly at constant temperature and pressure, the useful work is equal to the decrease in the free energy of the system. When the process occurs irreversibly at constant temperature and pressure, $T\,\Delta S > q$, and the useful work is always less than the decrease in the free energy of the system. Therefore, the decrease in free energy accompanying a process at constant temperature and pressure represents the maximum useful work which becomes available only when the process is reversible.

Most chemical reactions proceed under no restraint other than that of pressure and dissipate the energy of the system as heat with no production of useful work. Exceptional cases are found in the operation of electrolytic cells wherein a part of the energy content of the system is recovered as useful work. As an example of the significance of free energy 1 g-atom of zinc may be dissolved at constant temperature and pressure in excess hydrochloric acid of a given initial concentration, with the release of hydrogen gas and with no evaporation of water or hydrogen chloride. If it were possible to connect the zinc as anode through an external circuit with an insoluble cathode placed in the same solution and allow the reaction to proceed slowly and reversibly at constant pressure, the electric energy developed would equal the decrease in free energy of the system, $d'w_f = -dG$. In the absence of the electrolytic cell, this reaction would proceed irreversibly with no generation of electric energy. However, for the same initial and final conditions, the decrease in free energy of the system is independent of the path of the reaction and would be the same in both cases.

Another useful thermodynamic function is defined by the following relation and termed the *total work function* A,

$$A = U - TS \tag{8}$$

or, at constant temperature,

$$\Delta A = \Delta U - T\,\Delta S \tag{9}$$

From the first law, for a reversible nonflow process at constant temperature,

$$\Delta U = q - w = T\,\Delta S - w \tag{10}$$

Combining (9) and (10) for a reversible process at constant temperature gives

$$\Delta A = -w = -(w_e + w_f) \tag{11}$$

Therefore, the decrease in the total work function accompanying a process at constant temperature represents the maximum total work which becomes available only when the process is reversible. For a reaction proceeding at constant volume, $w_e = 0$, and $\Delta A = \Delta G$.

There has been considerable confusion between the concepts of free energy G and total work function A. Although the differences are usually slight, they are equal in a process at constant pressure only when there is no change in volume.

The Four Thermodynamic Energy Functions. The internal energy U, enthalpy H, free energy G and total work A are functions representing energies which are capable of performing useful work under certain conditions of restraint. For this reason these functions will be referred to as the four *energy functions.* These functions are frequently called the thermodynamic potentials, but this terminology leads to confusion with the usual concept of a potential as an intensive driving force. The significance of the four energy functions can be made tangible by again referring to the dissolution of zinc metal in a solution of hydrochloric acid and assuming that the system can be set up as a reversible electrolytic cell in a calorimeter. When the reaction proceeds isothermally at constant pressure and irreversibly with no generation of electric energy, the heat given up by the system and flowing into the calorimeter is equal to the decrease in the enthalpy of the system, or $-q_p = -\Delta H$. If the reaction proceeds isothermally at constant volume and irreversibly and with no generation of electric energy, the heat given up by the system to the calorimeter is equal to the decrease in internal energy of the system, or $-q_v = -\Delta U$.

If the zinc were connected as anode through an external electric circuit as previously described and the reaction allowed to proceed reversibly at constant pressure, the electric energy developed would be equal to the decrease in free energy, or $w_f = -\Delta G$. The work of expansion against the atmosphere w_e in the release of hydrogen would be $p\,\Delta V$ and the total work done $w_f + w_e$ would be equal to the decrease in the total work potential of the system, or $w_f + w_e = -\Delta A$.

In summary, for isothermal nonflow processes:

At constant volume	$\Delta U = q - w_f$
At constant pressure	$\Delta H = q - w_f$
In a reversible process at constant pressure	$\Delta G = -w_f$
In any reversible process	$\Delta A = -w_f - w_e$

Although in any reaction the values of ΔH, ΔU, ΔG, and ΔA depend only upon initial and final conditions and are independent of the path pursued, the direct calorimetric and electrical measurements of these quantities must be made under conditions of change similar to those previously described.

EQUILIBRIUM IN A CLOSED SYSTEM

A system is in equilibrium when its state is such that it can undergo no spontaneous or unaided changes. Such a condition can result only when all forces or potentials which tend to promote change are absent or are exactly balanced against similar opposing forces or potentials. An iron ball resting at the bottom of a spherical bowl is in mechanical equilibrium with the bowl, since the mechanical forces acting on the ball are balanced and the ball is at rest. No alteration in the position of the ball can possibly take place except at the expenditure of mechanical energy received from the outside.

Although the iron ball resting at the bottom of the bowl is in a state of mechanical equilibrium, it is not in a state of chemical or thermodynamic equilibrium with the surrounding air. Spontaneous, unaided and irreversible vaporization and oxidation of the iron can occur to form vapor and rust. This spontaneous process will proceed, although very slowly, until all the iron has been vaporized and oxidized to a condition of complete thermodynamic equilibrium. Thus many conditions of apparent equilibrium actually represent only partial equilibrium with respect to all possible changes. In dealing with such partial equilibria it is necessary that the changes with respect to which equilibrium is established are clearly recognized.

Criteria of Equilibrium. Since a principal objective of thermodynamic theory is the development of mathematical expressions for equilibrium conditions it is important to derive fundamental specifications which must be fulfilled by any system at equilibrium. Such specifications, termed *criteria of equilibrium*, are the foundation upon which complete relationships among the various properties of a system at equilibrium are established.

From a thermodynamic standpoint it is necessary that in a system at equilibrium every possible change which might take place to an infinitesimal extent shall be reversible since any irreversible change would result in a displacement which would destroy the original equilibrium. As was previously pointed out, reversible processes are accompanied by no change in total entropy of the combined system and its surroundings, whereas every spontaneous process is accompanied by an increase in total entropy. Thus, a universal thermodynamic criterion of equilibrium is that for any change which takes place the total entropy of the system *and its surroundings* shall be constant. In a completely isolated system the entropy of the system itself is constant at equilibrium. From Equation (1), defining entropy, it follows that where heat is added to a system in which all changes of state are reversible, $d'q = T\,dS$, where S is the entropy of the system itself, not including its surroundings. This expression may also be taken as a criterion of reversibility and equilibrium. Since for all irreversible changes of state $dS > d'q/T$, if any incremental addition of heat to the system is accompanied by an entropy increase equal to and not greater than $d'q/T$, all thermodynamic processes within the system must be reversible, and it follows that the system is in equilibrium.

A system is in *stable* equilibrium if after a finite displacement it spontaneously returns to its original state when the displacing force is returned to its original value. A round pencil lying in the bottom of a cylindrical trough is a mechanical example of this type of equilibrium. If, however, this pencil is carefully balanced on its sharpened point, it will be in a state of *unstable* equilibrium such that finite displacement does not lead to a spontaneous return to its original conditions. Thermodynamically, a system is in unstable equilibrium if, although any infinitesimal change is accompanied by no change in the total entropy of the system and its surroundings, a finite displacement involves an increase in total entropy. For example, a finite displacement of the pencil balanced upon its point results in an irreversible process whereby heat is developed and the entropy of the pencil and its surroundings increases. In a system in stable equilibrium no finite change can be accompanied by an increase in total entropy. Thus, at stable equilibrium, for any change resulting from a temporarily applied extraneous force, not associated with the system or its normal surroundings,

$$dS_t = 0, \qquad \Delta S_t \leqq 0 \tag{12}$$

where S_t = total entropy of the system and its normal surroundings. On the other hand, in unstable equilibrium, for any change

$$dS_t = 0, \qquad \Delta S_t > 0 \tag{13}$$

In addition to the general criteria of equilibrium just described, modified criteria are useful to define equilibrium in a system under different types of specified restraint. For example, liquid ether might be permitted to evaporate in a container initially filled with nitrogen at atmospheric pressure. If the walls of the container are rigid and do not conduct heat (and magnetic and electrical effects are absent), the system is completely isolated, and the process is restrained to conditions of constant volume and constant internal energy. Thus, $dV = 0$, $dU = 0$, $q = 0$, and $w = 0$. As vaporization of the ether proceeds spontaneously, the enthalpy, pressure, and entropy of the system all increase and the temperature decreases until equilibrium is reached at a condition where the entropy reaches a maximum and $dS = 0$.

Instead of restraining the system to conditions of constant volume and internal energy, the container might be provided with conducting walls in contact with surroundings of the same and constant temperature. The conditions of restraint would then be constant temperature and volume. Thus, $dT = 0$, $dV = 0$, and $p\,dV = 0$. By fitting this container with a frictionless piston the process could also be restrained to constant pressure. For all such changes of state under any conditions of restraint, equilibrium is reached when the fundamental requirement that $dq = T\,dS$ is satisfied for the system itself. A combination of this requirement with the first law yields the equation $T\,dS = dU + p\,dV + d'w_f$, which may therefore be taken as a criterion of equilibrium. This equation is applicable only to reversible changes, and, conversely, if it is applicable to all possible changes of a system these changes are all reversible and the system is at equilibrium. If the system approaches equilibrium under the conditions of restraint of constant entropy and volume,

$$dS = 0; \qquad dV = 0 \quad \text{and}$$

$$(dU)_{SV} = -d'w_f \tag{14}$$

or, if no means for performing useful work is present, $d'w_f = 0$

$$(dU)_{SV} = 0 \tag{15}$$

Equation (15) is a criterion of equilibrium, for a system restrained to constant entropy and volume and unable to perform useful work. Such a system may be visualized by considering ether evaporating in an atmosphere of nitrogen in a container with rigid conducting walls in contact with a reservoir of heat of variable temperature. As evaporation takes place, the temperature of the reservoir is so controlled that heat is withdrawn from the system in such a manner that the reduction in temperature offsets the increase in entropy resulting from the irreversible vaporization, and constant entropy is maintained in the system.

Under these conditions vaporization will proceed spontaneously until equilibrium is reached at a condition of minimum internal energy.

Since $U = H - pV = A + TS = G - pV + TS$, Equation (5a) may be written in terms of other thermodynamic energy functions. Thus, since $d'q = TdS$ and $d'w_e = p\,dV$,

$$T\,dS = dH - V\,dp + d'w_f \tag{16}$$

It follows that in a system restrained to conditions of constant pressure and entropy it is a criterion of equilibrium that $dH = -d'w_f$ or if no useful work is possible $dH = 0$.

$$(dH)_{pS} = 0 \tag{17}$$

Similarly, Equation (5a) may be written

$$dA + p\,dV + S\,dT + d'w_f = 0 \tag{18}$$

Thus, at conditions of equilibrium restrained to constant volume and temperature where no useful work is performed

$$(dA)_{TV} = 0 \tag{19}$$

Also, Equation (5a) may be written

$$dG - V\,dp + S\,dT + d'w_f = 0 \tag{20}$$

It follows that at conditions of equilibrium restrained to constant temperature and pressure it is a criterion of equilibrium that

$$(dG)_{Tp} = -d'w_f \tag{21}$$

or if no useful work can be performed

$$(dG)_{Tp} = 0 \tag{22}$$

Valid criteria of equilibrium under still other conditions of restraint might be similarly derived. However, that represented by Equations (21) and (22) is of the greatest interest because of the fact that the majority of chemical equilibria are restricted to conditions of constant temperature and pressure.

From consideration of Equations (14) through (22) it is evident that the proper criteria of equilibrium in a process is determined by the conditions of restraint under which it occurs. If the evaporation process used as an example is started from constant initial conditions but allowed to proceed under different conditions of restraint, the conditions at which equilibrium is reached will be different.

The various criteria of equilibrium form the basis for the prediction of equilibrium conditions in any process. For example, if a chemical reaction proceeds spontaneously under restraint of constant temperature and volume, the total work function A decreases as a function of n, the number of moles converted. Equilibrium will be reached at a minimum value of A where $dA/dn = 0$ if no useful work is performed or where $dA/dn = d'w_f/dn$ if useful work is performed. Since A is a property determined only by the state of the system, it follows that the state at which equilibrium is reached is dependent upon $d'w_f/dn$, the amount of useful work performed per mole of conversion at equilibrium conditions. Similarly, if a reaction proceeds under the restraint of a constant temperature and pressure, the change in free energy G is the basis of the criterion of equilibrium from Equations (21) and (22). The most valuable applications of these energy functions result from their fundamental relationships to the important criteria of equilibrium.

Thermodynamic Relationships. The thermodynamic properties of a system or a process may be classified into four groups as *reference properties, energy functions, derived properties*, and *path properties.*

The reference properties, temperature, pressure, volume, entropy, and composition, are those required to define completely the state of a system. Of these properties temperature, pressure, and composition are intensive; volume and entropy are extensive. The energy functions are extensive properties whose values are known only relative to some arbitrary reference state. Such properties as specific heat, coefficient of expansion, coefficient of compressibility, and Joule–Thomson coefficient are classified as derived properties.

All of these properties have intrinsic values which are determined by the existing state of a system and are independent of the course followed in arriving at that state. Properties such as these in which changes are dependent only upon the initial and final conditions of the system and are independent of the path followed in producing the changes are termed *point properties.*

As previously pointed out, heat and work including mechanical, radiant and electrical forms are not properties of a system but are manifestations of changes occurring within the system and as such may be considered properties of a process rather than of a system. From the previous discussions it is evident that these manifestations are dependent upon the particular course of the changes and may differ widely in two processes even though the initial and final properties of the systems are identical. For this reason these properties of a process are termed *path properties.*

When any two of the reference properties, temperature, pressure, volume, and entropy, of a pure substance or of a substance of fixed composition in all phases are specified, the relative values of other point properties are definitely established. With reference to one selected pair of properties, such as pressure and temperature, the magnitude of a third property such as volume may be represented as a surface in space. If extended to all conditions of pressure and temperature of a pure substance with its three states of aggregation, solid, liquid, and gas, the surface developed assumes a contour of great complexity. Upon such a contour surface, for example, with pressure and temperature as independent variables and volume as the dependent variable, equal values of all other point properties may be designated by lines extending over the contour surface. Separate sets of lines may be traced respectively for equal values of entropy, internal energy, specific heats, compressibilities, and so forth. The relationship of points on one path to those on another is definitely fixed by the contour of the surface.

If a mathematical equation can be established for the surface in question in terms of p, V, and T, then relative values of all related properties may be established by mathematical methods. If the equation is unknown or too complex for mathematical manipulation, the desired relationships may still be established by graphical methods. Obviously, the number of possible relations among the dozen or more different properties runs into millions even for first-order differential equations and becomes unlimited for all possible second-order differentials. Ordinarily, only the more important relations are developed in which some desired property is expressed in terms of properties which are easily measured, such as temperature and pressure.

In dealing with solutions and mixtures the foregoing discussion must be expanded to include the variable of composition. To represent the properties of a solution of fixed composition a contour surface can be constructed as for a pure compound, but in covering the entire range of compositions a graphical presentation by a contour surface is no longer adequate, and mathematical methods must be employed.

In developing the thermodynamic relations which follow, it will ordinarily be futile to attempt to visualize the relations involved.

Differential Energy Functions. If consideration is limited to systems of constant mass and composition and to reversible processes in which the only external force is pressure, differential equations for the four energy functions are developed as follows from their definitions and from Equation (1):

$$dU = d'q - d'w_e = T\,dS - p\,dV \tag{23}$$

$$dH = dU + d(pV) = dU + p\,dV + V\,dp \tag{24}$$

$$dG = dH - d(TS) = dH - T\,dS - S\,dT \tag{25}$$

$$dA = dU - d(TS) = dU - T\,dS - S\,dT \tag{26}$$

By combination of the foregoing, four basic differential equations are obtained which express the energy functions in terms of the four reference properties, p, V, T, and S. Thus,

$$dU = T\,dS - p\,dV \tag{27}$$

$$dH = T\,dS + V\,dp \tag{28}$$

$$dA = -S\,dT - p\,dV \tag{29}$$

$$dG = -S\,dT + V\,dp \tag{30}$$

Although these relations are developed by consideration of a highly restricted type of process, it may be noted that Equations (27–30) contain only terms which are intrinsic point properties of matter. Accordingly, these equations are generally applicable to a mass of matter of constant composition undergoing any type of physical change. The relationships between intrinsic point properties are not restricted to reversible operations and are unaffected by accompanying changes in external forms of energy.

Exact Differential Equations. Where u is a single-valued continuous function of several independent variables $x, y, z, \cdots$, the total differential of u can be expressed in terms of its partial derivatives with respect to its different independent variables, thus:

$$du = \left(\frac{\partial u}{\partial x}\right)_{y,z\ldots} dx + \left(\frac{\partial u}{\partial y}\right)_{x,z\ldots} dy + \left(\frac{\partial u}{\partial z}\right)_{x,y\ldots} dz + \cdots \tag{31}$$

The properties of a substance fulfill the requirements of u, and Equation (31) can be extended mathematically to include any number of variables. The value of the integral of du between any two conditions A and B of a relationship represented by Equation (31) is dependent only upon the location of these two points and is independent of the path followed between the two states. For a case involving only three variables, this situation is evident from consideration of Fig. 55 (page 286). Thus,

$$\int_A^B du = u_B - u_A = f(x, y) \tag{32}$$

Since u_A and u_B are wholly defined by the corresponding values of x and y, the value of the integral is unaffected by the path considered in the integration. An equation of the type of (31) is termed an exact differential equation.

If, however, any line is drawn on the surface of Fig. 55 connecting A and B, the length of the line will be dependent upon the path followed. A differential equation expressing the length of line will involve expressions in addition to those defining the surface and is termed an inexact differential equation. For the element of Fig. 55 the inexact differential is represented by the wavy line $d'\sigma$.

It is a property of an exact differential equation involving two independent variables x and y that

$$\frac{\partial}{\partial y}\left(\frac{\partial u}{\partial x}\right)_y = \frac{\partial}{\partial x}\left(\frac{\partial u}{\partial y}\right)_x \tag{33}$$

or

$$\frac{\partial^2 u}{\partial y\,\partial x} = \frac{\partial^2 u}{\partial x\,\partial y} \tag{34}$$

If more than two independent variables are involved, a relationship similar to Equation (33) may be written with respect to any pair of independent variables. Any exact differential equation must satisfy the relations of Equation (33).

For example, if du is expressed by the relation

$$du = M\,dx + N\,dy + P\,dz \tag{35}$$

where $M = f(x)$, $N = f(y)$, and $P = f(z)$, then Equation (35) is an exact differential equation only provided

$$M = \left(\frac{\partial u}{\partial x}\right)_{yz}; \quad N = \left(\frac{\partial u}{\partial y}\right)_{xz}; \quad P = \left(\frac{\partial u}{\partial z}\right)_{yx} \tag{36}$$

Under these conditions,

$$\left(\frac{\partial M}{\partial y}\right)_x = \frac{\partial^2 u}{\partial x\,\partial y}; \quad \left(\frac{\partial N}{\partial x}\right)_y = \frac{\partial^2 u}{\partial x\,\partial y} \quad \text{or} \quad \left(\frac{\partial M}{\partial y}\right)_x = \left(\frac{\partial N}{\partial x}\right)_y \tag{37}$$

Similarly,

$$\left(\frac{\partial M}{\partial z}\right)_x = \left(\frac{\partial P}{\partial x}\right)_z = \frac{\partial^2 u}{\partial x\,\partial z} \quad \text{and} \quad \left(\frac{\partial N}{\partial z}\right)_y = \left(\frac{\partial P}{\partial y}\right)_z = \frac{\partial^2 u}{\partial y\,\partial z} \tag{38}$$

If in the exact differential equation,

$$du = \left(\frac{\partial u}{\partial x}\right)_y dx + \left(\frac{\partial u}{\partial y}\right)_x dy \tag{39}$$

the condition is imposed that the property u is constant, the equation becomes

$$0 = \left(\frac{\partial u}{\partial x}\right)_y dx + \left(\frac{\partial u}{\partial y}\right)_x dy \tag{40}$$

or

$$\left(\frac{\partial y}{\partial x}\right)_u = -\left(\frac{\partial u}{\partial x}\right)_y \bigg/ \left(\frac{\partial u}{\partial y}\right)_x \tag{41}$$

Equation (39) may be differentiated with respect to x while any other property such as w is held constant. This gives, upon elimination of differentials of zero magnitude,

$$\left(\frac{\partial u}{\partial x}\right)_w = \left(\frac{\partial u}{\partial x}\right)_y + \left(\frac{\partial u}{\partial y}\right)_x \left(\frac{\partial y}{\partial x}\right)_w \tag{42}$$

Equations (37), (38), (41), and (42) represent mathematical principles commonly used in the development of thermodynamic relations.

Maxwell Relations. Since Equation (27) is an exact equation of the form of (31) the relationships of Equations (37) and (38) permit the derivation of a differential equation relating the coefficients T and p as shown in Equation (43). Similar equations result from equating the partial derivatives of the coefficients of Equations (28), (29), and (30). Thus,

$$\left(\frac{\partial T}{\partial V}\right)_S = -\left(\frac{\partial p}{\partial S}\right)_V \tag{43}$$

$$\left(\frac{\partial T}{\partial p}\right)_S = \left(\frac{\partial V}{\partial S}\right)_p \tag{44}$$

$$\left(\frac{\partial S}{\partial V}\right)_T = \left(\frac{\partial p}{\partial T}\right)_V \tag{45}$$

$$\left(\frac{\partial S}{\partial p}\right)_T = -\left(\frac{\partial V}{\partial T}\right)_p \tag{46}$$

Equations (43), (44), (45), and (46) are termed the Maxwell relations.

Equation (45) is the basis of the familiar Clapeyron equation. For the vaporization of a liquid the restriction of constant volume may be omitted, because the vapor pressure is independent of volume. Since vaporization is a reversible process at constant temperature, $(\partial S/\partial V)_T = \Lambda/T\,\Delta V$ and

$$\frac{dp_s}{dT} = \frac{\Lambda}{T\,\Delta V} = \frac{\Lambda}{T(V_g - V_l)} \tag{47}$$

which is the form of Equation (1), Chapter III, page 59.

Illustration 3. From a plot of the following properties per pound of superheated ammonia, show that

$$\left(\frac{\partial T}{\partial V}\right)_S = -\left(\frac{\partial p}{\partial S}\right)_V$$

using the reference conditions, $p = 173$ lb per sq in. abs, $t = 185°\text{F}$, $S = 1.300$, and $V = 2.20$ cu ft.

At $S = 1.300$

V	t	p
2.461	165.2	150
2.339	174.4	160
2.229	183.0	170
2.200	185.0	173
2.132	191.3	180
2.043	199.2	190
1.960	206.9	200

At $V = 2.20$

S	t	p
1.2466	114.8	150
1.2705	145.1	160
1.2935	180.3	170
1.3000	185.0	173
1.3154	208.0	180
1.3368	240.3	190
1.3575	283.1	200

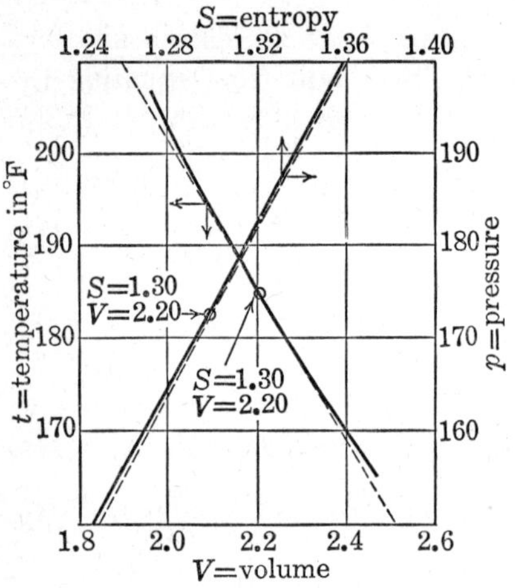

FIG 98. Evaluation of $(\partial T/\partial V)_S$ and $(\partial p/\partial S)_V$.

In Fig. 98, p is plotted against S at $V = 2.20$, and T is plotted against V at $S = 1.300$.

The value of $(\partial T/\partial V)_S$ is the slope of T plotted against V at $S = 1.300$, or

$$\left(\frac{\partial t}{\partial V}\right)_S = \frac{160 - 210}{2.50 - 1.90} = -83.3.$$

The value of $(\partial p/\partial S)_V$ is the slope of p plotted against S at $V = 2.20$. This slope is $(200 - 150)/(1.36 - 1.25) = 454$. To convert this slope into the same units as $(\partial T/\partial V)_S$ requires multiplication by 144/778, where 144 is the conversion factor for pounds per square inch to pounds per square foot and 778 represents the number of foot-pounds equivalent to 1 Btu. The result is 84, which is in agreement with the Maxwell relation.

Partial Derivatives of Energy Functions. By partial differentiation of the four differential energy functions Equations (27–30) while one of the terms appearing as a differential is held constant, there results:

$$\left(\frac{\partial U}{\partial V}\right)_S = -p \quad (48) \qquad \left(\frac{\partial H}{\partial p}\right)_S = V \quad (49)$$

$$\left(\frac{\partial A}{\partial V}\right)_T = -p \quad (50) \qquad \left(\frac{\partial G}{\partial p}\right)_T = V \quad (51)$$

$$\left(\frac{\partial U}{\partial S}\right)_V = T \quad (52) \qquad \left(\frac{\partial H}{\partial S}\right)_p = T \quad (53)$$

$$\left(\frac{\partial A}{\partial T}\right)_V = -S \quad (54) \qquad \left(\frac{\partial G}{\partial T}\right)_p = -S \quad (55)$$

By equating identical properties,

$$\left(\frac{\partial U}{\partial S}\right)_V = \left(\frac{\partial H}{\partial S}\right)_p = T \tag{56}$$

$$\left(\frac{\partial U}{\partial V}\right)_S = \left(\frac{\partial A}{\partial V}\right)_T = -p \tag{57}$$

$$\left(\frac{\partial H}{\partial p}\right)_S = \left(\frac{\partial G}{\partial p}\right)_T = V \tag{58}$$

$$\left(\frac{\partial G}{\partial T}\right)_p = \left(\frac{\partial A}{\partial T}\right)_V = -S \tag{59}$$

By again partially differentiating the four basic differential energy-function equations with respect to the coefficient terms, another set of eight relations results:

$$\left(\frac{\partial U}{\partial T}\right)_V = T\left(\frac{\partial S}{\partial T}\right)_V \tag{60}$$

$$\left(\frac{\partial U}{\partial p}\right)_S = -p\left(\frac{\partial V}{\partial p}\right)_S \tag{61}$$

$$\left(\frac{\partial H}{\partial T}\right)_p = T\left(\frac{\partial S}{\partial T}\right)_p \tag{62}$$

$$\left(\frac{\partial H}{\partial V}\right)_S = V\left(\frac{\partial p}{\partial V}\right)_S \tag{63}$$

$$\left(\frac{\partial A}{\partial p}\right)_T = -p\left(\frac{\partial V}{\partial p}\right)_T \tag{64}$$

$$\left(\frac{\partial A}{\partial S}\right)_V = -S\left(\frac{\partial T}{\partial S}\right)_V \tag{65}$$

$$\left(\frac{\partial G}{\partial V}\right)_T = V\left(\frac{\partial p}{\partial V}\right)_T \tag{66}$$

$$\left(\frac{\partial G}{\partial S}\right)_p = -S\left(\frac{\partial T}{\partial S}\right)_p \tag{67}$$

Thermal Capacity Relations. One of the most important applications of thermodynamics is to establish the relations of physical properties and energy changes from a minimum amount of easily determined experimental data. The thermal capacities of a substance are among those most frequently measured experimentally. There are four useful types of thermal capacities, designated mathematically as follows, for operations in which all changes of state are reversible:

$$C_p = \left(\frac{\partial' q}{\partial T}\right)_p = \text{heat capacity at constant pressure} \tag{68}$$

$$C_v = \left(\frac{\partial' q}{\partial T}\right)_V = \text{heat capacity at constant volume} \tag{69}$$

$$l_p = \left(\frac{\partial' q}{\partial p}\right)_T = \text{latent heat of pressure change at constant temperature} \tag{70}$$

$$l_v = \left(\frac{\partial' q}{\partial V}\right)_T = \text{latent heat of expansion at constant temperature} \tag{71}$$

The thermal capacities are independent of the manner in which heat is added but are restricted by definition to reversible changes of state.

The first two thermal capacities are those most commonly used. The latent heat of volume change at constant temperature includes the usual latent heats of fusion and evaporation, and l_p corresponds to the heat of compression.

Since entropy is a point function, it can be expressed by the following exact differential equations:

$$dS = \left(\frac{\partial S}{\partial T}\right)_p dT + \left(\frac{\partial S}{\partial p}\right)_T dp = \frac{d'q}{T} \tag{72}$$

$$dS = \left(\frac{\partial S}{\partial T}\right)_V dT + \left(\frac{\partial S}{\partial V}\right)_T dV = \frac{d'q}{T} \tag{73}$$

Combining (72) and (73) with Equations (68), (69), (70), and (71) gives

$$\left(\frac{\partial S}{\partial T}\right)_p = \frac{1}{T}\left(\frac{\partial' q}{\partial T}\right)_p = \frac{C_p}{T} \tag{74}$$

$$\left(\frac{\partial S}{\partial p}\right)_T = \frac{1}{T}\left(\frac{\partial' q}{\partial p}\right)_T = \frac{l_p}{T} \tag{75}$$

$$\left(\frac{\partial S}{\partial T}\right)_V = \frac{1}{T}\left(\frac{\partial' q}{\partial T}\right)_V = \frac{C_v}{T} \tag{76}$$

$$\left(\frac{\partial S}{\partial V}\right)_T = \frac{1}{T}\left(\frac{\partial' q}{\partial V}\right)_T = \frac{l_v}{T} \tag{77}$$

At constant temperature, from Equations (75) and (77),

$$l_p = T\left(\frac{\partial S}{\partial p}\right)_T \tag{78}$$

$$l_v = T\left(\frac{\partial S}{\partial V}\right)_T \tag{79}$$

From (46) and (78),

$$l_p = -T\left(\frac{\partial V}{\partial T}\right)_p \tag{80}$$

From (45) and (79),

$$l_v = T\left(\frac{\partial p}{\partial T}\right)_V \tag{81}$$

For an ideal gas, Equations (80) and (81) reduce to

$$l_p^* = -V \quad \text{and} \quad l_v^* = p$$

Substituting the values of Equations (74–77) into (72) and (73) yields

$$dS = \frac{C_p}{T}\,dT + \frac{l_p}{T}\,dp \tag{82}$$

$$dS = \frac{C_v}{T}\,dT + \frac{l_v}{T}\,dV \tag{83}$$

If Equation (82) is applied to conditions of constant entropy and Equation (80) is substituted

$$\frac{C_p}{T} = -\frac{l_p}{T}\left(\frac{\partial p}{\partial T}\right)_S = \left(\frac{\partial V}{\partial T}\right)_p\left(\frac{\partial p}{\partial T}\right)_S \tag{84}$$

Similarly, from Equations (83) and (81),

$$\frac{C_v}{T} = -\frac{l_v}{T}\left(\frac{\partial V}{\partial T}\right)_S = -\left(\frac{\partial p}{\partial T}\right)_V\left(\frac{\partial V}{\partial T}\right)_S \tag{85}$$

$(C_p - C_v)$ and (C_p/C_v). By equating Equations (82) and (83), an expression for the difference between the heat capacities at constant pressure and volume may be derived. Thus,

$$C_p - C_v = l_v\frac{dV}{dT} - l_p\frac{dp}{dT} \tag{86}$$

or, in terms of partial derivatives, restricted first to constant pressure and then to constant volume,

$$C_p - C_v = l_v\left(\frac{\partial V}{\partial T}\right)_p = -l_p\left(\frac{\partial p}{\partial T}\right)_V \tag{87}$$

Combining (80) or (81) with (87) gives

$$\boxed{C_p - C_v = T\left(\frac{\partial p}{\partial T}\right)_V\left(\frac{\partial V}{\partial T}\right)_p} \tag{88}$$

By method of Equation (41),

$$\left(\frac{\partial V}{\partial T}\right)_p = -\left(\frac{\partial p}{\partial T}\right)_V \Big/ \left(\frac{\partial p}{\partial V}\right)_T \quad \text{and}$$

$$\left(\frac{\partial p}{\partial T}\right)_V = -\left(\frac{\partial V}{\partial T}\right)_p \Big/ \left(\frac{\partial V}{\partial p}\right)_T \tag{89}$$

Combining (88) and (89) gives

$$C_p - C_v = -T\left(\frac{\partial p}{\partial T}\right)_V^2\left(\frac{\partial V}{\partial p}\right)_T = -T\left(\frac{\partial V}{\partial T}\right)_p^2\left(\frac{\partial p}{\partial V}\right)_T \tag{90}$$

The ratio of the heat capacities κ at constant pressure is obtained by combining Equations (84) and (85):

$$\kappa = \frac{C_p}{C_v} = -\frac{\left(\frac{\partial V}{\partial T}\right)_p \left(\frac{\partial p}{\partial T}\right)_S}{\left(\frac{\partial p}{\partial T}\right)_V \left(\frac{\partial V}{\partial T}\right)_S} = -\left(\frac{\partial V}{\partial T}\right)_p \left(\frac{\partial T}{\partial p}\right)_V \left(\frac{\partial p}{\partial V}\right)_S \tag{91}$$

Effects of Pressure and Volume on Heat Capacity. The variation with pressure of the heat capacities C_p and C_v at constant temperature is of special importance because of the difficulties of calorimetric measurements at pressures other than atmospheric. Application of the principle of Equations (35–38) to (82) results in

$$\left(\frac{\partial \frac{C_p}{T}}{\partial p}\right)_T = \left(\frac{\partial \frac{l_p}{T}}{\partial T}\right)_p \tag{92}$$

Substituting (80) in (92) gives

$$\left(\frac{\partial C_p}{\partial p}\right)_T = -T\left(\frac{\partial^2 V}{\partial T^2}\right)_p \tag{93}$$

Similarly, from (83) and (81),

$$\left(\frac{\partial C_v}{\partial V}\right)_T = T\left(\frac{\partial^2 p}{\partial T^2}\right)_V \tag{94}$$

***PVT* Relationships of the Energy Functions.** The effects of pressure, volume, and temperature upon the four thermodynamic energy functions are among the most useful of relations. The more important relations are developed herewith. Others may be developed in a similar manner or more readily by Shaw's method as discussed on page 465.

A differential expression for internal energy is obtained by combining Equations (27), (83), and (81):

$$dU = C_v\, dT + \left[T\left(\frac{\partial p}{\partial T}\right)_V - p\right] dV \tag{95}$$

Application of Equation (95) at constant temperature and at constant volume, respectively, results in

$$\left(\frac{\partial U}{\partial V}\right)_T = T\left(\frac{\partial p}{\partial T}\right)_V - p \tag{96}$$

$$\left(\frac{\partial U}{\partial T}\right)_V = C_v \tag{97}$$

Similarly, combining Equations (28), (80), and (82) gives

$$dH = C_p\,dT + \left[-T\left(\frac{\partial V}{\partial T}\right)_p + V\right]dp \tag{98}$$

Applying Equation (98) to constant temperature and constant pressure operations, respectively, results in

$$\left(\frac{\partial H}{\partial p}\right)_T = V - T\left(\frac{\partial V}{\partial T}\right)_p \tag{99}$$

$$\left(\frac{\partial H}{\partial T}\right)_p = C_p \tag{100}$$

One of the most important relations expresses the variation in free-energy change with change in temperature at constant pressure. From Equation (5),

$$\frac{\Delta G}{T} = \frac{\Delta H}{T} - \Delta S \tag{101}$$

$$\left(\frac{\partial \frac{\Delta G}{T}}{\partial T}\right)_p = \frac{1}{T}\left(\frac{\partial \Delta H}{\partial T}\right)_p - \frac{\Delta H}{T^2} - \left(\frac{\partial \Delta S}{\partial T}\right)_p \tag{102}$$

By Equation (62),

$$\left(\frac{\partial \Delta H}{\partial T}\right)_p = T\left(\frac{\partial \Delta S}{\partial T}\right)_p \tag{103}$$

combining (102) and (103) gives

$$\left(\frac{\partial \frac{\Delta G}{T}}{\partial T}\right)_p = -\frac{\Delta H}{T^2} \tag{104}$$

Illustration 4. From van der Waals' equation of state calculate the increase in c_p of CO_2 where the pressure is increased from 1 to 100 atm at 100°C. The van der Waals equation is

$$\left(p + \frac{a}{v^2}\right)(v - b) = RT \tag{a}$$

where v = molal volume
a, b = van der Waals' constants, characteristic of the gas

For 1 lb-mole, Equation (93) is written

$$\left(\frac{\partial c_p}{\partial p}\right)_T = -T\left(\frac{\partial^2 v}{\partial T^2}\right)_p \tag{b}$$

Differentiating the van der Waals equation gives

$$\left(\frac{\partial v}{\partial T}\right)_p = \frac{R}{p - \dfrac{a}{v^2} + \dfrac{2ab}{v^3}} \tag{c}$$

$$\left(\frac{\partial^2 v}{\partial T^2}\right)_p = -R\left[\frac{2av^{-3} - 6abv^{-4}}{(p - av^{-2} + 2abv^{-3})^2}\right]\left(\frac{\partial v}{\partial T}\right)_p \tag{d}$$

Integration of (b) gives

$$\Delta c_p = -\int_1^{100} T\left(\frac{\partial^2 v}{\partial T^2}\right)_p dp \tag{e}$$

Equation (e) is solved by substitution of (d) and graphical integration where,

$$a = 3.60(10^6); \quad b = 42.8; \quad R = 82.1 \text{ (cc-atm)/(g-mole)(°K)}$$

Volume v is first obtained from Equation (a) at various values of p. Equations (c) and (d) are then solved in turn at various values of p.

p atm	v cc	$\left(\frac{\partial v}{\partial T}\right)_p$	$-T\left(\frac{\partial^2 v}{\partial T^2}\right)_p$
0	∞	∞	0
1	30,590	82.7	0.64
10	2,985	8.55	0.734
25	1,150	3.65	0.931
50	530	2.09	1.52
75	320	1.67	2.8
100	213	1.56	5.07

From graphical integration,

$$\Delta c_p = 194.20 \text{ (cc-atm)/(g-mole)(°C)} = 4.7 \text{ cal/(g-mole)(°C)}$$

or

$$c_p = 9.6 + 4.7 = 14.3$$

Illustration 5. Prove that $C_p - C_v = \dfrac{TV\beta^2}{\alpha}$ where

$$\beta = \frac{1}{V}\left(\frac{\partial V}{\partial T}\right)_p = \text{coefficient of expansion}$$

$$\alpha = -\frac{1}{V}\left(\frac{\partial V}{\partial p}\right)_T = \text{coefficient of compressibility}$$

From this relation calculate C_v for mercury, where at 0°C, $\beta = 0.00018$ per °C, $\alpha = 0.000\,003\,9$ per atm, $C_p = 0.0333$ cal/(g)(°C), and the density = 13.596 g per cc. From (90),

$$C_p - C_v = -T\left(\frac{\partial V}{\partial T}\right)_p^2\left(\frac{\partial p}{\partial V}\right)_T$$

Substituting α and β gives

$$C_p - C_v = \frac{-T\beta^2V^2}{-\alpha V} = \frac{T\beta^2 V}{\alpha}$$

For mercury at 0°C,

$$C_p - C_v = \frac{(273)(0.000\ 18)^2}{(13.596)(0.000\ 003\ 9)(41.3)} = 0.0040$$

where 41.3 is the conversion factor from cubic-centimeter-atmospheres to calories. Hence,

$$C_v = 0.0333 - 0.0040 = 0.0293 \text{ cal/(g)(°K)}$$

Shaw's Method of Derivation. The usual procedure in making thermodynamic transformations to arrive at a usable or desirable relation is time-consuming and often results in circuitous procedures which return to the starting point. A. N. Shaw[2] has devised a means of circumventing these difficulties and arriving at the desired result in a systematic manner. This method employs a system of abbreviated mathematical notations termed *Jacobians*. A Jacobian is a determinant, the elements of which are partial derivatives.

It will be recalled that the solution of simultaneous algebraic equations can be accomplished by the use of determinants. For example, consider the two simultaneous equations:

$$\begin{aligned} a_1x + b_1y &= c_1 \\ a_2x + b_2y &= c_2 \end{aligned} \tag{105}$$

The values of x and y satisfying these two equations will be found to be

$$\begin{aligned} x &= \frac{c_1b_2 - c_2b_1}{a_1b_2 - a_2b_1} \\ y &= \frac{a_1c_2 - a_2c_1}{a_1b_2 - a_2b_1} \end{aligned} \tag{106}$$

These solutions can be arrived at systematically by expressing the results in determinant form, thus:

$$\begin{aligned} x &= \frac{\begin{vmatrix} c_1 & b_1 \\ c_2 & b_2 \end{vmatrix}}{\begin{vmatrix} a_1 & b_1 \\ a_2 & b_2 \end{vmatrix}} \\ y &= \frac{\begin{vmatrix} a_1 & c_1 \\ a_2 & c_2 \end{vmatrix}}{\begin{vmatrix} a_1 & b_1 \\ a_2 & b_2 \end{vmatrix}} \end{aligned} \tag{107}$$

The lower determinant in each case is the principal determinant con-

[2] A. N. Shaw, *Phil. Trans. Roy. Soc. A*, **334**, 299–328 (1935).

sisting of the four coefficients of the unknowns arranged in a square

$$\begin{vmatrix} a_1 b_1 \\ a_2 b_2 \end{vmatrix}$$

the value of which is obtained by cross-multiplying terms, thus, $a_1b_2 - a_2b_1$. The determinant in the numerator is derived from the principal determinant by replacing the coefficient of the unknown sought with the constant c; thus, for x the a terms are replaced by c terms.

The Jacobian $\dfrac{\partial(x, y)}{\partial(\alpha, \beta)}$ represents the following determinant,

$$\frac{\partial(x, y)}{\partial(\alpha, \beta)} = \begin{vmatrix} \left(\dfrac{\partial x}{\partial \alpha}\right)_\beta & \left(\dfrac{\partial x}{\partial \beta}\right)_\alpha \\ \left(\dfrac{\partial y}{\partial \alpha}\right)_\beta & \left(\dfrac{\partial y}{\partial \beta}\right)_\alpha \end{vmatrix} = \left(\frac{\partial x}{\partial \alpha}\right)_\beta \left(\frac{\partial y}{\partial \beta}\right)_\alpha - \left(\frac{\partial x}{\partial \beta}\right)_\alpha \left(\frac{\partial y}{\partial \alpha}\right)_\beta \tag{108}$$

where x and y are two dependent variables and α and β are any two independent variables of which x and y are functions. For example, x might be enthalpy; y, entropy; α, temperature; and β, pressure. The values of α and β may equal x and y. The foregoing Jacobian is also written in the form $J(x, y)$ where $J(x, y) = \dfrac{\partial(x, y)}{\partial(\alpha, \beta)}\cdot$ Where this abbreviated notation is used without designation of α and β these variables may be any independent variables of which x and y are functions. However, if several Jacobians, $J(x, y)$, $J(x, z)$, and $J(z, w)$ appear together in a single expression the values of α and β must be the same for all. With this restriction Jacobians may be combined algebraically in any desired manner.

The following four properties of Jacobians are the basis of their use in thermodynamic transformations:

1. Interchanging the order of any two terms in a Jacobian changes its sign, thus,

$$\frac{\partial(x, y)}{\partial(\alpha, \beta)} = -\frac{\partial(y, x)}{\partial(\alpha, \beta)} = \frac{\partial(y, x)}{\partial(\beta, \alpha)} = -\frac{\partial(x, y)}{\partial(\beta, \alpha)} \tag{109}$$

or

$$J(x, y) = -J(y, x) \tag{110}$$

2. $$\frac{\partial(x, x)}{\partial(\alpha, \beta)} = 0 \quad \text{or} \quad J(x, x) = 0 \tag{111}$$

3. $$\left(\frac{\partial y}{\partial x}\right)_z = \frac{J(y, z)}{J(x, z)} = \frac{J(z, y)}{J(z, x)} = -\frac{J(y, z)}{J(z, x)} = -\frac{J(z, y)}{J(x, z)} \tag{112}$$

4. The exact differential equation,

$$dx = M\,dy + N\,dz$$

may be written as

$$J(x, \alpha) = MJ(y, \alpha) + NJ(z, \alpha) \tag{113}$$

The first three relationships can be established readily by setting up in determinant form with α and β as the independent variables and expanding in the usual manner. The fourth relationship may be proved readily by differentiating the exact differential equation with respect to β holding α constant, thus,

$$\left(\frac{\partial x}{\partial \beta}\right)_\alpha = M\left(\frac{\partial y}{\partial \beta}\right)_\alpha + N\left(\frac{\partial z}{\partial \beta}\right)_\alpha \tag{114}$$

The principle of Equation (112) combined with (114) gives

$$\frac{J(x, \alpha)}{J(\beta, \alpha)} = \frac{MJ(y, \alpha)}{J(\beta, \alpha)} + \frac{NJ(z, \alpha)}{J(\beta, \alpha)} \tag{115}$$

Equation (115) is identical with Equation (113).

Thermodynamic equations are most frequently expressed in terms of p, V, T, and S as the independent variables. Any of the six possible pairs can be chosen as the independent variables represented by α and β. The Jacobians of the six possible pairs of independent variables are designated as follows:

$$\left.\begin{aligned} J(V, T) &= a \\ J(p, V) &= b \\ J(p, S) &= c \\ J(p, T) &= l \\ J(V, S) &= n \\ J(T, S) = J(p, V) &= b \end{aligned}\right\} \tag{116}$$

The proof of this last relationship is derived by writing Equation (43) in Jacobian form in accordance with Equation (112), thus,

$$\frac{J(T, S)}{J(V, S)} = -\frac{J(p, V)}{J(S, V)} = \frac{J(p, V)}{J(V, S)} \tag{117}$$

or

$$J(T, S) = J(p, V) \tag{118}$$

Using the symbols of Equation (116) Shaw developed Table XXIII from which the Jacobian $J(x, y)$ of any pair of thermodynamic properties x and y may be obtained in terms of the six possible Jacobians of the four independent variables p, V, T, and S. The properties considered are p, V, T, S, U, H, A, and G. In addition, w_r the work of expansion

TABLE XXIII. VALUES OF JACOBIANS, $J(x, y)$

$y \rightarrow$ / $x \downarrow$	p	V	T	S	U	H	A	G	q (rev)	w (rev)
p	0	b	l	c	$Tc-pb$	Tc	$-Sl-pb$	$-Sl$	Tc	pb
V	$-b$	0	a	n	Tn	$Tn-Vb$	$-Sa$	$-Sa-Vb$	Tn	0
T	$-l$	$-a$	0	b	$Tb+pa$	$Tb-Vl$	pa	$-Vl$	Tb	$-pa$
S	$-c$	$-n$	$-b$	0	pn	$-Vc$	$Sb+pn$	$Sb-Vc$	0	$-pn$
U	$-Tc+pb$	$-Tn$	$-Tb-pa$	$-pn$	0	$-TVc$ $-p(Tn-Vb)$	$T(Sb+pn)$ $+pSa$	$T(Sb-Vc)$ $+p(Sa+Vb)$	$-pTn$	$-Tpn$
H	$-Tc$	$-Tn+Vb$	$-Tb+Vl$	Vc	TVc $+p(Tn-Vb)$	0	$T(Sb+pn)$ $-V(Sl+pb)$	$T(Sb-Vc)$ $-VSl$	VTc	$-Tpn$ $+Vpb$
A	$Sl+pb$	Sa	$-pa$	$-Sb-pn$	$-T(Sb+pn)$ $-pSa$	$-T(Sb+pn)$ $+V(Sl+pb)$	0	SVl $+p(Sa+Vb)$	$-STb$ $-pTn$	Spa
G	Sl	$Sa+Vb$	Vl	$-Sb+Vc$	$-T(Sb-Vc)$ $-p(Sa+Vb)$	$-T(Sb-Vc)$ $+VSl$	$-SVl$ $-p(Sa+Vb)$	0	$-STb$ $+VTc$	$Spa+Vpb$
q (rev)	$-Tc$	$-Tn$	$-Tb$	0	pTn	$-VTc$	$STb+pTn$	$STb-VTc$	0	$-Tpn$
w (rev)	$-pb$	0	pa	pn	Tpn	$Tpn-Vpb$	$-Spa$	$-Spa-Vpb$	Tpn	0

$$a = J(V, T) = -J(T, V)$$
$$b = J(p, V) = J(T, S) = -J(V, p) = -J(S, T)$$
$$c = J(p, S) = -J(S, p)$$
$$l = J(p, T) = -J(T, p)$$
$$n = J(V, S) = -J(S, V)$$

$$\left(\frac{\partial x}{\partial y}\right)_z = \frac{J(x, z)}{J(y, z)}$$
$$J(T, S) - J(p, V) = 0$$
$$J(x, y) \cdot J(z, w) + J(y, z) \cdot J(w, x) + J(z, x) \cdot J(y, w) = 0$$
$$b^2 + ac - ln = 0$$

$$J(x, y) = \frac{\left(\frac{\partial x}{\partial \alpha}\right)_\beta \left(\frac{\partial x}{\partial \beta}\right)_\alpha}{\left(\frac{\partial y}{\partial \alpha}\right)_\beta \left(\frac{\partial y}{\partial \beta}\right)_\alpha} = \left(\frac{\partial x}{\partial \alpha}\right)_\beta \left(\frac{\partial y}{\partial \beta}\right)_\alpha - \left(\frac{\partial x}{\partial \beta}\right)_\alpha \left(\frac{\partial y}{\partial \alpha}\right)_\beta$$

done by and q_r the heat added to a system undergoing a reversible process are included. The Jacobian $J(x, y)$ of any pair of properties is obtained by entering the table at the values of x in the left-hand column and y in the top row. For example, if $x = H$ and $y = T$,

$$J(x, y) = J(H, T) = -Tb + Vl = -TJ(p, V) + VJ(p, T) \quad (119)$$

Table XXIII was constructed from the differential energy functions, Equations (27) to (30). For example, by the fourth Jacobian rule Equation (28) may be written as follows with T replacing α as the independent variable:

$$J(H, T) = TJ(S, T) + VJ(p, T) = -Tb + Vl \quad (120)$$

The other Jacobians are developed in a similar manner.

The Jacobians taken from Table XXIII may be modified to include any selected pair of independent variables α and β in accordance with Equation (109), or may be transformed according to the four rules of Equations (109–113) and expanded according to Equation (108) to obtain any desired relationship. For example, Equation (119) may be modified to include p and V as independent variables α and β. Thus, applying Equation (109) to Equation (119) gives

$$\frac{\partial(H, T)}{\partial(p, V)} = -T + V\frac{\partial(p, T)}{\partial(p, V)} \quad (121)$$

If desired, Equation (121) might be expanded in accordance with Equation (108) to yield a complete relationship among H, T, p, and V. A similar relationship among the same variables but of different form results if p and T are chosen as the independent variables instead of p and V. In this manner a limitless number of relationships may be developed. Where a specific relationship is sought in terms of specified variables, it is desirable to simplify the expressions and eliminate unwanted variables before expanding the Jacobians.

Illustration 6. From Table XXIII evaluate $(\partial G/\partial T)_p$. This expression may be written in Jacobian form in accordance with Equation (112) and the Jacobians evaluated from the table.

$$\left(\frac{\partial G}{\partial T}\right)_p = \frac{J(G, p)}{J(T, p)} = \frac{Sl}{-l} = -S$$

This relationship may be transformed into other variables by replacing S by its equivalent in Jacobians from Table XXIII. Thus,

$$S = \frac{J(A, V)}{a} = \frac{J(A, V)}{J(V, T)} = -\left(\frac{\partial A}{\partial T}\right)_V$$

or

$$\left(\frac{\partial G}{\partial T}\right)_p = \left(\frac{\partial A}{\partial T}\right)_V$$

In order to eliminate an undesirable independent variable, use is made of the following important relationship:

$$b^2 + ac - ln = 0 \tag{122}$$

This relationship can be proved as follows: From Equation (72),

$$dS = \left(\frac{\partial S}{\partial p}\right)_T dp + \left(\frac{\partial S}{\partial T}\right)_p dT \tag{123}$$

or

$$dS = \frac{J(S, T)}{J(p, T)} dp + \frac{J(S, p)}{J(T, p)} dT \tag{124}$$

$$J(p, T)\, dS = J(S, T)\, dp - J(S, p)\, dT \tag{125}$$

Differentiation with respect to any variable w, with V held constant, gives

$$J(p, T)\left(\frac{\partial S}{\partial w}\right)_V = J(S, T)\left(\frac{\partial p}{\partial w}\right)_V - J(S, p)\left(\frac{\partial T}{\partial w}\right)_V \tag{126}$$

Expansion in accordance with Equation (112) results in

$$J(p, T)J(S, V) = J(S, T)J(p, V) - J(S, p)J(T, V) \tag{127}$$

or, substituting symbols for the Jacobians gives

$$l(-n) = (-b)b - (-c)(-a)$$

or

$$b^2 + ac - ln = 0 \tag{128}$$

Illustration 7. From Jacobians, prove that

$$T\left(\frac{\partial p}{\partial T}\right)_V \left(\frac{\partial V}{\partial T}\right)_p = -T\left(\frac{\partial V}{\partial T}\right)_p^2 \left(\frac{\partial p}{\partial V}\right)_T \tag{a}$$

Rewriting in Jacobians gives

$$\frac{TJ(p, V)J(V, p)}{J(T, V)J(T, p)} = -\frac{TJ(V, p)^2J(p, T)}{J(T, p)^2J(V, T)} \tag{b}$$

Substitution of values from Table XXIII gives

$$\frac{Tb(-b)}{(-a)(-l)} = \frac{-T(-b)^2l}{(-l)^2a} \tag{c}$$

or

$$\frac{b^2T}{al} = \frac{b^2T}{al} \tag{d}$$

Illustration 8. It is desired to evaluate $(\partial p/\partial V)_H$ in terms of the independent variables p and T and if possible to eliminate S. Equation (112) and Table XXIII give

$$\left(\frac{\partial p}{\partial V}\right)_H = \frac{J(p, H)}{J(V, H)} = \frac{Tc}{Tn - Vb} \tag{a}$$

The Jacobians c and b each include one of the independent variables and may be expanded directly into a partial derivative in terms of p and T. However, $n = J(V, S)$ contains neither of the independent variables. It is therefore desirable to eliminate this quantity by means of Equation (128). Thus,

$$\frac{Tc}{Tn - Vb} = \frac{Tcl}{Tb^2 + Tac - Vbl} \tag{b}$$

The symbols may now be replaced by their corresponding Jacobians, thus,

$$\left(\frac{\partial p}{\partial V}\right)_H = \frac{TJ(p, S)J(p, T)}{TJ(p, V)^2 + TJ(V, T)J(p, S) - VJ(p, V)J(p, T)} \tag{c}$$

Since p and T are the independent variables, this last equation becomes, by (109),

$$\left(\frac{\partial p}{\partial V}\right)_H = \frac{T\left(\frac{\partial(p, S)}{\partial(p, T)}\right)\left(\frac{\partial(p, T)}{\partial(p, T)}\right)}{T\left(\frac{\partial(p, V)}{\partial(p, T)}\right)^2 + T\left(\frac{\partial(V, T)}{\partial(p, T)}\right)\left(\frac{\partial(p, S)}{\partial(p, T)}\right) - V\left(\frac{\partial(p, V)}{\partial(p, T)}\right)\left(\frac{\partial(p, T)}{\partial(p, T)}\right)} \tag{d}$$

The like terms in (d) are canceled, and the remaining Jacobians written as partial derivatives according to Equation (112):

$$\left(\frac{\partial p}{\partial V}\right)_H = \frac{T\left(\frac{\partial S}{\partial T}\right)_p}{T\left(\frac{\partial V}{\partial T}\right)_p^2 + T\left(\frac{\partial V}{\partial p}\right)_T\left(\frac{\partial S}{\partial T}\right)_p - V\left(\frac{\partial V}{\partial T}\right)_p} \tag{e}$$

Since $\left(\frac{\partial S}{\partial T}\right)_p = \frac{C_p}{T}$, Equation (e) reduces to

$$\left(\frac{\partial p}{\partial V}\right)_H = \frac{C_p}{T\left(\frac{\partial V}{\partial T}\right)_p^2 + \left(\frac{\partial V}{\partial p}\right)_T C_p - V\left(\frac{\partial V}{\partial T}\right)_p} \tag{f}$$

For 1 mole of an ideal gas $\left(\frac{\partial v}{\partial T}\right)_p = \frac{R}{p}$; $\quad \left(\frac{\partial v}{\partial p}\right)_T = -\frac{v}{p}$

hence,

$$\left(\frac{\partial p}{\partial v}\right)_H = \frac{c_p}{T\left(\frac{R}{p}\right)^2 - c_p\left(\frac{v}{p}\right) - v\left(\frac{R}{p}\right)} = -\frac{p}{v} \tag{g}$$

Equation (g) is in agreement with the requirement that for an ideal gas H is a function of temperature only. For this requirement

$$\left(\frac{\partial p}{\partial v}\right)_H = \left(\frac{\partial p}{\partial v}\right)_T = -\frac{p}{v}$$

Summary of the Most Useful Relations. It cannot be said that one thermodynamic relation is more significant or more important than another, but it is found that relatively few relations suffice for most thermodynamic problems. A brief selection of these more useful relations is presented in Table XXIV.

TABLE XXIV

THERMODYNAMIC RELATIONS

	Equation Number in Text
Differential Energy Functions	
(a) $dU = T\,dS - p\,dV$	(27)
(b) $dH = T\,dS + V\,dp$	(28)
(c) $dA = -S\,dT - p\,dV$	(29)
(d) $dG = -S\,dT + V\,dp$	(30)
Maxwell Relations	
(e) $\left(\frac{\partial T}{\partial V}\right)_S = -\left(\frac{\partial p}{\partial S}\right)_V$	(43)
(f) $\left(\frac{\partial T}{\partial p}\right)_S = \left(\frac{\partial V}{\partial S}\right)_p$	(44)
(g) $\left(\frac{\partial S}{\partial V}\right)_T = \left(\frac{\partial p}{\partial T}\right)_V$	(45)
(h) $\left(\frac{\partial S}{\partial p}\right)_T = -\left(\frac{\partial V}{\partial T}\right)_p$	(46)
Energy-Function Derivatives	
(i) $\left(\frac{\partial U}{\partial S}\right)_V = \left(\frac{\partial H}{\partial S}\right)_p = T$	(56)
(j) $\left(\frac{\partial U}{\partial V}\right)_S = \left(\frac{\partial A}{\partial V}\right)_T = -p$	(57)
(k) $\left(\frac{\partial H}{\partial p}\right)_S = \left(\frac{\partial G}{\partial p}\right)_T = V$	(58)
(l) $\left(\frac{\partial G}{\partial T}\right)_p = \left(\frac{\partial A}{\partial T}\right)_V = -S$	(59)
Thermal-Capacity Relations	
(m) $C_p = \left(\frac{\partial' q}{\partial T}\right)_p = T\left(\frac{\partial S}{\partial T}\right)_p$	(74)
(n) $l_p = \left(\frac{\partial' q}{\partial p}\right)_T = T\left(\frac{\partial S}{\partial p}\right)_T = -T\left(\frac{\partial V}{\partial T}\right)_p$	(75)(80)
(o) $C_v = \left(\frac{\partial' q}{\partial T}\right)_V = T\left(\frac{\partial S}{\partial T}\right)_V$	(76)
(p) $l_v = \left(\frac{\partial' q}{\partial V}\right)_T = T\left(\frac{\partial S}{\partial V}\right)_T = T\left(\frac{\partial p}{\partial T}\right)_V$	(77)(81)
(q) $dS = \frac{C_p\,dT}{T} + \frac{l_p}{T}\,dp$	(82)

TABLE XXIV (*cont.*)

	Equation Number in Text
(r) $dS = \frac{C_v\,dT}{T} + \frac{l_v\,dV}{T}$	(83)
(s) $C_p - C_V = -T\left(\frac{\partial p}{\partial T}\right)_V^2\left(\frac{\partial V}{\partial p}\right)_T = -T\left(\frac{\partial V}{\partial T}\right)_p^2\left(\frac{\partial p}{\partial V}\right)_T$	(90)
(t) $\frac{C_p}{C_v} = -\left(\frac{\partial V}{\partial T}\right)_p\left(\frac{\partial T}{\partial p}\right)_V\left(\frac{\partial p}{\partial V}\right)_S = \kappa$	(91)
(u) $\left(\frac{\partial C_p}{\partial p}\right)_T = -T\left(\frac{\partial^2 V}{\partial T^2}\right)_p$	(93)
(v) $\left(\frac{\partial C_v}{\partial V}\right)_T = T\left(\frac{\partial^2 p}{\partial T^2}\right)_V$	(94)
Effect of p, V, T on U, H, and G	
(w) $\left(\frac{\partial U}{\partial V}\right)_T = T\left(\frac{\partial p}{\partial T}\right)_V - p$	(96)
(x) $\left(\frac{\partial U}{\partial T}\right)_V = C_v$	(97)
(y) $\left(\frac{\partial H}{\partial p}\right)_T = V - T\left(\frac{\partial V}{\partial T}\right)_p$	(99)
(z) $\left(\frac{\partial H}{\partial T}\right)_p = C_p$	(100)
(α) $\left(\frac{\partial \frac{\Delta G}{T}}{\partial T}\right)_p = -\frac{\Delta H}{T^2}$	(104)

The Thermodynamic Temperature Scale. The properties of an ideal gas are defined by the requirements that $pv = RT$ and that the internal energy is a function of temperature only and is independent of volume. It has been experimentally demonstrated that real gases approach this behavior at low pressures. The gas-temperature scale is defined from the behavior of an ideal gas as $T = pv/R$ and has been numerically evaluated by the extrapolation to zero pressure of observations with a constant-volume hydrogen thermometer.

All of the general thermodynamic relations developed in the preceding sections were defined without reference to the character of the temperature scale on which the definition of entropy is based. If it can be shown that any of these relations lead to an expression involving temperature which can also be derived from the ideal-gas law, this fact will constitute proof of the identity of the thermodynamic and ideal-gas temperature scales.

From Equation (w) of Table XXIV,

$$\left(\frac{\partial U}{\partial V}\right)_T = T\left(\frac{\partial p}{\partial T}\right)_V - p \qquad (129)$$

Since Equation (129) relates p, V, and T, it may be termed a thermodynamic equation of state. If this equation is applied to an ideal gas, for which $(dU/dV)_T = 0$

$$T = p^* \bigg/ \left(\frac{\partial p}{\partial T}\right)_V \tag{130*}$$

Differentiation of the ideal-gas law gives

$$\left(\frac{\partial p^*}{\partial T}\right)_V = \frac{R}{v} = \frac{p}{T}$$

or

$$T = p^* \bigg/ \left(\frac{\partial p}{\partial T}\right)_V \tag{131*}$$

Since Equations (130) and (131) are identical, it follows that the thermodynamic and the ideal-gas temperature scales are similar functions, both having the same zero point.

Thermodynamic Properties of an Ideal Gas. As previously mentioned, the properties of an ideal gas are defined by the requirements that the pVT relations are exactly expressed by the equation $pv = RT$ and that the internal energy is dependent only upon temperature and is independent of pressure and volume. Equations whose applicability is restricted to ideal gases are designated by asterisks (*) following the equation numbers. Similarly an asterisk following any thermodynamic symbol indicates restriction to the ideal-gaseous state.

By differentiation of the ideal-gas law, it follows that

$$p\,dv + v\,dp = R\,dT \tag{132*}$$

or

$$\left(\frac{\partial p}{\partial T}\right)_V = \frac{R}{v} = \frac{p}{T} \quad \text{and} \quad \left(\frac{\partial v}{\partial p}\right)_T = -\frac{v}{p} \quad \text{and} \quad \left(\frac{\partial v}{\partial T}\right)_p = \frac{R}{p} = \frac{v}{T} \tag{133*}$$

From (s), Table XXIV,

$$c_p - c_v = -T\left(\frac{R}{v}\right)^2\left(-\frac{v}{p}\right) = \frac{R^2T}{pv} = R \tag{134*}$$

The differential change in internal energy may be written as

$$dU = \left(\frac{\partial U}{\partial T}\right)_V dT + \left(\frac{\partial U}{\partial V}\right)_T dV \tag{135}$$

By substitution of Equations (w) and (x) of Table XXIV, there results

$$dU = C_v\,dT + \left[T\left(\frac{\partial p}{\partial T}\right)_V - p\right] dV \tag{136}$$

For an ideal gas $p = T\left(\frac{\partial p}{\partial T}\right)_V$, and Equation (136) becomes for 1 mole

$$d\mathrm{U} = c_v\, dT \tag{137*}$$

or

$$\mathrm{U} = \int_0^T c_v\, dT + \mathrm{U}_0 \tag{138*}$$

Since for an ideal gas $(\partial^2 p/\partial T^2)_V = 0$, it follows from Equation (v), Table XXIV, that c_v^* is independent of volume and pressure.

By combination of (98) with (133) and (134) on the basis of 1 mole,

$$d\mathrm{H} = c_p\, dT = c_v\, dT + R\, dT = (c_v + R)\, dT \tag{139*}$$

or

$$\mathrm{H} = \int_0^T c_v\, dT + RT + \mathrm{H}_0 \tag{140*}$$

From (n), Table XXIV, and (133),

$$l_p = -\frac{RT}{p} = -v \tag{141*}$$

Combining (141) with (q) of Table XXIV gives

$$d\mathrm{s} = c_p\, d\ln T - R\, d\ln p$$

or

$$\Delta\mathrm{s} = \int_{T_1}^{T_2} c_p\, d\ln T - R \ln \frac{p_2}{p_1} \tag{142*}$$

It is evident that Equation (142) cannot be directly integrated with lower limits of either $T = 0$ or $p = 0$, since the integral of each term at these limits becomes negative infinity. In order to evaluate directly the entropy of an actual gas in the ideal state from heat-capacity measurements it is necessary to start with the crystalline solid at a temperature near the absolute zero, heat it to its fusion point and then to its boiling point T' under a pressure p' sufficiently low that ideal-gas behavior is obtained. The entropy s′ of the saturated vapor at this temperature and pressure is calculated from the low-temperature heat capacities of the crystalline and liquid state and the heats of fusion and vaporization. Equation (142) is then integrated from this limit. Thus,

$$\mathrm{s} = \int_{T'}^{T} c_p\, d\ln T - R \ln \frac{p}{p'} + \mathrm{s}' \tag{143*}$$

A *perfect gas* may be defined as one which behaves ideally at all conditions and hence is incapable of condensation to either a liquid or a solid phase. It follows that for the internal energy of such a gas to be independent of volume at all conditions of pressure and temperature it can possess only translational energy of motion and its heat capacity must be independent of temperature as well as of pressure, as shown by Equation (VII-19), page 211. Thus, for a perfect gas, $c_v = \frac{3}{2}R$ and $c_p = \frac{5}{2}R$, whereas the heat capacity of what has been termed an ideal gas may vary with temperature. It is evident that no substance fulfills the properties of a perfect gas but that perfect behavior is approached by monatomic gases at low pressures.

For a perfect gas, Equation (142) may be integrated as follows:

$$\text{s} = \tfrac{5}{2}R \ln T - R \ln p + b \qquad (144)^*$$

The integration constant b has been evaluated by methods of statistical mechanics as $(\frac{3}{2}R \ln M - 2.298)$ where M is the molecular weight.

Illustration 9. Assuming that nitrogen behaves ideally over the range indicated, calculate the values of s*, u*, h*, a*, and g* for 1 lb-mole at 10 atm, and 100°F, relative to conditions at 1 atm, and 60°F. Assume $c_v = 5.0$.

$$\Delta\text{u}^* = c_v\,\Delta T = 5.0(40) = 200 \text{ Btu}$$

$$\Delta\text{h}^* = (c_v + R)\,\Delta T = (5.0 + 1.99)(40) = 279.6 \text{ Btu}$$

$$\Delta\text{s}^* = c_p \ln \frac{T_2}{T_1} - R \ln \frac{p_2}{p_1} = 6.99 \ln \frac{560}{520} - 1.99 \ln \frac{10}{1} = -4.07 \text{ entropy units}$$

It will be evident that numerical values cannot be secured for Δa and Δg unless absolute values of entropy are known.

The absolute entropy of nitrogen at 60°F and 1 atm pressure is 45.56 Btu/(lb-mole)(°R). The final entropy after heating and compressing to 100°F and 10 atm, from Illustration 8, is 45.56 + Δs* = 41.49.

Then, since $\Delta\text{a} = \Delta\text{u} - \Delta(T\text{s}) = \Delta\text{u} - T_2\text{s}_2 + T_1\text{s}_1$,

$$\Delta\text{a}^* = 200 - (560)(41.49) + (520)(45.56) = 666$$

Similarly,

$$\Delta\text{g}^* = \Delta\text{h}^* - T_2\text{s}_2^* + T_1\text{s}_1^* = 279.6 - 23\,234 + 23\,700 = 746$$

PROBLEMS

1. (*a*) Calculate the increase in entropy when 1 lb of ice at 0°F and atmospheric pressure is converted into steam at 220°F and 1 atm pressure.

(*b*) The absolute entropy of water vapor as an ideal gas at 25°C and 1.0 atm is 45.13 cal/(°K)(g-mole). Using the data of part (*a*), calculate the absolute entropy of ice at 32°F in Btu/(°R)(lb). Steam at 220°F and 1 atm may be assumed to behave ideally.

2. The volume coefficient of expansion $(\partial V/\partial T)/V$ of water at 100°C is 0.00078 per °C. Calculate the change in entropy in cal/(gram)(°C) when the pressure is reduced from 100 to 1 atm at 100°C, neglecting the effect of pressure on volume.

3. Derive the differential equations for the following relations:

(*a*) Change of total work function with temperature at constant volume.

(*b*) Change of enthalpy with entropy at constant pressure.

(*c*) Change of internal energy with volume at constant entropy.

(*d*) Change of temperature with volume at constant entropy.

4. Calculate the latent heat of vaporization per pound of water at 160°F from the following data:

$$\frac{dp}{dT} = 0.113 \text{ lb/(sq in.)(°R)}$$

Volume of saturated vapor = 77.29 cu ft per lb and of liquid = 0.017 cu ft per lb.

5. From selected data on the properties of superheated steam taken from Keenan and Keyes, "Thermodynamic Properties of Steam" (John Wiley & Sons, New York, 1936),

(*a*) Verify the relation $dU = T\,dS - p\,dV$.

(*b*) Verify any one or more of the four Maxwell relations.

6. Assuming ideal behavior, calculate the entropy of 10 lb of hydrogen gas at 60°F and 10 atm relative to 0°F and 1 atm.

7. Evaluate the following coefficients per pound-mole of an ideal gas:

(*a*) $\left(\frac{\partial p}{\partial T}\right)_V$ (*b*) $\left(\frac{\partial^2 p}{\partial T^2}\right)_V$ (*c*) $\left(\frac{\partial p}{\partial v}\right)_T$ (*d*) $\left(\frac{\partial v}{\partial T}\right)_p$

(*e*) $\left(\frac{\partial^2 v}{\partial T^2}\right)_p$ (*f*) $\left(\frac{\partial \mathrm{S}}{\partial p}\right)_T$ (*g*) $\left(\frac{\partial \mathrm{S}}{\partial v}\right)_T$ (*h*) $\left(\frac{\partial \mathrm{G}}{\partial T}\right)_V$

(*i*) $\left(\frac{\partial \mathrm{U}}{\partial v}\right)_T$ (*j*) $\left(\frac{\partial \mathrm{U}}{\partial p}\right)_T$ (*k*) $\left(\frac{\partial c_p}{\partial p}\right)_T$ (*l*) $\left(\frac{\partial c_v}{\partial v}\right)_T$

(*m*) $\left(\frac{\partial p}{\partial v}\right)_S$ (*n*) $\left(\frac{\partial T}{\partial p}\right)_H$ (*o*) $c_p - c_v$ (*p*) l_p (*q*) l_v

8. For a van der Waals gas (see Illustration 4, page 463), evaluate the following coefficients:

(*a*) $\left(\frac{\partial p}{\partial T}\right)_V$ (*b*) $\left(\frac{\partial^2 p}{\partial T^2}\right)_V$ (*c*) $\left(\frac{\partial v}{\partial T}\right)_p$ (*d*) $\left(\frac{\partial^2 v}{\partial T^2}\right)_p$

(*e*) $\left(\frac{\partial \mathrm{U}}{\partial v}\right)_T$ (*f*) $\left(\frac{\partial \mathrm{U}}{\partial p}\right)_T$ (*g*) $\left(\frac{\partial c_p}{\partial p}\right)_T$ (*h*) $\left(\frac{\partial p}{\partial v}\right)_S$

9. By combination of Equation (47) with Equation (III-16), page 73, evaluate the heat of vaporization of methylamine at its normal boiling point, −6.5°C, assuming that the vapor obeys the ideal-gas law and neglecting the volume of the liquid.

10. From Table XXIII, express

(*a*) $C_p - C_v$ in terms of p, V, and T.

(*b*) $\left(\frac{\partial H}{\partial p}\right)_S$ in terms of V.

(*c*) $\left(\frac{\partial H}{\partial p}\right)_S$ in terms of G and p.

(*d*) $\left(\frac{\partial A}{\partial p}\right)_T$ in terms of p and V.

(*e*) $\left(\frac{\partial S}{\partial T}\right)_p$ in terms of C_p and T.

11. From Table XXIII derive equations for the following quantities in terms of the indicated variables:

Quantities	*Variables*
(*a*) $\left(\frac{\partial T}{\partial V}\right)_S$	S, p
(*b*) $\left(\frac{\partial G}{\partial T}\right)_p$	S
(*c*) $\left(\frac{\partial G}{\partial T}\right)_p$	A, T
(*d*) $\left(\frac{\partial A}{\partial S}\right)_V$	S, T

12. From Table XXIII verify the following equations:

(*a*) $C_p - C_v = T\left(\frac{\partial p}{\partial T}\right)_V\left(\frac{\partial V}{\partial T}\right)_p$

(*b*) $(C_p - C_v) = -T\left(\frac{\partial V}{\partial T}\right)_p^2\left(\frac{\partial p}{\partial V}\right)_T$

(*c*) $\frac{C_p}{C_v} = -\left(\frac{\partial V}{\partial T}\right)_p\left(\frac{\partial T}{\partial p}\right)_V\left(\frac{\partial p}{\partial V}\right)_S$

13. Calculate the values of ΔS, ΔU, ΔH, ΔA, and ΔG when 1 lb-mole of H_2 gas at 77°F and 1 atm is heated and compressed to 500°F and 100 atm. The initial absolute entropy of hydrogen is 31.23 Btu/(lb-mole) (°R). Assume ideal behavior.

14. From the relationships of problem 8 and Table XXIV, calculate the following changes in the properties of 1 lb-mole of CO_2 gas when the pressure is increased isothermally from 1.0 to 100 atm at 100°C. $a = 3.60\ (10^6)\ (\text{atm})(\text{cc})^2/(\text{g-mole})^2$; $b = 42.8$ cc per g-mole.

(*a*) ΔH, (*b*) ΔU, (*c*) ΔS, (*d*) Δc_p, (*e*) Δc_v.

CHAPTER XII

THERMODYNAMIC PROPERTIES OF FLUIDS

The ideal-gas law permits satisfactory calculations of pressure–volume–temperature relationships at low pressures where the volumes per mole are relatively large and the distances between molecules great or where the temperatures are relatively high. However, under conditions of small molal volumes, corresponding to high pressures, the errors in assuming ideal-gas behavior may be as great as 500 per cent.

Actual Behavior of Gases. Several hundred equations of state have been proposed to express the pVT relationship of gases, but none has been found universally satisfactory, and most are applicable only to a single gas over a limited range of temperatures and pressures. The most generally satisfactory equation of state is that of Beattie and Bridgman wherein the various constants must be determined uniquely for each gas. The van der Waals equation is presented because of its value in giving an elementary understanding of the reasons for departure of gases from ideal behavior and because of its support of the theory of corresponding states which serves as the basis of the most useful generalized graphical method. However, the van der Waals equation is less accurate than the generalized graphical method, and its application is rarely warranted.

The actual pVT relationships of carbon dioxide are shown graphically in Fig. 99. Each curve represents the relationship between the pressure and the molal volume of carbon dioxide at the indicated temperature. The critical point is indicated by C. The double-crosshatched area represents the region of the liquid state. The plain area is the region of a homogeneous fluid state which at low pressures is recognized as a gas but at high pressures has continuity with the liquid. The single-crosshatched area represents a region in which both the liquid and gaseous states are present in equilibrium with each other. Thus, following along the 21.5°C experimental isothermal line from right to left, an increasing pressure is required to cause a reduction in volume until the *saturation curve* CB is reached. At this point the attractive forces between the molecules become sufficiently great to start condensation. The volume may then be diminished without further increase in pressure until the curve CA is reached. This curve represents the completion of condensation into the relatively incompressible liquid state. A further decrease in volume must be accompanied by a large increase in pressure.

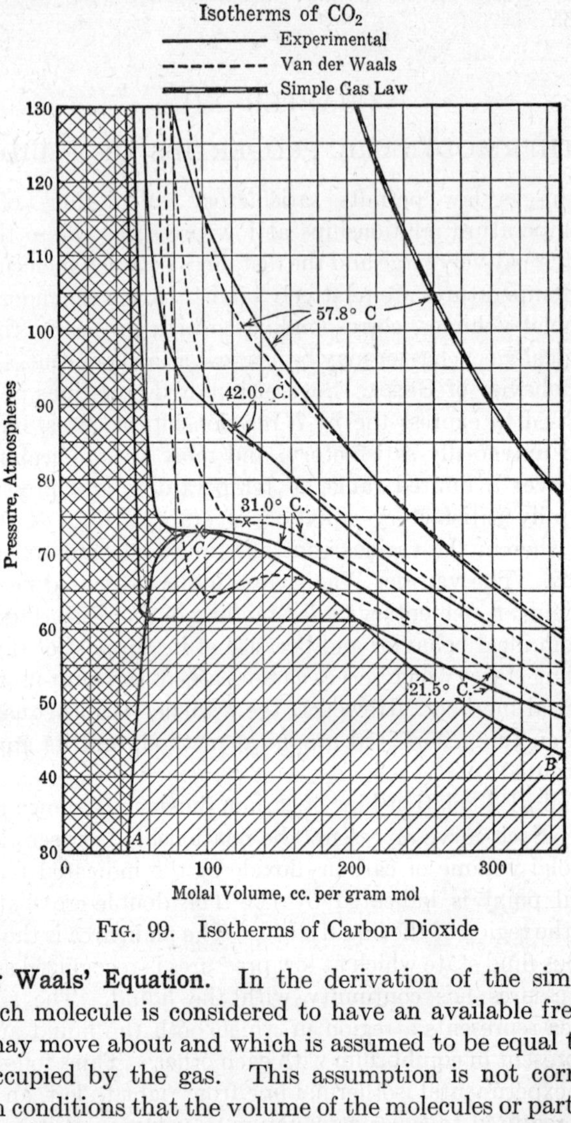

FIG. 99. Isotherms of Carbon Dioxide

Van der Waals' Equation. In the derivation of the simple kinetic theory, each molecule is considered to have an available free space, in which it may move about and which is assumed to be equal to the total volume occupied by the gas. This assumption is not correct except under such conditions that the volume of the molecules or particles themselves is negligible as compared to the total volume occupied by them. Actually, in each mole of gas there is a space of volume $(v - b)$ available for free motion, somewhat less than the total volume. The correction b by which the available free volume per mole differs from the total volume per mole is dependent on the volume actually represented by the N molecules themselves. It is this free volume which must be used in the more rigorous derivation of the kinetic gas laws.

TABLE XXV

VAN DER WAALS' AND CRITICAL CONSTANTS OF GASES
From the data of the International Critical Tables

a, b = van der Waals' constants $\quad a = (\text{atm})\left(\frac{\text{cc}}{\text{g-mole}}\right)^2; \quad b = \frac{\text{cc}}{\text{g-mole}}$

t_c = critical temperature, degrees centigrade
p_c = critical pressure, atmospheres
d_c = critical vapor density, grams per cubic centimeter
v_c = critical volume, cubic centimeters

To convert values of a to $\frac{\text{lb}}{(\text{in.})^2}\left(\frac{\text{ft}^3}{\text{lb-mole}}\right)^2$ multiply values in table by 0.003776

To convert values of b to $\frac{\text{ft}^3}{\text{lb-mole}}$ multiply values in table by 0.0160

	a	b	t_c	p_c	d_c	v_c
Argon	1.35×10^6	32.3	−122	48	0.531	75.2
Acetylene	4.37×10^6	51.2	36	62	0.231	113
Air	1.33×10^6	36.6	−140.7	37.2	—	—
Ammonia	4.19×10^6	37.3	132.4	111.5	0.235	72.4
Carbon dioxide	3.60×10^6	42.8	31.1	73.0	0.460	95.5
Carbon monoxide	1.46×10^6	39.4	−139	35	0.311	90.0
Chlorine	6.50×10^6	56.2	144	76.1	0.573	124
Ethylene	4.48×10^6	57.2	9.7	50.9	0.22	127
Hydrogen chloride	3.65×10^6	40.8	51.4	81.6	0.42	87
Hydrogen	0.245×10^6	26.6	−239.9	12.8	0.0310	64.5
Methane	2.25×10^6	42.8	−82.5	45.8	0.162	99
Methyl chloride	7.50×10^6	65.1	143.1	65.8	0.37	137
Nitrogen	1.347×10^6	38.6	−147.1	33.5	0.3110	90
Oxygen	1.36×10^6	31.9	−118.8	49.7	0.430	74
Sulfur dioxide	6.80×10^6	57.2	157.2	77.7	0.52	123
Water	5.48×10^6	30.6	374	217.7	0.4	45

Another factor, neglected in the simple kinetic treatment of a gas, is the attraction existing between molecules, known as van der Waals' forces.

These forces tend to draw the molecules together and diminish the pressure exerted below the value corresponding to ideal behavior. It may be demonstrated from kinetic theory that this reduction is inversely proportional to the square of the molal volume. Thus

$$p = p' - \frac{a}{v^2} \quad \text{or} \quad p' = p + \frac{a}{v^2}$$

where p' = pressure calculated from the simple kinetic theory
p = actual pressure

The term p' is the *internal pressure* of the gas. The corrected equation of state then becomes

$$\left(p + \frac{a}{v^2}\right)(v - b) = \tfrac{1}{3}Nmu^2 = RT \quad (1)$$

This is the equation of van der Waals, which represents the general form of the pressure–volume relationships of a gas, even when compressed to the region of liquefaction. Although this equation is an improvement over the ideal-gas law, its numerical results represent only a fair approximation where molal volumes are small. The factors a and b, characteristic of each gas, are termed the van der Waals constants. In the first two columns of Table XXV are listed values of a and b corresponding to pressure in atmospheres, volumes per gram-mole in cubic centimeters, and temperatures in degrees Kelvin.

Beattie–Bridgeman Equation of State. One of the most widely used equations of state is that of Beattie and Bridgeman.[1,2]

$$pv^2 = RT\left[v + B_0\left(1 - \frac{b}{v}\right)\right]\left[1 - \frac{c}{vT^3}\right] - A_0\left(1 - \frac{a}{v}\right) \quad (2)$$

where a, b, A_0, B_0, c = empirically determined constants
v = molal volume, liters per gram-mole
p = pressure, atmospheres
T = temperature, degrees Kelvin
R = gas-law constant = 0.08206

This equation contains five constants which must be determined for each particular system. The methods for evaluating the constants are discussed by Beattie and Bridgeman in the papers cited and also by Deming and Shupe.[3]

In Table XXVI are values of the constants of Equation (2) for several common gases. In the ranges of temperature and molal volume indicated in Table XXVI, the equation was found to yield an average deviation of only 0.18 per cent from the accepted experimental values.

Equation (2) and the constants of Table XXVI may be used to calculate accurately pressure, volume, temperature relationships over wide ranges of conditions. Such an equation is of value for the extrapolation of limited experimental data and for highly precise interpolation between observed values. It is also valuable in thermodynamic calculations

[1] *J. Am. Chem. Soc.*, **49**, 1665 (1927).
[2] *J. Am. Chem. Soc.*, **50**, 3133 (1928).
[3] *J. Am. Chem. Soc.*, **52**, 1382 (1930); **53**, 843, 860 (1931).

TABLE XXVI

BEATTIE–BRIDGEMAN CONSTANTS

Gas	A_0	*a*	B_0	*b*	*c*	*Temperature Range,* °C	*Minimum v* cc/-g mole
He	0.0216	0.05984	0.01400	0.0	0.004×10^4	400 to −252	100
H_2	0.1975	−0.00506	0.02096	−0.04359	0.0504×10^4	200 to −244	100
N_2	1.3445	0.02617	0.05046	−0.00691	4.20×10^4	400 to −149	180
O_2	1.4911	0.02562	0.04624	0.004208	4.80×10^4	100 to −117	110
Air	1.3012	0.01931	0.04611	−0.01101	4.34×10^4	200 to −145	125
CO_2	5.0065	0.07132	0.10476	0.07235	66.00×10^4	100 to 0	180
CH_4	2.2769	0.01855	0.05587	−0.01587	12.83×10^4	200 to 0	166

because it expresses compressibility data in a rigorous equation which may be incorporated in any desired mathematical operations. Its use in compressibility calculations is cumbersome, but once the constants are evaluated for a particular gas a complete set of compressibility factors may be calculated and used in the manner described in the following sections.

Benedict–Webb–Rubin Equation of State. An empirical equation of state with eight constants has been formulated by Benedict, Webb, and Rubin[4] for the lighter hydrocarbons from the experimental data for methane, ethane, propane, and *n*-butane; thus, if *d* is density, *g*-moles/liter,

$$p = RTd + \left(B_0RT - A_0 - \frac{C_0}{T^2}\right)d^2 + (bRT - a)d^3 + a\alpha d^6 + cd^3\,\frac{(1 + \gamma d^2)}{T^2}\,e^{-\gamma d^2} \tag{2a}$$

TABLE XXVII

CONSTANTS OF THE BENEDICT–WEBB–RUBIN EQUATION OF STATE

Units: Atmospheres, Liters, Gram-Moles, Degrees Kelvin; $R = 0.08207$.

	Methane	*Ethane*	*Propane*	*n-Butane*
B_0	0.0426000	0.0627724	0.0973130	0.124361
A_0	1.85500	4.15556	6.87225	10.0847
$C_0\ 10^{-6}$	0.0225700	0.179592	0.508256	0.992830
b	0.00338004	0.0111220	0.0225000	0.0399983
a	0.0494000	0.345160	0.947700	1.88231
$c\ 10^{-6}$	0.00254500	0.0327670	0.129000	0.316400
γ	0.0060000	0.0118000	0.0220000	0.0340000
α	0.000124359	0.000243389	0.000607175	0.00110132

[4] M. Benedict, G. W. Webb, and L. C. Rubin, *J. Chem. Phys.*, **8**, 334–45 (1940).

Values of the constants for these four gases are recorded in Table XXVII. This equation holds within 0.34 per cent, even at gas densities twice the critical value.

COMPRESSIBILITY FACTOR

The equation of state may be written

$$pV = znRT \tag{3}$$

where z is termed the compressibility factor and is a function of pressure, temperature, and the nature of the gas.

If values of the compressibility factor of a gas are known, all calculations involving its pVT relationships may be carried out by simple proportionalities derived from Equation (3). Thus, applying Equation (3) to a given mass of gas at two different conditions and combining, gives

$$\frac{p_1V_1}{p_2V_2} = \frac{z_1T_1}{z_2T_2} \tag{4}$$

The correct normal molal volume at 0°C and 1 atm is equal to $359z_s$ cu ft per lb-mole or $22.41z_s$ liters per g-mole, where z_s is the compressibility factor at standard conditions. With a knowledge of compressibility factors it is thus possible to extend the entire system of calculation which was used with the ideal-gas law to apply at any desired conditions of temperature or pressure. The ideal-gas law may be considered as representing a special case in which the compressibility factor is equal to unity.

The nature of the variation of the compressibility factor z with pressure, molal volume, and temperature is shown by the curves of Figs. 100, 101, 102 for nitrogen. In the range covered by these three charts, the compressibility factor of nitrogen may be found at any specified conditions by interpolating on the proper chart. If the pressure and temperature are specified, the compressibility factor is read from Fig. 100. If the molal volume and temperature or the pressure and molal volume are specified, Fig. 101 or Fig. 102 is used.

The use of compressibility-factor data is demonstrated by the following illustrations, dealing with nitrogen.

Illustration 1. One cubic foot of nitrogen at 50°C and 30 atm is compressed to 60 atm and cooled to −50°C. Calculate the final volume.

The compressibility factors at the initial and final conditions may be obtained from Fig. 100.

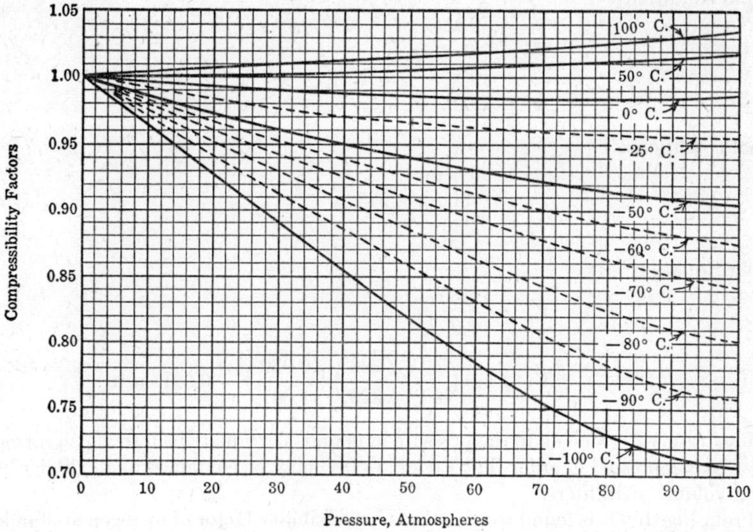

FIG. 100. Compressibility Factors of Nitrogen at Specified Pressures and Temperatures.

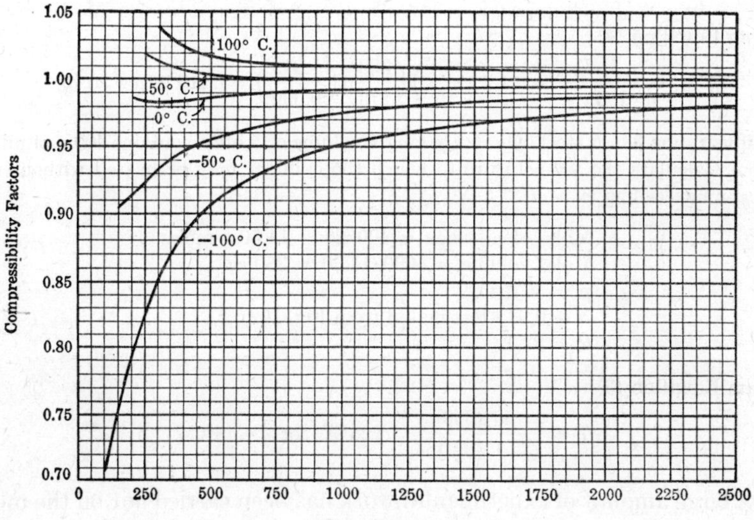

FIG. 101. Compressibility Factors of Nitrogen at Specified Molal Volumes and Temperatures.

Initial conditions:

$$p_1 = 30 \text{ atm}$$
$$V_1 = 1.0 \text{ cu ft}$$
$$T_1 = 50°\text{C} = 323°\text{K}$$
$$z_1 = 1.001$$

Final conditions:

$$p_2 = 60 \text{ atm}$$
$$T_2 = -50°\text{C} = 223°\text{K}$$
$$z_2 = 0.930$$

From Equation (4):

$$V_2 = \frac{V_1 p_1 T_2 z_2}{p_2 T_1 z_1}$$

$$V_2 = 1.0\left(\frac{30}{60}\right)\left(\frac{223}{323}\right)\left(\frac{0.930}{1.001}\right) = 0.321 \text{ cu ft}$$

Illustration 2. Calculate the pressure in pounds per square inch to which nitrogen must be compressed in order that 1.0 kg at 50°C may be contained in a cylinder having a volume of 10 liters.

From Fig. 100 it is found that z_s, the compressibility factor of nitrogen at standard conditions, is equal to 0.999. The volume of 1.0 kg or 35.7 g-moles of nitrogen at standard conditions is then (0.999)(35.7)(22.41) or 801 liters.

Final conditions:

$$v = \text{molal volume} = 10/35.7 = 0.28 \text{ liter}$$
$$T = 323°\text{K}$$
$$z = 1.015 \text{ (from Fig. 101)}$$

From Equation (4):

$$p_2 = \frac{p_1 V_1 T_2 z_2}{V_2 T_1 z_1} = 14.7\left(\frac{801}{10}\right)\left(\frac{323}{273}\right)\left(\frac{1.015}{0.999}\right) = 1415 \text{ lb per sq in.}$$

Illustration 3. A steel cylinder having a volume of 5 liters contains 400 g of nitrogen. Calculate the temperature to which the cylinder may be heated without the pressure exceeding 50 atm.

$$\text{Moles of nitrogen} = 400/28 = 14.30 \text{ g-moles}$$
$$\text{Molal volume} = 5000/14.30 = 350 \text{ cc}$$
$$p = 50 \text{ atm}$$
$$z = 0.945 \text{ (estimated from Fig. 102)}$$
$$R = 82.1$$

From Equation (3):

$$T = \frac{(50)(350)}{(82.1)(0.945)} = 225°\text{K} \quad \text{or} \quad -48°\text{C}$$

A large amount of experimental work has been carried out on the more common and industrially important gases. In the International Critical Tables are experimental data for many gases and extensive references to the literature.

Experimental-compressibility data are commonly presented in tables or curves showing values of the product pV at various values of p for constant values of T where the product is taken as 1.0 at 1 atm pressure and 0°C. These data are arranged to show the isothermal variation of pV with pressure. If z_s is the compressibility factor of the gas at 0°C

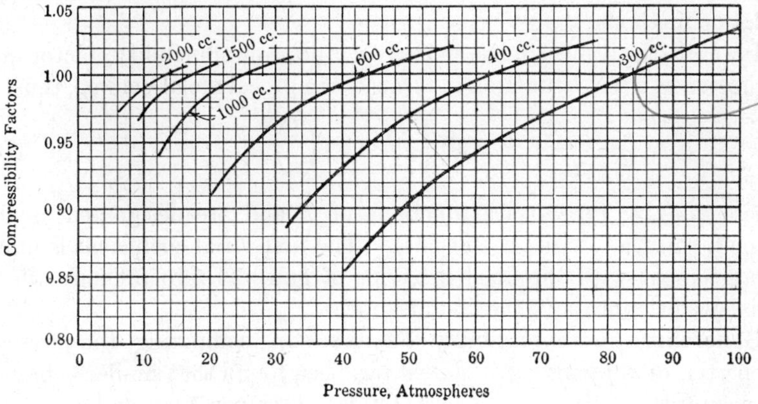

FIG. 102. Compressibility Factors of Nitrogen at Specified Pressures and Molal Volumes.

and 1 atm pressure, then the compressibility factor of the gas for any other condition may be obtained from the following:

$$z = z_s(pV)\frac{273}{T} \tag{5}$$

The value of z_s is nearly unity for the common gases. It may be obtained accurately from density measurements made at standard conditions from the relation,

$$z_s = \frac{M}{\rho_s 22.41} \tag{6}$$

where M is the molecular weight and ρ_s is the density in grams per liter at 0°C and 1 atm pressure.

The compressibility factor of all gases is unity when the pressure is zero. Thus, values of z_s may also be obtained by plotting the values of pV against p at 0°C and extrapolating the graph to $p = 0$. Then

$$z_s = \frac{p_s V_s}{p_0 V_0} \tag{7}$$

where p_0V_0 is the extrapolated value of pV at $p = 0$.

Experimental-compressibility measurements are also frequently expressed in the form of isometric data showing the variation of pressure with change in temperature of a known weight of gas confined in a constant volume. Such isometrics are determined at various selected volumes. These data yield curves of the type of Fig. 102 directly. Curves of the types of Figs. 100 and 101 may then be derived by interpolation methods.

In statistical mechanics values of the compressibility factor are expressed as a series function in terms of reciprocal molal volume, thus,

$$z = 1 + \beta(T)\frac{1}{v} + \gamma(T)\left(\frac{1}{v}\right)^2 + \cdots \tag{7a}$$

where $\beta(T)$, $\gamma(T)$, $\cdots$ are functions of T and are designated as the second, third, $\cdots$ virial coefficients. The first virial coefficient is unity in agreement with the ideal behavior of gases at zero pressure where virial coefficients above the first disappear.

Generalized Compressibility Factors. In comparing the physical properties of different substances it has been found that similar behavior is encountered at the same values of reduced temperatures and pressures. The reduced properties are related to the critical properties of a substance as defined on page 54. Thus,

$$T_r = \frac{T}{T_c}, \quad p_r = \frac{p}{p_c} \quad \text{and} \quad v_r = \frac{v}{v_c} \tag{8}$$

where the subscript r indicates the reduced and c the critical properties. This similarity in the behavior of a substance at equal values of reduced temperature and pressure is referred to as the theory of corresponding states and forms the basis of the generalized procedure described in the following pages for establishing the deviations of thermodynamic properties from ideal behavior. The van der Waals and other equations of state may be shown to be consistent with and to support this theory.

If the theory of corresponding states were rigorous, a single equation of state would suffice for all gases if expressed in terms of reduced properties instead of in absolute values. In Fig. 103 are plotted values of compressibility factors as a function of reduced temperature and reduced pressure. This chart was derived as an average of data reported in the literature for carbon dioxide, nitrogen, hydrogen, ammonia, methane, propane, and pentane, and is not in rigorous agreement with all the data on any one of these substances. However, these discrepancies, which result from the fact that the theory of corresponding states is only an approximation, are small enough so that for many purposes the chart may be taken as applicable to all pure gases. It then constitutes a con-

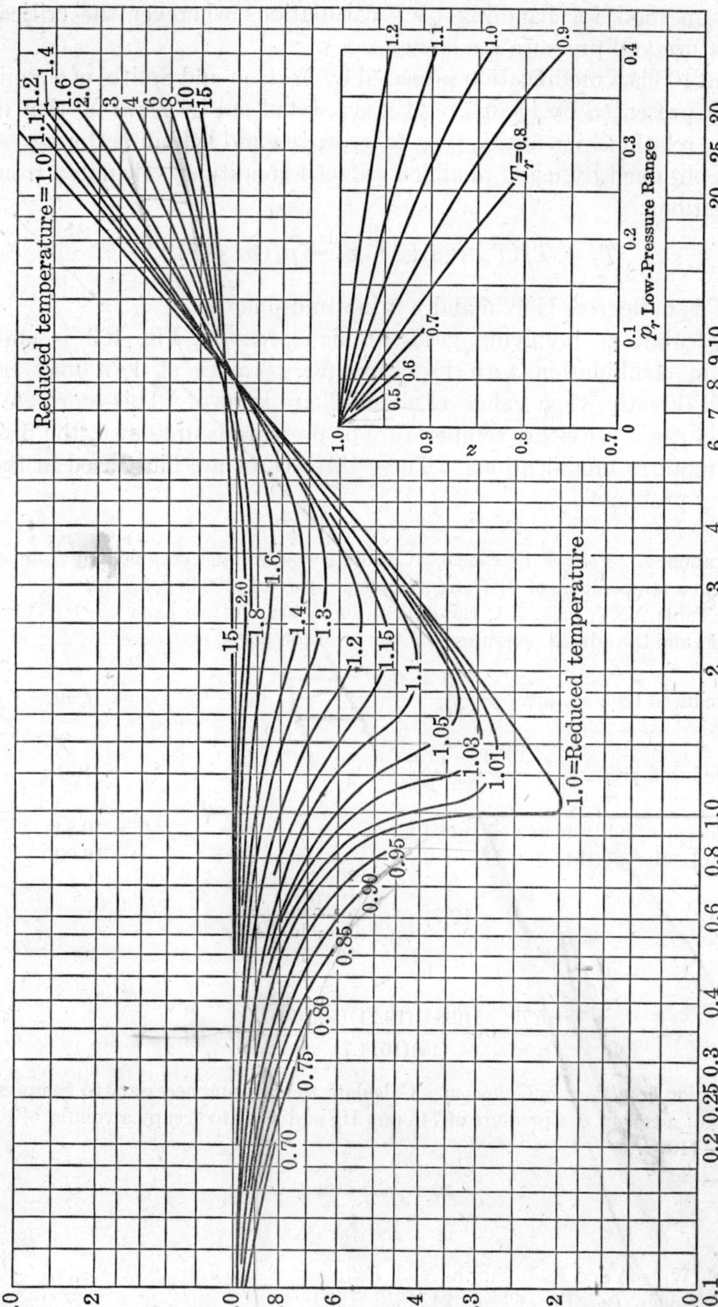

FIG. 103. Compressibility Factors of Gases and Vapors.

(Reproduced in "C.P.P. Charts")

venient method for handling pVT calculations wherever the critical temperature and pressure are known.

Figure 103 is a modification prepared by Watson and Smith[5] of a similar chart presented by Dodge.[6] It was pointed out by Newton[7] that in applying relationships of this type to hydrogen and helium better agreement is obtained by using modified reduced conditions calculated from the equations:

$$T_r = T/(T_c + 8); \qquad p_r = p/(p_c + 8) \tag{9}$$

where T is in degrees Kelvin and p is in atmospheres.

In calculations involving gases at high pressure, Fig. 103 is employed in combination with Equation (3). Values of V/n may be obtained directly when values of p and T are known. This represents the usual case. If either temperature or pressure is unknown, the procedure requires modifications. These three cases are illustrated in the following problems.

Illustration 4. *Volume Unknown.* Calculate the volume occupied by 1 lb of methane at a temperature of 40°F and a gauge pressure of 1000 lb per sq in.

From Table XXV, page 481, the critical temperature of methane is −82.5°C or −116.5°F, and the critical pressures 45.8 atm or 673 lb per sq in.

$$\text{Reduced temperature} = \frac{40 + 460}{-116.5 + 460} = 1.40$$

$$\text{Reduced pressure} = \frac{1000 + 14.7}{673} = 1.51$$

Compressibility factor (Figure 103) = 0.84
Molecular weight = 16.00

$$R = 10.71 \frac{(\text{lb})(\text{ft})^3}{(\text{in.})^2(\text{lb-mole})°(\text{R})}$$

hence

$$V = \frac{nzRT}{p} = \frac{1(0.84)(10.71)(500)}{(16)(1014.7)} = 0.278 \text{ cu ft}$$

Illustration 5. *Pressure Unknown.* Calculate the pressure necessary to compress 100 liters of nitrogen at a pressure of 745 mm Hg and 23°C to occupy a volume of 1.0 liter at −110°C.

$$p_c = 33.5 \text{ atm}$$

$$T_c = 126°\text{K}$$

[5] K. M. Watson and R. L. Smith, *Natl. Petroleum News*, July 1936.
[6] B. F. Dodge, *Ind. Eng. Chem.*, **24**, 1353 (1932).
[7] R. H. Newton, *Ind. Eng. Chem.*, **27**, 302 (1935).

$$\text{Initial } p_r = \frac{0.982}{33.5} = 0.0293$$

$$\text{Initial } T_r = \frac{296}{126} = 2.35$$

$$\text{Initial } z = 1.00$$

$$\text{Final } T_r = \frac{163}{126} = 1.29$$

$$\text{Final } p_r = \frac{p_2}{33.5}$$

From Equation (4):

$$\frac{p_1V_1}{p_2V_2} = \frac{z_1T_1}{z_2T_2}$$

Substituting the known values gives

$$\frac{(0.982)(100)}{p_2(1)} = \frac{1.00}{z_2}\left(\frac{296}{163}\right)$$

$$z_2 = 0.0185p_2 = (0.0185)(33.5)p_{r_2} = 0.621p_{r_2}$$

The straight line corresponding to the equation, $z_2 = 0.621p_r$, is extended across the compressibility chart, Fig. 103. Since Fig. 103 is plotted on double logarithmic scales, the line must have a positive slope of 45° and pass through the point $z = 0.621$ where $p_r = 1.0$. Where the 45° line intersects $T_r = 1.29$, the value of z_2 is found to be 0.80, and the value of $p_r = 1.30$. Hence

$$p = (1.30)(33.5) = 43.6 \text{ atm}$$

Illustration 6. *Temperature Unknown.* The volume occupied by 1 lb of *n*-octane at 27 atm is 0.20 cu ft. Calculate the temperature.

$$t_c = 565°\text{F } (1025°\text{R}) \qquad p_c = 24.6 \text{ atm}$$

$$p_r = \frac{27}{24.6} = 1.10 \qquad T_r = \frac{T}{1025}$$

$$R = \frac{0.730 \text{ (atm)(cu ft)}}{\text{(lb-mole)(°R)}}; \qquad \text{molecular weight} = 114$$

From Equation (3):

$$(27)(0.20) = \frac{z(0.730T)}{114}$$

or

$$z = \frac{845}{T} = \frac{845}{1025T_r} = \frac{0.825}{T_r}$$

To solve the problem graphically z is plotted in Fig. 104 against T_r for a value of $p_r = 1.10$ from Fig. 103. On the same scale is plotted the equation $z = \dfrac{0.825}{T_r}$. Where these two curves intersect, $z = 0.72$ and $T_r = 1.15$.

Hence $T = 1025T_r = (1025)(1.15) = 1180°\text{R}$ or 720°F.

The generalized compressibility-factor correlation offers the most convenient method available for estimating pVT relationships whenever critical-temperature and critical-pressure values are available. Several remarkable properties in the behavior of gases are shown by Fig. 103. The compressibility factor is less than unity for all values of T_r below 2 at reduced pressures below 8. In this range actual gases are more compressible than ideal. At values of p_r above 8 the compressibility factor of all gases exceeds unity regardless of temperature. In this range all

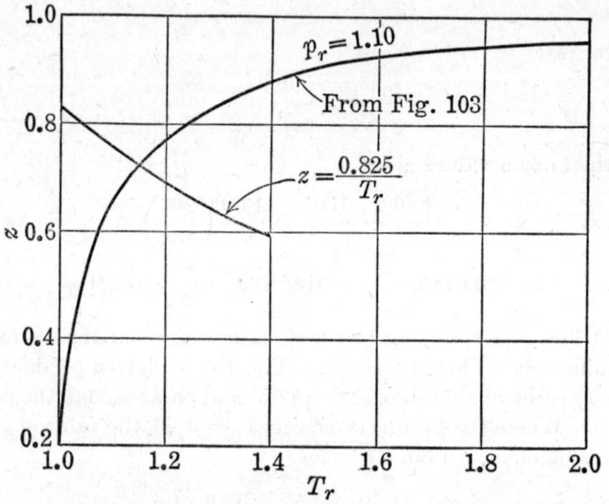

FIG. 104. Graphical Solution for T_r.

gases are less compressible than the ideal. At a value of $p_r = 8$, at all values of T_r the compressibility factor is approximately unity. The greatest deviation from ideal behavior occurs near the critical state where the gas becomes almost five times as compressible as in the ideal state.

Care should be exercised to avoid use of gaseous compressibility factors at pressures higher than the vapor pressure at reduced temperatures less than 1.0. Such conditions produce liquefaction to which Fig. 103 is not applicable.

Based upon the generalized chart for compressibility factors corresponding charts have been constructed for deviation from ideal behavior of other thermodynamic properties such as enthalpy, entropy, heat capacity, and the Joule–Thomson effect.

Effect of Pressure on Enthalpy of Gases. It has been shown that the enthalpy of an ideal gas is independent of pressure and a function only of temperature. However, at elevated pressures all actual gases deviate from ideal behavior and enthalpy changes with change in pressure, particularly in the region of the critical point.

The enthalpy of a real gas relative to the enthalpy of an ideal gas when both are at the same temperature may be obtained by integration of Equation (y) of Table XXIV, between the limits of the existing pressure and zero pressure where all gases behave ideally and the enthalpy becomes independent of pressure. Thus,

$$(H^* - H)_T = \int_0^p \left[T\left(\frac{\partial V}{\partial T}\right)_p - V \right]_T dp \tag{10}$$

Where the equation of state is known, the values of V and $(\partial V/\partial T)_p$ may be expressed as functions of pressure and the integration of (10) may be carried out analytically or graphically.

The value of $(H^* - H)_T$ may also be expressed in terms of the compressibility factors as a variable and thus obtained from these factors for a specific material or from an approximate, generalized chart such as Fig. 103. Differentiation of Equation (3) gives

$$\left(\frac{\partial v}{\partial T}\right)_p = \frac{R}{p}\left[z + T\left(\frac{\partial z}{\partial T}\right)_p \right] \tag{11}$$

Combining Equations (y) of Table XXIV and (11) results in

$$\left(\frac{\partial \text{H}}{\partial p}\right)_T = -\frac{RT^2}{p}\left(\frac{\partial z}{\partial T}\right)_p \tag{12}$$

If experimental-compressibility data are available, values of $(\partial z/\partial T)_p$ may be calculated and Equation (12) integrated graphically.

Expressed in reduced properties,

$$\left(\frac{\partial \text{H}}{\partial p_r}\right)_T \frac{1}{p_c} = -\frac{RT_r^2 T_c^2}{p_r p_c T_c}\left(\frac{\partial z}{\partial T_r}\right)_p \tag{13}$$

Integration, at constant temperature, gives

$$\left(\frac{\text{H}^* - \text{H}}{T_c}\right)_T = RT_r^2 \int_0^{p_r} \left(\frac{\partial z}{\partial T_r}\right)_p d \ln p_r \tag{14}$$

The derivative $(\partial z/\partial T_r)_p$ is obtained by graphical differentiation of a cross-plot of Fig. 103, on regular coordinates, and Equation (14) is then integrated over the pressure range required at various constant temperatures. Values of $(\text{H}^* - \text{H})/T_c$ are plotted in Figs. 105 and 106 in terms of reduced conditions. These charts permit corrections for enthalpy to be made readily and with usually sufficient accuracy. An enthalpy-correction chart similar to Fig. 106 was first developed by Watson and Nelson.[8] Figures 105 and 106 have been corrected to repre-

[8] K. M. Watson and E. F. Nelson, *Ind. Eng. Chem.*, **25**, 880 (1933).

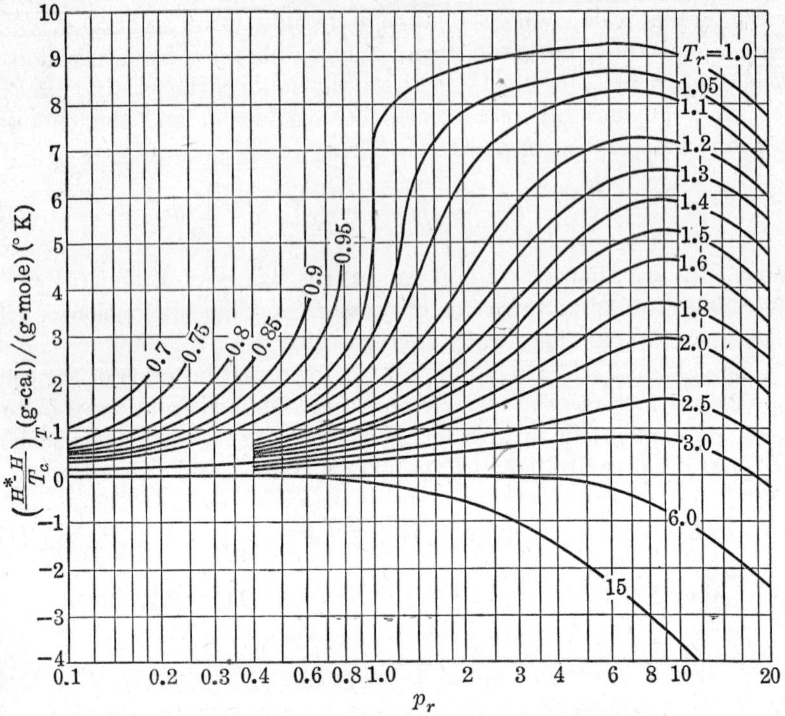

FIG. 105. Enthalpy Correction for Gases (High Range)

sent an average of the data more recently calculated by Edmister[9] and York and Weber.[10] Data for the Joule–Thomson coefficient of water vapor were used to extend the low-temperature ranges of the charts.

York and Weber[10] found that in the temperature range above $T_r = 1$ the enthalpy corrections vary systematically with the critical temperature of the substance when correlated as in Figs. 105 and 106. Accordingly they proposed that the values read from such charts should be multiplied by a correction factor ϕ defined as follows for use with Figs. 105 and 106, based upon pentane, $T_c = 470$°K:

$$\phi = \left(\frac{T_c}{470}\right)^n \tag{15}$$

where T_c is the critical temperature of the substance in degrees Kelvin, and n is a function of reduced temperature as follows:

T_r	1.0	1.05	1.1	1.2	1.3	1.4	1.5	1.6
n	0.37	0.28	0.25	0.20	0.18	0.16	0.15	0.14

[9] W. C. Edmister, *Ind. Eng. Chem.*, **30**, 352 (1938).
[10] R. York and H. C. Weber, *Ind. Eng. Chem.*, **32**, 388 (1940).

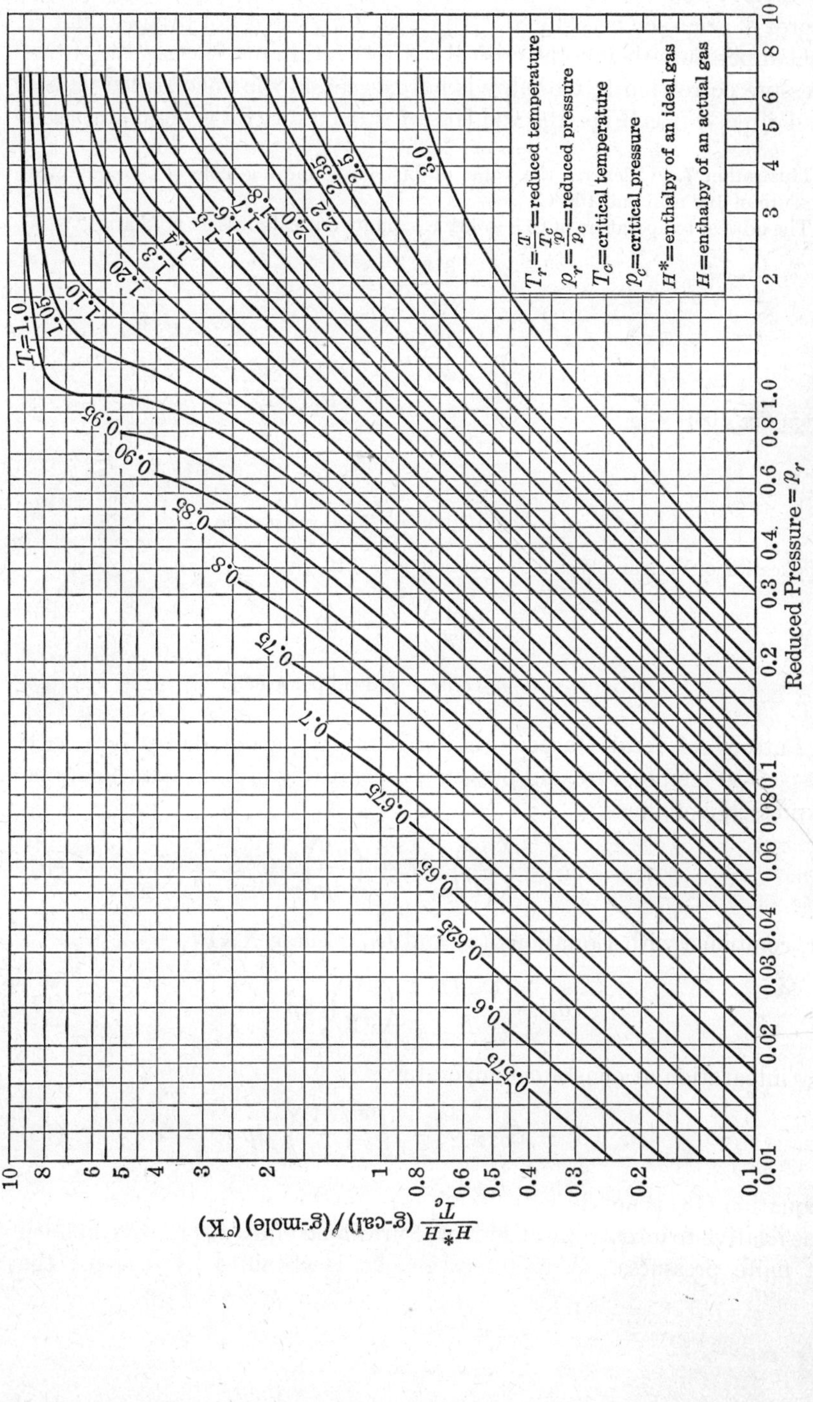

Fig. 106. Enthalpy Correction for Gases (Low Range).

(Reproduced in "C.P.P. Charts")

Use of the correction factor ϕ results in excellent correlation of data for several hydrocarbons and steam, and it should be used where a high degree of accuracy is required. Unfortunately ϕ has not been evaluated for temperatures below the critical. However, in many cases the entire pressure correction to enthalpy is not large, and the uncorrected values read directly from Figs. 105 and 106 are satisfactory for general purposes.

Illustration 7. Calculate the value of $(\text{H}^* - \text{H})_T$ for 1 g-mole of CO_2 gas at a pressure of 100 atm and 100°C.

The critical temperature of CO_2 is 304°K, and its critical pressure is 73 atm.

$$T_r = \frac{373}{304} = 1.23$$

$$p_r = \frac{100}{73} = 1.37$$

From Fig. 106:

$$\frac{\text{H}^* - \text{H}}{T_c} = 2.55$$

$$\text{H}^* - \text{H} = (2.55)(304) = 775 \text{ cal per g-mole}$$

Application of York and Weber's correction, $n = 0.19$, gives

$$\phi = \left(\frac{304}{470}\right)^{0.19} = 0.92$$

$$\text{H}^* - \text{H} = (0.92)(775) = 713 \text{ cal per g-mole}$$

Entropy of Real Gases. In terms of its partial derivatives with respect to temperature and pressure, the entropy of a substance may be expressed as follows:

$$dS = \left(\frac{\partial S}{\partial T}\right)_p dT + \left(\frac{\partial S}{\partial p}\right)_T dp \tag{16}$$

or, combined with Equations (n) and (m), Table XXIV,

$$dS = \frac{C_p\, dT}{T} - \left(\frac{\partial V}{\partial T}\right)_p dp \tag{17}$$

By integration at constant temperature,

$$(S_p - S_0^*)_T = -\int_0^p \left(\frac{\partial V}{\partial T}\right)_p dp \tag{18}$$

Equation (18) is not directly useful, because it expresses the entropy of a gas relative to its entropy at zero pressure and is equal to negative infinity at finite pressures. A useful expression is obtained by deriving the entropy correction due to deviation from ideal behavior at a given pressure. For an ideal gas,

$$(S_p^* - S_0^*)_T = -\int_0^p \left(\frac{\partial V}{\partial T}\right)_p^* dp \tag{19}$$

The desired entropy correction is obtained by subtracting (18) from (19):

$$(S_p^* - S_p)_T = \int_p^0 \left[\left(\frac{\partial V}{\partial T}\right)_p^* - \left(\frac{\partial V}{\partial T}\right)_p\right] dp \tag{20}$$

From the ideal-gas law:

$$\left(\frac{\partial v}{\partial T}\right)_p^* = \frac{R}{p} \tag{21}$$

Hence, for 1 mole:

$$(\text{s}_p^* - \text{s}_p)_T = \int_p^0 \left[\frac{R}{p} - \left(\frac{\partial v}{\partial T}\right)_p\right] dp \tag{22}$$

Equation (22) may be integrated if $(\partial v/\partial T)_p$ is evaluated from an equation or from Fig. 103. Thus, from Equation (3),

$$\left(\frac{\partial v}{\partial T}\right)_p = \frac{zR}{p} + \frac{RT}{p}\left(\frac{\partial z}{\partial T}\right)_p \tag{23}$$

Substitution of (23) in (22) gives

$$(\text{s}_p^* - \text{s}_p)_T = R\int_p^0 \left[\frac{(1-z)}{p} - \frac{T}{p}\left(\frac{\partial z}{\partial T}\right)_p\right] dp \tag{24}$$

Writing (24) in terms of reduced properties gives

$$(\text{s}_p^* - \text{s}_p)_T = R\int_{p_r}^0 (1-z)\, d\ln p_r + RT_r \int_0^{p_r} \left(\frac{\partial z}{\partial T_r}\right)_p d\ln p_r \tag{25}$$

Combining (14) and (25) gives

$$(\text{s}_p^* - \text{s}_p)_T = R\int_{p_r}^0 (1-z)\, d\ln p_r + \frac{\text{H}^* - \text{H}}{T_r T_c} \tag{26}$$

Equation (26) may be evaluated by graphical integration of a plot of $(1 - z)$ against $\ln p_r$ derived from Fig. 103. Figure 107 was plotted from values determined in this manner and added to the corresponding values of $(\text{H}^* - \text{H})/T_c T_r$ obtained from Fig. 106.

Because of the large effect of pressure on entropy and the complex nature of the deviations from ideal behavior, it is customary to tabulate

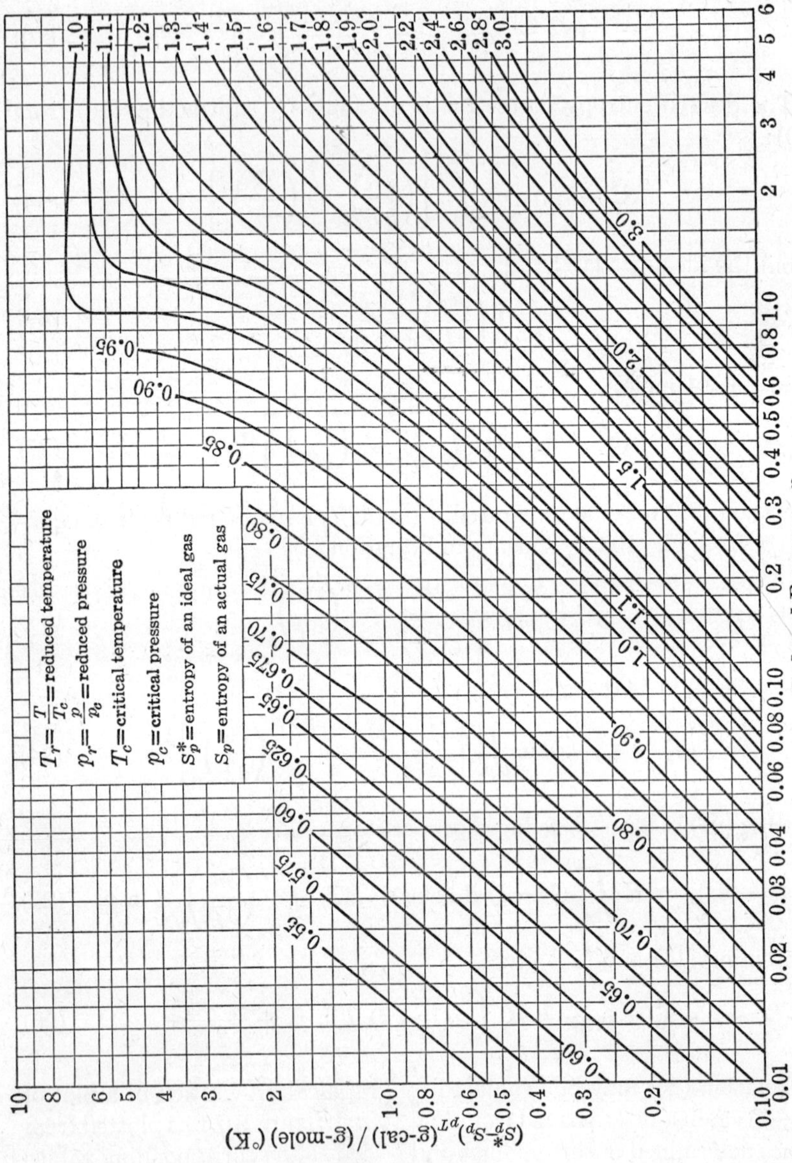

FIG. 107. Entropy Correction for Gases.

(Reproduced in "C.P.P. Charts")

absolute entropy data for gases at a pressure of 1 atm in a hypothetical ideal state. These values express the entropy $s^*_{p=1.0}$ which the gas would have if it behaved ideally at pressures up to 1 atm. The entropy of the actual gas at 1 atm pressure must be obtained from Equation (26).

Illustration 8. The absolute entropy of carbon dioxide in the hypothetical ideal state at a pressure of 1 atm and 25°C is 51.08 entropy units per mole. Calculate the molal entropy at a temperature of 100°C and a pressure of 100 atm. The entropy at 100°C in the ideal state at 1 atm is first calculated from heat-capacity data. The entropy at 100°C and 100 atm, if ideal behavior is assumed, is then determined, and from this value is subtracted the correction for deviation from ideality read from Fig. 107.

From Table V, page 214, the molal heat capacity at 1 atm is

$$c_p^* = 6.85 + (8.533)(10^{-3})T - (2.475)(10^{-6})T^2$$

By substitution in Equation (17) and integration at constant pressure,

$$s^*_{T_2} - s^*_{T_1} = 6.85 \ln \frac{T_2}{T_1} + (8.533)(10^{-3})(T_2 - T_1) - 1.237(10^{-6})(T_2^2 - T_1^2) = 2.11$$

$$s^*_{(100°\text{C}, 1 \text{ atm})} = 51.08 + 2.11 = 53.19$$

In cases such as this where relatively small temperature changes are involved, a satisfactory solution may be obtained by assuming the heat capacity to be its value at the average temperature. From Equation (XI–142) at constant temperature,

$$s^*_{p_2} - s^*_{p_1} = -R \ln \frac{p_2}{p_1}$$

or

$$s^*_{(100°\text{C}, 100 \text{ atm})} = 53.19 - 1.99 \ln 100 = 44.00$$

From Fig. 107, at values of $T_r = 1.23$ and $p_r = 1.37$,

$$s^* - s = 1.50$$

hence,

$s = 44.00 - 1.50 = 42.50$, the entropy of the actual gas at 100°C and 100 atm.

Effect of Pressure upon C_p. The heat capacity at constant pressure of any substance is represented by the following differential equation:

$$dC_p = \left(\frac{\partial C_p}{\partial T}\right)_p dT + \left(\frac{\partial C_p}{\partial p}\right)_T dp \tag{27}$$

From Equation (u), Table XXIV:

$$\left(\frac{\partial C_p}{\partial p}\right)_T = -T\left(\frac{\partial^2 V}{\partial T^2}\right)_p$$

and

$$dC_p = \left(\frac{\partial C_p}{\partial T}\right)_p dT - T\left(\frac{\partial^2 V}{\partial T^2}\right)_p dp \tag{28}$$

By integration at constant temperature

$$(C_p - C_{p0})_T = -\int_0^p T\left(\frac{\partial^2 V}{\partial T^2}\right)_p dp \tag{29}$$

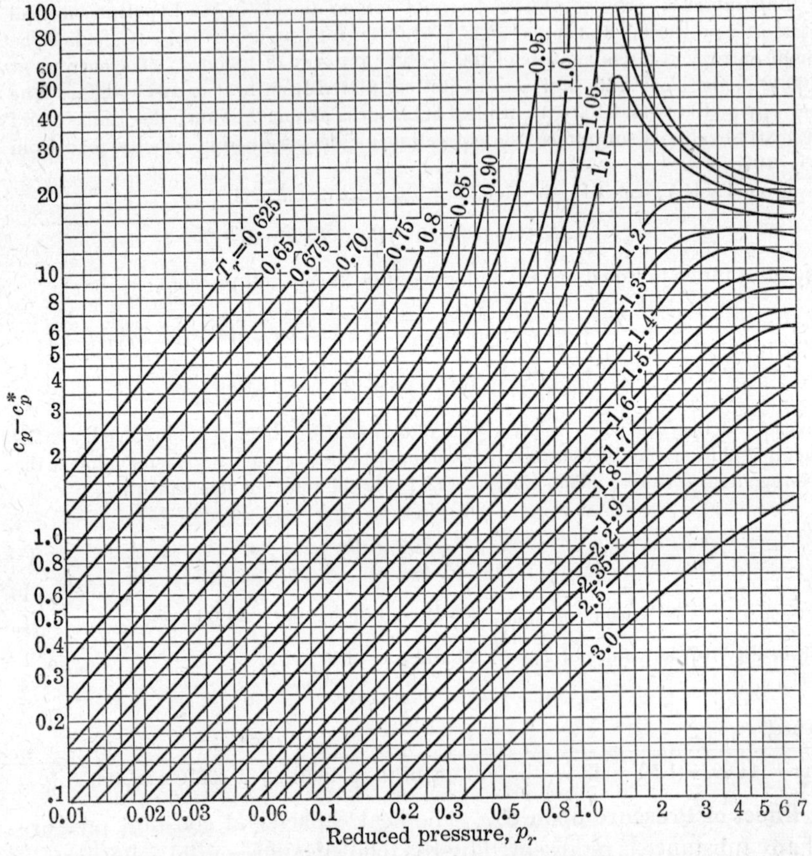

FIG. 108. Molal Heat Capacity Correction for Gases.

(Reproduced in "C.P.P. Charts")

Since for an ideal gas $(\partial^2 V/\partial T^2)_p = 0$, the heat capacity of any gas is independent of pressure at low pressures and $C_{p0} = C_p^*$. Thus, for the molal heat capacity of any gas,

$$(c_p - c_p^*)_T = -T\int_0^p \left(\frac{\partial^2 v}{\partial T^2}\right)_p dp = -\frac{T_r}{T_c}\, p_c \int_0^{p_r}\left(\frac{\partial^2 v}{\partial T_r^2}\right)_p dp_r \tag{30}$$

If an equation of state is known, values of $c_p - c_p^*$ can be calculated by integration of Equation (30). The pressure correction to heat capacity at constant pressure can also be expressed in terms of the compressibility factor. Writing Equation (3) in terms of reduced temperature and pressure and differentiating, gives

$$\frac{p_c}{T_c}\left(\frac{\partial^2 v}{\partial T_r}\right)_p = \frac{R}{p_r}\left[T_r\left(\frac{\partial^2 z}{\partial T_r^2}\right)_p + 2\left(\frac{\partial z}{\partial T_r}\right)_p\right] = \theta \tag{31}$$

Substituting (31) in (30) gives

$$(c_p - c_p^*)_T = -T_r \int_0^{p_r} \theta \, dp_r \tag{32}$$

The function θ can be evaluated in terms of reduced temperatures and pressures, by graphical differentiation of Fig. 103. Graphical integration of this function in Equation (32) then gives values of $(c_p - c_p^*)_T$ as a function of reduced conditions.

In Fig. 108 values of $(c_p - c_p^*)_T$ are plotted against reduced properties. This is a modification of a chart developed by Watson and Smith.[11]

Illustration 9. Calculate the heat capacity of CO_2 gas at 100°C and 100 atm pressure.

From Table IV, page 212, c_p^*, the heat capacity in the ideal state and at 100°C, is 9.7. From Fig. 108, at $T_r = 1.23$ and $p_r = 1.37$,

$$c_p - c_p^* = 8.5$$
$$c_p = 8.5 + 9.7 = 18.2$$

Joule–Thomson Effect. The differential temperature change resulting when a gas is expanded freely and adiabatically is known as the Joule–Thomson coefficient μ and is designated mathematically as

$$\mu = (\partial T/\partial p)_H \tag{33}$$

This coefficient can be expressed in terms of pVT relations by writing Equation (XI-98) for conditions of constant enthalpy. Thus,

$$C_p \, dT = \left[T\left(\frac{\partial V}{\partial T}\right)_p - V\right] dp \tag{34}$$

or

$$\left(\frac{\partial T}{\partial p}\right)_H = \frac{1}{C_p}\left[T\left(\frac{\partial V}{\partial T}\right)_p - V\right] = \mu \tag{35}$$

For an ideal gas $T(\partial V/\partial T)_p = V$, and from Equation (35) $\mu = 0$. From Equation (35) and Fig. 103 a relationship between μc_p and reduced temperature and pressure can be developed.

[11] K. M. Watson and R. L. Smith, *Natl. Petroleum News*, July 1936.

Integration of (35) applied to 1 mole gives

$$(T_1 - T_2)_H = \int_{p_2}^{p_1} \frac{1}{c_p} \left[T \left(\frac{\partial v}{\partial T} \right)_p - v \right] dp \tag{36}$$

Combining (36) and (10) gives

$$(T_1 - T_2)_H = \frac{(\text{H}^* - \text{H}_{p1})_{T(\text{avg})} - (\text{H}^* - \text{H}_{p2})_{T(\text{avg})}}{c_{p(\text{avg})}} \tag{37}$$

where $c_{p(\text{avg})}$ is the average molal heat capacity over the range of temperature and pressure involved. Equation (37) is useful only for moderate changes in temperature. For other cases a rigorous integration of Equation (36) is necessary. Problems involving estimation of such temperature changes are most conveniently solved from temperature–enthalpy charts.

THE LIQUID STATE

All the exact thermodynamic relations developed in the preceding sections for the gaseous state are equally applicable to the liquid state and may be used for the calculation of thermodynamic properties and their variations. Thus, from data relating the volume of a liquid to pressure and temperature together with heat capacity data, it is possible to calculate the effects of pressure and temperature on enthalpy, entropy, and heat capacity by the same methods used for the gaseous state.

At reduced temperatures below 0.8 the molal volumes of liquids are small and change little with pressure. As a result, the effects of pressure on the enthalpy, entropy, and heat capacity of liquids are small and frequently may be neglected. However, at high pressures, particularly at temperatures approaching the critical, such effects become large and must be carefully evaluated.

Generalized Liquid Densities. It was pointed out by Watson[12] that the compressibility and thermal expansion data of a variety of liquids which were investigated can be satisfactorily represented by the following relation:

$$\rho = \left(\frac{\rho_1}{\omega_1} \right) \omega \tag{38}$$

or

$$v = (v_1 \omega_1)/\omega \tag{39}$$

[12] K. M. Watson, *Ind. Eng. Chem.*, **35**, 398 (1943).

where ρ or v = density or molal volume, respectively, of liquid at p_r and T_r

ρ_1 or v_1 = density or molal volume, respectively, of liquid at p_{r_1} and T_{r_1}

ω = expansion factor of liquid at p_r and T_r

ω_1 = expansion factor of liquid at p_{r_1} and T_{r_1}

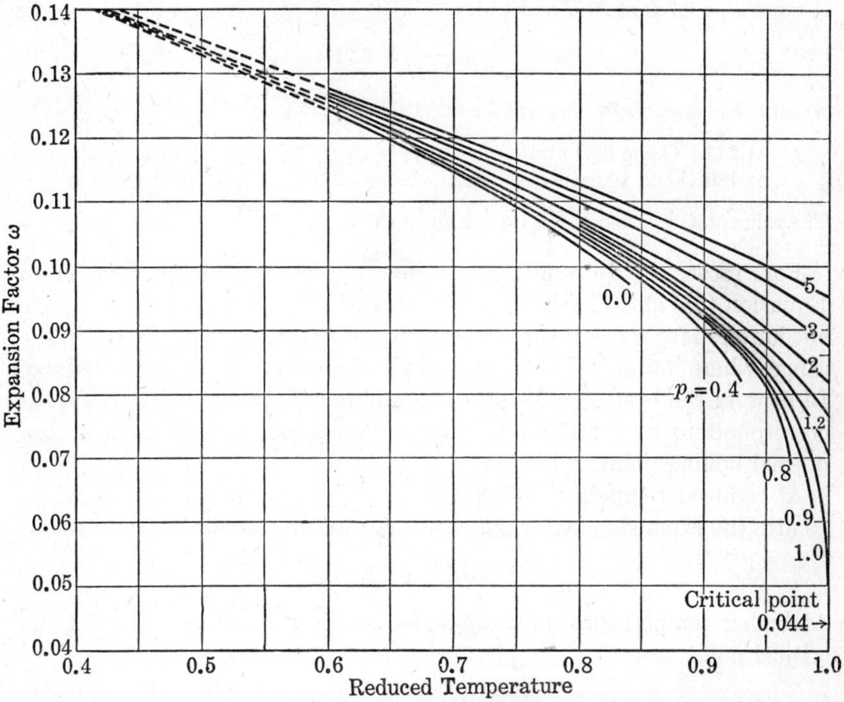

FIG. 109. Thermal Expansion and Compression of Liquids.
(Reproduced in " C.P.P. Charts ")

The factor ω is a dimensionless quantity termed the *expansion factor* and when expressed as a function of reduced temperature and pressure is approximately the same for all liquids. Values of ω are plotted in Fig. 109.

The terms (ρ_1/ω_1) or $(v_1\omega_1)$ are characteristic constants for any one liquid which are established by a single density measurement and the corresponding value of ω_1. Thus, from one known value of density together with the ω chart, Fig. 109, the density of that liquid can be estimated at any other temperature and pressure by means of Equation (38).

Illustration 10. The density of ethyl alcohol at 20°C and 1 atm is 0.789 g per cc. Estimate the critical density and the density at 180°C and 100 atm pressure.

For ethyl alcohol, $t_c = 243.1°C$, and $p_c = 63.1$ atm.

Temperature, °C....	20	243.1	180
Pressure, atm.......	1.0	63.1	100
T_r	0.568	1.0	0.878
p_r	0.0159	1.0	1.59
ω	0.127	0.044	0.0984

Since $\rho_1 = 0.789$ at 20°C and 1 atm

$$\frac{\rho_1}{\omega_1} = \frac{0.789}{0.127} = 6.21$$

which is a characteristic constant for ethyl alcohol.

At 243.1°C and 63.1 atm $\quad \rho_c = (6.21)(0.044) = 0.273$ g per cc
At 180°C and 100 atm $\quad \rho = (6.21)(0.0984) = 0.611$ g per cc

The value of ρ_c reported in the International Critical Tables is 0.275 g per cc.

For most of the substances investigated the deviations from Equation (38) and Fig. 109 were less than 5 per cent. Larger errors are encountered for water if the reference value is at a low temperature in the range where water exhibits anomalous density–temperature relations. In general the best over-all results are obtained from a reference value corresponding to a reasonably high temperature, preferably near the normal boiling point or higher.

At reduced temperatures below 0.65 and at pressures not exceeding 10 atm the expansion factor ω is satisfactorily expressed by the equation:

$$\omega = 0.1745 - 0.0838T_r \tag{40}$$

At higher temperatures and pressures Fig. 109 must be used. At the critical point,

$$v_c = (v_1\omega_1)/0.0440 \tag{41}$$

Critical volumes or densities are calculated from Equation (41) with considerably more accuracy than from Fig. 103 which is based on gas-phase data.

In Table XXVIII are values of (ρ_1/ω_1) and $(v_1\omega_1)$ for the normal paraffins as developed by Gamson and Watson.[13] The corresponding critical properties, boiling points, and the constants of the vapor pressure equation (III-16, page 73) are also included. It was noted by these investigators that the values of (ρ_1/ω_1) and $(v_1\omega_1)$ are substantially independent of isomerization. The lower paraffins, both normal and isomeric, are satisfactorily represented by the following equation:

[13] B. W. Gamson and K. M. Watson, *Natl. Petroleum News*, Tech. Sec., May 1944. Also "Process Engineering Data," National Petroleum Publishing Company, Cleveland (1944).

Paraffins:

$$(v_1\omega_1) = 1.88 + 2.44n_C; \quad \text{cc per g-mole} \tag{42}$$

where

$$n_C = \text{number of carbon atoms}$$

The observed densities for the normal paraffins above tetradecane tend to be lower than calculated from Equation (42), but it is believed that these deviations may result from the fact that the densities were measured close to the melting points. On this basis densities calculated from the equation should be more reliable for high-temperature applications than those based on the experimental values.

It was found that the density data of other hydrocarbon series may be represented by similar equations in which the addition of a CH_2 group is accompanied by an increase of 2.44 cc per g-mole in the value of $(v_1\omega_1)$. Thus,

Mono-olefins:

$$(v_1\omega_1) = 1.1 + 2.44n_C \tag{43}$$

Monocyclic aromatics:

$$(v_1\omega_1) = -3.0 + 2.44n_C \tag{44}$$

For homologous series of polar compounds the increment in $(v_1\omega_1)$ for the addition of a CH_2 group tends to vary in the lower members of the series and approach a value of 2.44 with increased molecular weight. This behavior is expressed by the general relationship,

$$(v_1\omega_1) = a + 2.44n_C + m \log n_C \tag{45}$$

where a and m are constants characteristic of the series. For the hydrocarbons m is equal to zero. For other series it may have a finite value, either positive or negative. Thus, for the alcohols, both normal and isomeric,

$$(v_1\omega_1) = 2.89 + 2.44n_C - 1.23 \log n_C \tag{46}$$

Similar equations may be developed for other homologous series.

It was pointed out by Gamson and Watson that equations of the type of (45) may be used together with the methods discussed in Chapter III (pages 68–72) for estimating critical temperatures as a means of predicting liquid densities at all conditions without direct experimental-density data. Conversely, if a single experimental-density value is available, such equations may be used to estimate the critical temperature of a homolog. Thus, combination of Equations (39) and (40) and solving for

T_c give

$$T_c = \frac{0.0838T}{0.1745 - (v_1\omega_1)/v} \tag{47}$$

where

v = the molal volume at temperature T under conditions such that Equation (40) is applicable

Equation (47) may be used to estimate the critical temperature from density data when the boiling point is unknown. The boiling point may then be estimated from the relationships between boiling and critical points in Chapter III.

By means of Equation (III-9), page 70, it is possible to calculate the critical volume from no data other than the structural formula. This value may be substituted in Equation (41) and $(v_1\omega_1)$ evaluated. Then the critical temperature may be calculated from Equation (47) if a single liquid-density measurement is available. In this manner it is possible to estimate densities at all conditions and, with the relationships of Chapter III, boiling points and vapor pressures from only the structural formula and one liquid-density measurement.

These methods for estimating physicochemical properties are of particular importance in dealing with unstable materials under conditions which preclude direct measurements. However, it must be recognized that all such methods are approximations to be used only in the absence of more reliable data. In many cases several alternate relationships may be used for estimating a property. In the present state of the data it is best to compare the various methods with what direct measurements are available on the compound in question and similar compounds, and in this way select the most consistent and logical value for each particular problem. The relative reliability of the various methods is to a large extent dependent upon the type of fundamental data available.

Pressure Correction to Enthalpy of Liquids. Combination of Equation (38) with (XI-99) written in terms of reduced temperature and pressure gives

$$\frac{1}{p_c}\left(\frac{\partial H}{\partial p_r}\right)_T = \frac{1}{\rho} - T_r\left(\frac{\partial \frac{1}{\rho}}{\partial T_r}\right)_p = \frac{\omega_1}{\rho_1}\left[\frac{1}{\omega} - T_r\left(\frac{\partial \frac{1}{\omega}}{\partial T_r}\right)_p\right] \tag{48}$$

Equation (48) is conveniently integrated between p_r and the critical pressure, $p_r = 1.0$,

$$\frac{\rho_1}{p_c\omega_1}(H_{cp} - H)_T = \int_{p_r}^{1.0}\left[\frac{1}{\omega} + \frac{T_r}{\omega^2}\left(\frac{\partial \omega}{\partial T_r}\right)_p\right]dp_r = \psi \tag{49}$$

where H_{cp} = enthalpy at temperature T and the critical pressure p_c. The value of $(\partial\omega/\partial T)_p$ is obtained by graphically differentiating Fig. 109. The results of the integration of Equation (49) are plotted against reduced temperature and pressure in Fig. 110.

TABLE XXVIII

PHYSICAL PROPERTIES OF NORMAL PARAFFINS

1	2	3	4	5	6	7	8	9	10	11
C Atoms	Name	Mol. Wgt.	T_b°R	T_c°R	p_c lb per sq in.	ρ_1/ω_1, g per cc	$v_1\omega_1$, cc per g-mole	b	A	B lb per sq in.
1	Methane	16.04	201.2	343	673	3.679	4.36	0.000	2.338	5.166
2	Ethane	30.07	331.8	550	717	4.429	6.81	0.088	2.573	5.428
3	Propane	44.09	416.2	666	642	4.803	9.18	0.125	2.661	5.468
4	Butane	58.12	491.1	766	544	5.002	11.62	0.166	2.767	5.502
5	Pentane	72.15	557	847	482	5.128	14.07	0.190	2.887	5.570
6	Hexane	86.17	615.7	915	433	5.216	16.52	0.209	2.994	5.631
7	Heptane	100.20	669	972	394	5.285	18.96	0.224	3.124	5.719
8	Octane	114.22	718	1025	362	5.340	21.39	0.236	3.232	5.790
9	Nonane	128.25	763	1073	332	5.382	23.83	0.248	3.298	5.819
10	Decane	142.28	805	1114	308	5.414	26.28	0.258	3.407	5.896
12	Dodecane	170.32	881	1185	272	5.459	31.20	0.275	3.637	6.071
14	Tetradecane	198.38	948	1248	244	5.483	36.18	0.289	3.818	6.205
16	Hexadecane	226.43	1007	1300	221	*5.48	*41.3	0.300	4.006	6.350
18	Octadecane	254.48	1060	1345	202	*5.49	*46.3	0.308	4.195	6.500
20	Eicosane	282.54	1105	1380	187	*5.5	*51.	0.315	4.403	6.675
25	Pentacosane	352.67	1210	1465	156	*5.5	*64.	0.329	4.834	7.027
30	Triacontane	422.80	1305	1545	133	*5.5	*76.	0.339	5.169	7.293
35	Pentatriacontane	492.93	1385	1610	120	*5.5	*88.	0.347	5.580	7.660
40	Tetracontane	563.06	1465	1675	108	*5.5	*103.	0.353	6.011	8.045
45	Pentatetracontane	633.19	1540	1740	100	*5.4	*118.	0.358	6.358	8.358

* Based on experimental density, questionable because near melting point.

Equation (49) is dimensionless, and any consistent set of units may be used directly with Fig. 110. However, care must be used to recognize the dimensions of the resultant enthalpy term and apply the proper conversion factor in order to obtain conventional thermal units. For example, if ρ is in pounds per cubic foot, and p is in pounds per square foot, H will be expressed in foot-pounds per pound and must be divided by 778 in order to obtain Btu per pound.

Illustration 11. The enthalpy of saturated liquid propane at 190°F and an absolute pressure of 522 lb per sq in. is 102.3 Btu per lb.[14] Calculate the enthalpy of the liquid at this same temperature under critical pressure and also at a pressure of 1500 lb per sq in. At 70°F and 200 lb per sq in. pressure the density of liquid propane is 31.5 lb per cu ft.

$$T_c = 666°\text{R}$$

$$p_c = 642 \text{ lb per sq in.} = 9{,}450 \text{ lb per sq ft}$$

$$T_{r1} = 530/666 = 0.796$$

$$p_{r1} = 200/642 = 0.312$$

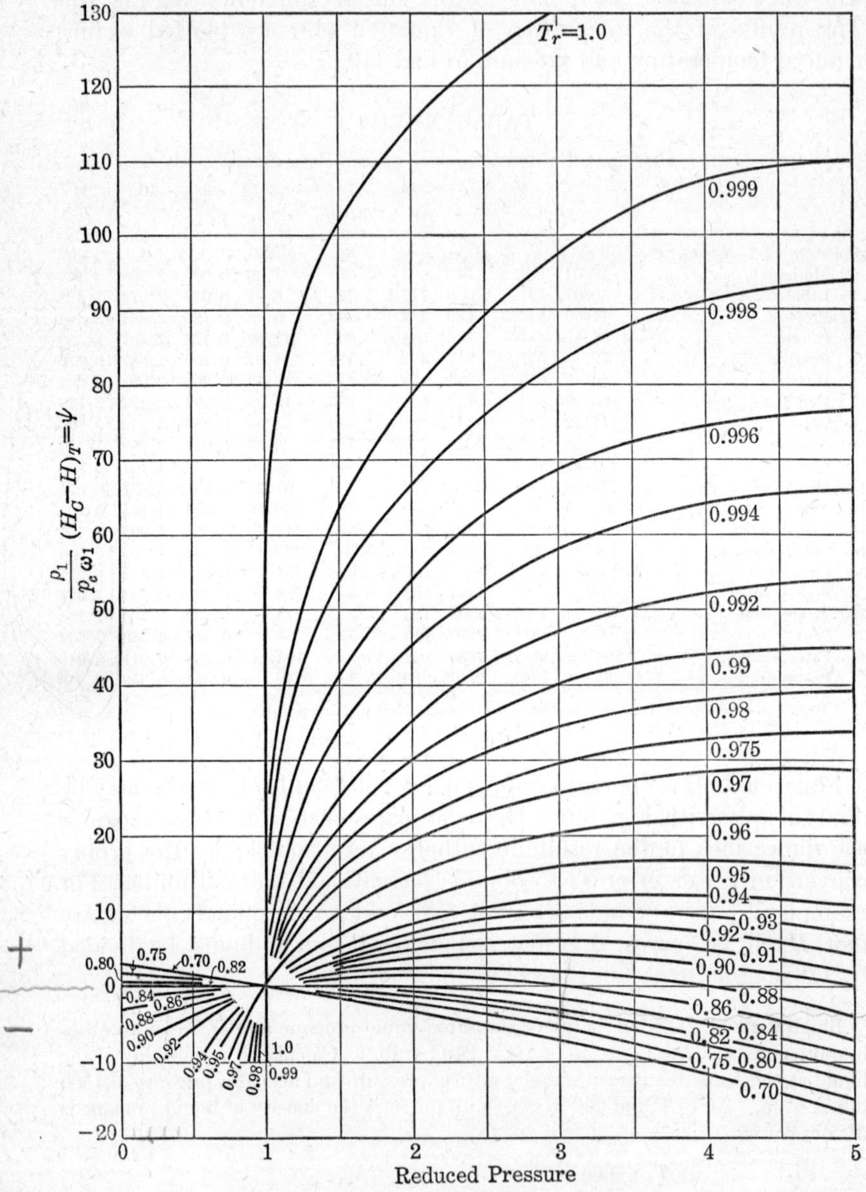

Fig. 110. Enthalpy Correction for Liquids

(Reproduced in "C.P.P. Charts")

From Fig. 109, $\omega_1 = 0.1045$, and $\rho_1/\omega_1 = 301$ lb per cu ft

$$\frac{\rho_1}{p_c\omega_1} = 301/92{,}450 = 0.00326 \text{ lb per ft-lb}$$

At a temperature of 190°F and 522 lb per sq in.

$$T_{r2} = 0.976 \qquad p_{r2} = 0.813 \qquad H_2 = 102.3°$$

From Fig. 110,

$$\frac{\rho_1}{p_c\omega_1}(H_{cp} - H_2)_T = -11.0$$

$$(H_{cp} - H_2)_T = -11.0/0.00326 = -3370 \text{ ft-lb per lb}$$

$$= -3370/778 = -4.3 \text{ Btu per lb}$$

$$H_{cp} = 102.3 - 4.3 = 98.0 \text{ Btu per lb}$$

At a temperature of 190°F and 1500 lb per sq in.

$$T_{r3} = 0.976; \qquad p_{r3} = 2.34$$

From Fig. 110,

$$\frac{\rho_1}{p_c\omega_1}(H_{cp} - H_3)_T = 25$$

$$(H_{cp} - H_3)_T = \frac{25}{(0.00326)(778)} = 9.9 \text{ Btu per lb}$$

$$H_3 = 98.0 - 9.9 = 88.1 \text{ Btu per lb}$$

A value of 90.4 was calculated from direct experimental data.[14]

Pressure Correction to Entropy of Liquids. Combination of Equation (38) with (h) Table XXIV written in terms of reduced temperature gives

$$\frac{1}{p_c}\left(\frac{\partial S}{\partial p_r}\right)_T = -\frac{1}{T_c}\left(\frac{\partial(1/\rho)}{\partial T_r}\right)_p = -\frac{\omega_1}{T_c\rho_1}\left(\frac{\partial\left(\frac{1}{\omega}\right)}{\partial T_r}\right)_p \tag{50}$$

Integrating between p_r and the critical pressure, $p_r = 1.0$, gives

$$\frac{T_c\rho_1}{p_c\omega_1}(S_{cp} - S)_T = +\int_{p_r}^{1.0}\frac{1}{\omega^2}\left(\frac{\partial\omega}{\partial T_r}\right)_p dp_r \tag{51}$$

where S_{cp} = entropy at temperature T and the critical pressure p_c. The results of this integration are plotted against reduced temperature and pressure in Fig. 111.

[14] B. H. Sage, W. N. Lacey and J. G. Schaafsma, *Ind. Eng. Chem.*, **26**, 1218 (1934).

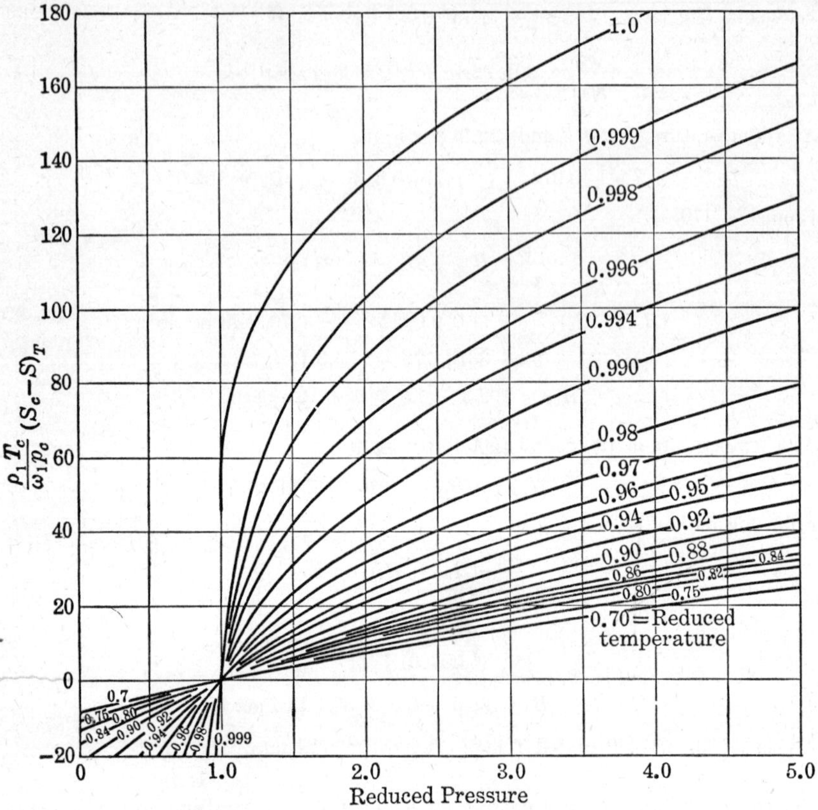

FIG. 111. Entropy Correction for Liquids.
(Reproduced in "C.P.P. Charts")

Figure 111 may be used by a procedure similar to that of Illustration 10 to determine the change of entropy accompanying any change of pressure at constant temperature. Similar care must be exercised with regard to units.

Pressure Correction to Heat Capacity at Constant Pressure of Liquids. By differentiating Equation (49) with respect to temperature at constant pressure, an expression relating heat capacity at constant pressure to heat capacity at the critical pressure is found.

$$\frac{\rho_1}{p_c\omega_1}\left[\left(\frac{\partial H_{cp}}{\partial T}\right)_p - \left(\frac{\partial H}{\partial T}\right)_p\right]_T = \frac{1}{T_c}\left(\frac{\partial \psi}{\partial T_r}\right)_p \tag{52}$$

$$\frac{T_c\rho_1}{p_c\omega_1}(C_{pc} - C_p)_T = \left(\frac{\partial \psi}{\partial T_r}\right)_p \tag{53}$$

where C_{pc} = heat capacity at constant pressure at temperature T and

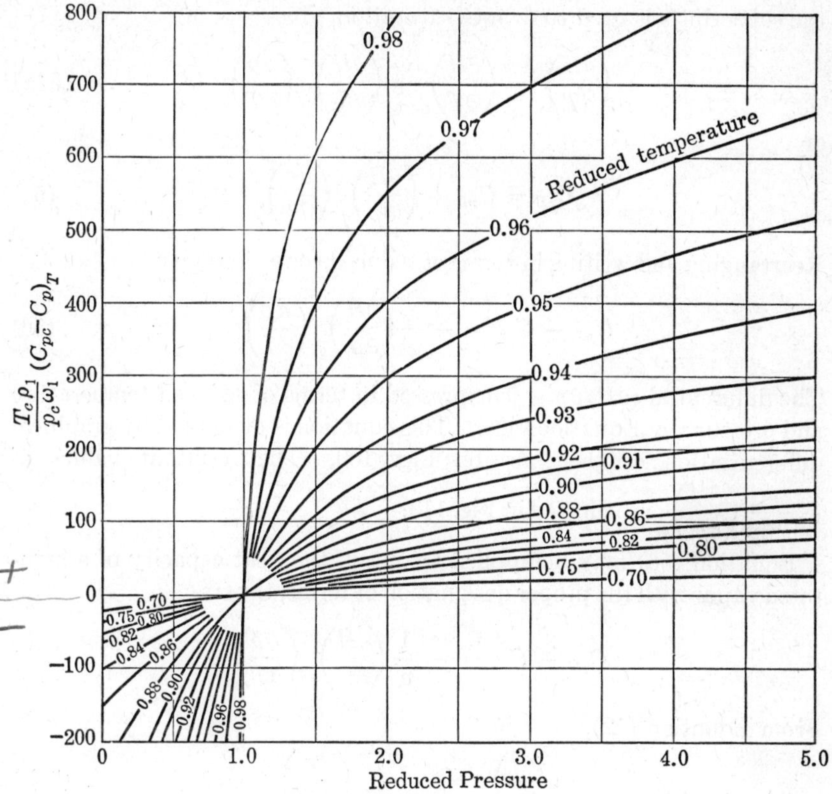

FIG. 112. Heat-Capacity Correction for Liquids.
(Reproduced in "C.P.P. Charts")

the critical pressure. A generalized evaluation of Equation (53) is obtained by differentiating Fig. 110. The results are plotted in Fig. 112.

By the general procedure of Illustration 10, Fig. 112 may be used to estimate the change in heat capacity at constant pressure accompanying any change in pressure at constant temperature.

Heat Capacity of a Saturated Liquid or Saturated Vapor. When the temperature of a liquid at its boiling point is increased while saturation is maintained, an accompanying increase in pressure results. The heat capacity of such a saturated liquid involves the enthalpy change resulting from both the temperature and pressure changes. Thus, from Equation (XI–39),

$$dH = \left(\frac{\partial H}{\partial T}\right)_p dT + \left(\frac{\partial H}{\partial p}\right)_T dp \tag{54}$$

Differentiation of Equation (54) with respect to T at saturation, if the subscript s is used to denote saturation, gives

$$\left(\frac{dH}{dT}\right)_s = \left(\frac{\partial H}{\partial T}\right)_p + \left(\frac{\partial H}{\partial p}\right)_T \left(\frac{dp}{dT}\right)_s \tag{54a}$$

or

$$C_{sL} = C_{pL} + \left(\frac{\partial H}{\partial p}\right)_T \left(\frac{dp}{dT}\right)_s \tag{55}$$

Rearranging and writing in terms of reduced conditions give

$$C_{pL} - C_{sL} = -\frac{1}{p_c}\left(\frac{\partial H}{\partial p_r}\right)_T \left(\frac{dp}{dT}\right)_s \tag{56}$$

The differential $(\partial H/\partial p_r)_T$ is expressed in terms of reduced temperature and pressure by Equation (48). This equation is evaluated by graphical differentiation of the lines on Fig. 109. The resultant values of $-\frac{\rho_1}{\omega_1 p_c}\left(\frac{\partial H}{\partial p_r}\right)_T$ are plotted in Fig. 113.

Equation (56) may be applied to obtain the heat capacity of a saturated vapor with the proper designation of terms, thus:

$$C_{pG} - C_{sG} = -\frac{1}{p_c}\left(\frac{\partial H}{\partial p_r}\right)_T \left(\frac{dp}{dT}\right)_s \tag{56a}$$

From Equation (12),

$$\left(\frac{\partial H}{\partial p_r}\right)_T = -\frac{RT^2}{p_r T_c}\left(\frac{\partial z}{\partial T_r}\right)_p \tag{56b}$$

Values of $(\partial z/\partial T_r)_p$ are obtained from a cross-plot of the compressibility chart (Fig. 103) on regular co-ordinates. The heat capacity of a saturated gas assumes a negative value as the critical temperature is approached.

An approximate value of $(dp/dT)_s$ which is satisfactory for many purposes is obtained by differentiating the Calingaert–Davis vapor pressure Equation (III-7),

$$\ln p = A - \frac{B}{T - 43} \tag{56c}$$

Thus,

$$\left(\frac{dp}{dT}\right)_s = \frac{p_s B}{(T - 43)^2} \tag{57}$$

where

T = temperature, °K
B = constant of Calingaert–Davis equation

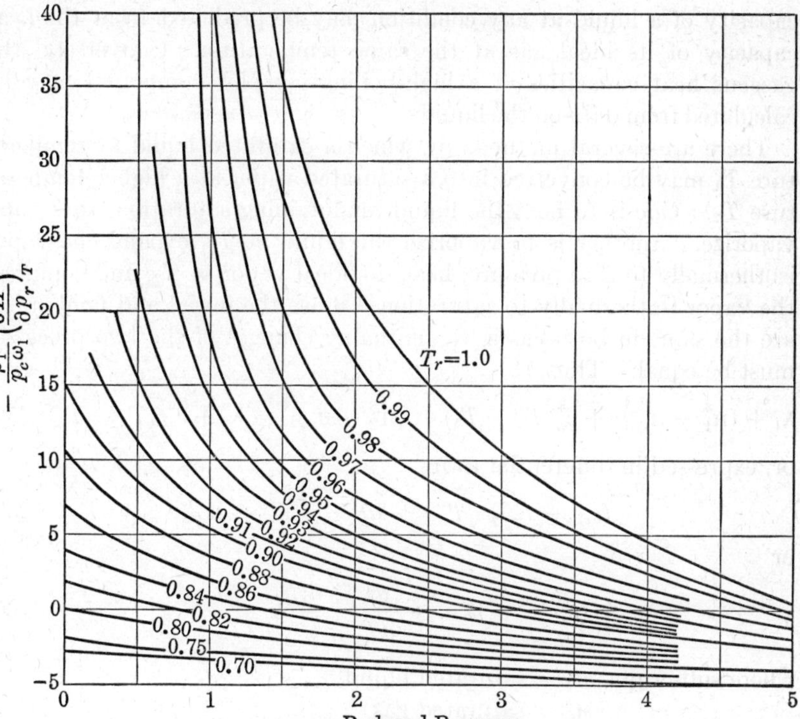

FIG. 113. Differential Effect of Pressure upon the Enthalpy of Liquids.
(Reproduced in " C.P.P. Charts ")

The constant B may be evaluated from the boiling point and critical data using Equation (VII-30), page 232. More nearly accurate values of $(dp/dT)_s$ are obtained by differentiating Equation (III-16), page 73.

By the use of Fig. 113 in combination with Equation (57) the heat capacity at constant pressure of a saturated liquid is calculated from a known value of the heat capacity of the saturated liquid. The heat capacity at any other pressure is then calculated by means of Fig. 112. This procedure is demonstrated in Illustration 12.

Heat Capacity of a Saturated Liquid from C_p of Its Ideal Gas. Data on the heat capacities of liquids, particularly at temperatures other than atmospheric are scanty and frequently unreliable. However, generalizations of statistical data discussed in Chapter XVI permit the estimation of heat capacities of ideal gases over wide temperature ranges. A thermodynamic relation exists between the heat capacity of the ideal gas and the heat capacity of the saturated liquid involving the heat of vaporization and the effect of pressure on the enthalpy of the gas. From this relation in generalized form together with Figs. 112 and 113 the heat capacity of a liquid at any condition may be predicted from the heat capacity of its ideal gas at the same temperature. Conversely, the gaseous heat capacities of a high-boiling complex compound may be calculated from data on the liquid.

There are several methods by which a saturated liquid at temperature T_1 may be converted into a saturated vapor at a higher temperature T_2. One is to heat the liquid, maintaining saturation, to T_2 and vaporize. Another is to vaporize the liquid at T_1, expand the vapor isothermally to zero pressure, heat the ideal vapor to T_2, and compress the vapor isothermally to saturation. Since the initial and final states are the same in both cases, the enthalpy changes of the two processes must be equal. Thus,

$$\lambda_1 + (\text{H}_1^* - \text{H}_{sG1}) + c_p^*(T_2 - T_1) - (\text{H}_2^* - \text{H}_{sG2}) = \lambda_2 + c_{sL}(T_2 - T_1) \quad (58)$$

or, expressed in differential form,

$$(c_{sL} - c_p^*)_T \, dT = -d\lambda - d(\text{H}^* - \text{H}_{sG}) \quad (59)$$

or

$$(c_{sL} - c_p^*)_T = -\frac{d\lambda}{dT} - \frac{d(\text{H}^* - \text{H}_{sG})}{dT} \quad (60)$$

where subscripts, sL = saturated liquid
sG = saturated gas

The total derivative $\dfrac{d(\text{H}^* - \text{H}_{sG})}{dT}$ can be expressed in terms of its partials in the usual manner, thus,

$$\left[\frac{d(\text{H}^* - \text{H}_{sG})}{dT}\right]_s = \left[\frac{\partial(\text{H}^* - \text{H}_{sG})}{\partial T}\right]_p + \left[\frac{\partial(\text{H}^* - \text{H}_{sG})}{\partial p}\right]_T \left(\frac{dp}{dT}\right)_s \quad (61)$$

Substituting Equation (61) into (60) and writing in terms of reduced conditions, we have

$$(c_{sL} - c_p^*)_T = -\frac{1}{T_c}\left(\frac{d\lambda}{dT_r}\right) - \frac{1}{T_c}\left[\frac{\partial(\text{H}^* - \text{H}_{sG})}{\partial T_r}\right]_p - \frac{T_c}{p_c T_c}\left[\frac{\partial(\text{H}^* - \text{H}_{sG})}{\partial p_r}\right]_T \left(\frac{dp}{dT}\right)_s \quad (62)$$

The value of $d\lambda/dT_r$ is obtained by combining Equations (VII-29 and 32), page 231, and differentiating. Thus,

$$\frac{d\lambda}{dT_r} = -0.361RB\left(\frac{T_b}{T_b - 43}\right)^2 \frac{(1 - T_{rb})^{-0.38}}{(1 - T_r)^{0.62}} \quad (63)$$

where B = constant of Calingaert–Davis equation
T_b = normal boiling point, degrees Kelvin

The derivatives in Equation (62) which contain enthalpy difference are evaluated by differentiation of Fig. 106. The results of this operation are plotted in Figs. 114 and 115 as functions of reduced temperature and pressure. These figures, together with Equations (57) and (63), permit evaluation of all terms of Equation (62) and calculation of

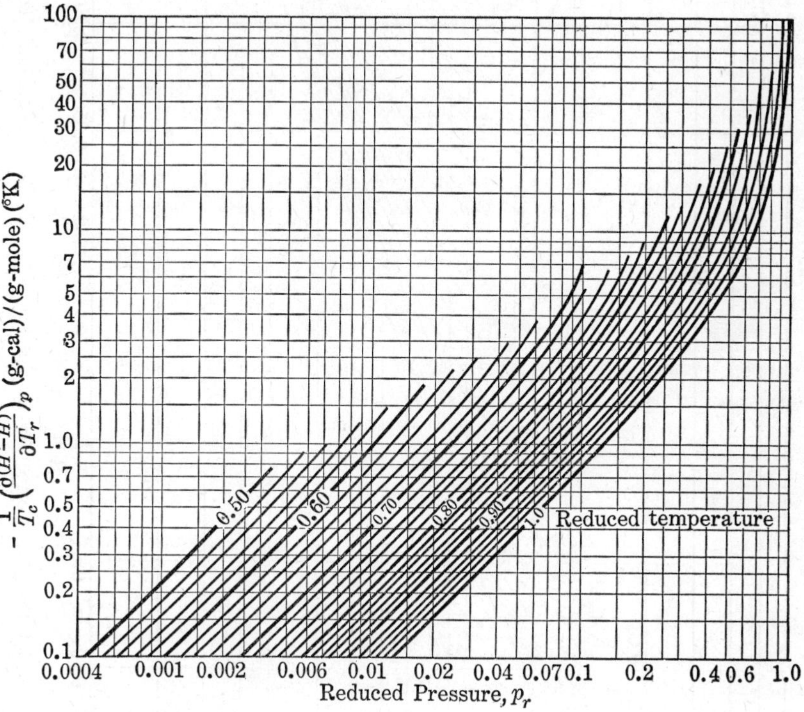

FIG. 114. Differential Pressure Correction to the Enthalpy of Gases (Constant Pressure).
(Reproduced in " C.P.P. Charts ")

$(c_{sL} - c_p^*)_T$. For use in Equation (62) it is recommended that vapor pressures be calculated from the Calingaert–Davis Equation (56c). However, this equation is not recommended for calculation of vapor pressures as such where accuracy is required.

Use of Equation (63) and Figs. 114 and 115 is not recommended at reduced temperatures below 0.55 or above 0.96. At reduced temperatures of 0.55 and below it may be assumed that $(c_{sL} - c_p^*)_T$ is independent of temperature.

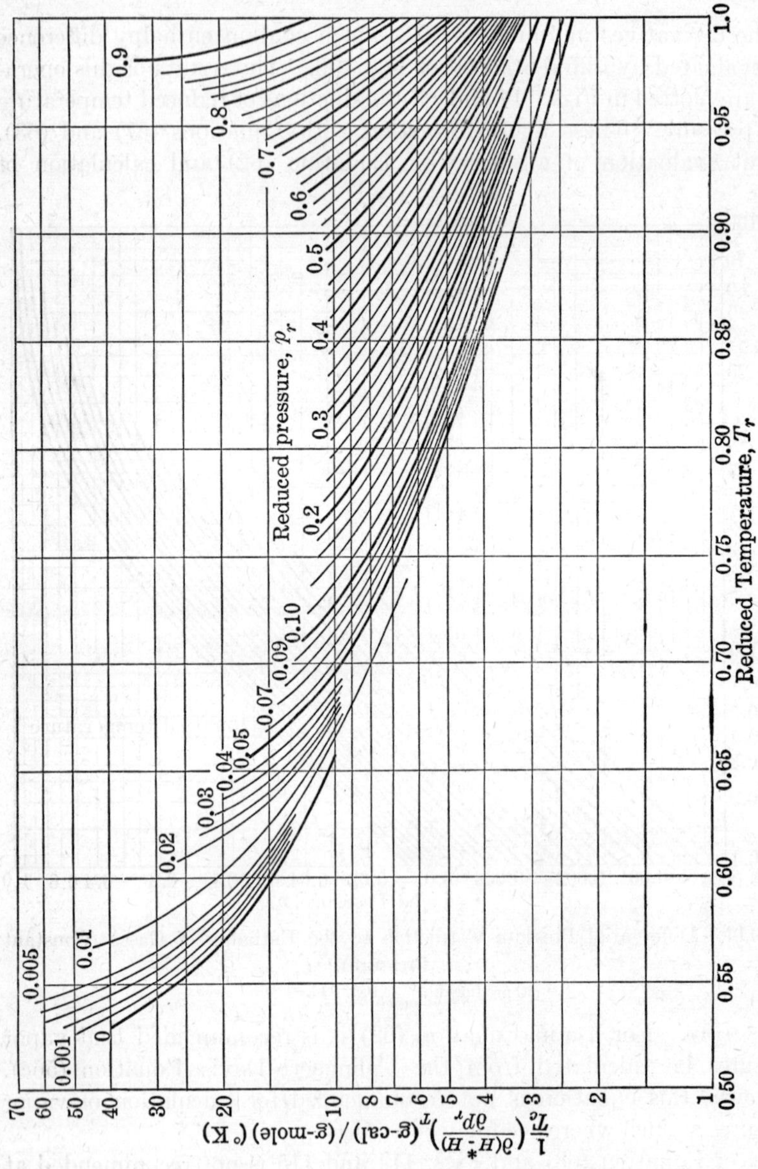

Fig. 115. Differential Pressure Correction to the Enthalpy of Gases (Constant Temperature).

(Reproduced in "C.P.P. Charts")

Illustration 12. Propane gas at a temperature of 63°C and atmospheric pressure has a molal heat capacity[15] of 18.8. Calculate for this temperature:

(*a*) Heat capacity of the saturated liquid.
(*b*) Heat capacity at constant pressure of the saturated liquid.
(*c*) Heat capacity of the liquid at the critical pressure.
(*d*) Heat capacity of the liquid at a pressure of 200 atm.

Physical constants of propane:

$$T_c = 370°\text{K}; \qquad p_c = 43.7 \text{ atm}; \qquad T_b = 228.6°\text{K}$$

Solution: (*a*) From Equation (VII–30), page 232,

$$B = \ln 43.7 \bigg/ \left[\frac{1}{228.6 - 43} - \frac{1}{370 - 43}\right] = 1620$$

From Equation (56c), where $p = 1.0$,

$$A = \frac{B}{T_b - 43} = \frac{1620}{185.6} = 8.73$$

$$\ln p_s = 8.73 - \frac{1620}{336 - 43} = 3.20$$

$$p_s = 24.5 \text{ atm}; \qquad p_r = 24.5/43.7 = 0.56$$

$$T_r = 336/370 = 0.91; \qquad T_{r_b} = 0.618$$

Evaluation of Equation (63) gives

$$\frac{d\lambda}{dT_r} = -(0.361)(1.987)(1620)\left(\frac{228.6}{185.6}\right)^2 \frac{(1 - 0.618)^{-0.38}}{(1 - 0.91)^{0.62}} = -11{,}290 \text{ cal/(g-mole)}$$

From Fig. 114,

$$-\frac{1}{T_c}\left(\frac{\partial(\text{H}^* - \text{H})}{\partial T_r}\right)_p = 25.5 \text{ cal/(g-mole)(°K)}$$

From Fig. 115,

$$\frac{1}{T_c}\left(\frac{\partial(\text{H}^* - \text{H})}{\partial p_r}\right)_T = 11.0 \text{ cal/(g-mole)(°K)}$$

From Equation (57),

$$\left(\frac{dp}{dT}\right)_s = \frac{(24.5)(1620)}{(336 - 43)^2} = 0.462 \text{ atm per °K}$$

Substitution in Equation (62) gives

$$c_{sL} - c_p^* = \frac{11{,}290}{370} + 25.5 - \frac{(0.462)(370)(11.0)}{43.7}$$

$$= 30.5 + 25.5 - 41.9 = 14.1$$

$$c_{sL} = 18.8 + 14.1 = 32.9$$

Direct experimental measurements[16] give a value of 33.0.

[15] B. H. Sage, D. C. Webster, and W. N. Lacey, *Ind. Eng. Chem.*, **29**, 1309 (1937).
[16] B. H. Sage and W. N. Lacey, *Ind. Eng. Chem.*, **27**, 1484 (1935).

(*b*) From the data of Illustration 11, $\rho_1/\omega_1 = 301$ lb per cu ft or 4.84 g per cc. From Fig. 113,

$$-\frac{1}{p_c}\left(\frac{\partial \text{H}}{\partial p_r}\right)_T = \frac{(7.8)(44)}{(41.3)(4.84)} = 1.73 \text{ cal/(g-mole)(atm)}$$

The factor 41.3 converts the energy units from cc-atm to calories while 44 is the molecular weight of the gas. From Equation (56) and the value of $(dp/dT)_s$ previously determined, $(c_p - c_s)_T = (1.73)(0.438) = 0.755$ cal/(g-mole)(°K).

$c_p = 31.2 + 0.755 = 32.0$, the heat capacity at constant pressure of the saturated liquid at 63°C.

(*c*) From Fig. 112,

$$(c_{p_c} - c_p) = -\frac{(110)(43.7)(44)}{(41.3)(373)(4.8)} = -2.9$$

$c_{p_c} = 32.0 - 2.9 = 29$, the molal heat capacity of the liquid at 63°C and 43.7 atm.

(*d*) At 200 atm, $p_r = 4.58$.

From Fig. 112,

$$(c_{p_c} - c_p) = \frac{(185)(43.7)(44)}{(41.3)(373)(4.8)} = 4.8$$

$c_p = 29.1 - 4.8 = 24.3$, the molal heat capacity of the liquid at 63°C and 200 atm.

DETERMINATION OF THERMODYNAMIC DATA

Considerable attention has been devoted to generalized methods for estimating thermodynamic properties from a minimum of direct data. These methods are important because of the frequency with which process design calculations must be carried out in the absence of direct thermodynamic data of any type. The generalized relations derived from the theory of corresponding states are also of value for correlating the behavior of solutions and mixtures. However, it must be recognized that all such generalizations are approximations, not based on rigorous principles, and they should be used only in the absence of reliable direct data. The errors incurred may not be serious for many purposes, but for some requirements they render the application of generalized methods unsatisfactory.

It is difficult to predict the probable errors in the use of the generalized relations of the preceding sections. For gases the greatest errors are encountered for the saturated vapors approaching the critical temperature. The following estimates represent the orders of magnitude of the maximum errors which may be encountered.

	Per Cent
Compressibility of gases, Fig. 103	15
Pressure correction for enthalpy of gas, Figs. 105–106:	
Without York and Weber's correction	30

	Per Cent
In the range of $T_r = 1.0$ to 3.0 and $p_r = 0$ to 10 with York and Weber's correction	5
Nonideality correction for entropy of gas, Fig. 107	30
Pressure correction for heat capacity, Fig. 108	30
Expansion factors of liquids, Fig. 109	5
Pressure correction to enthalpy of liquids, Fig. 110	25
Pressure correction to entropy of liquids, Fig. 111	15
Pressure correction to heat capacity of liquids, Fig. 112	20
Difference between heat capacity at constant pressure and heat capacity of saturated liquid, Fig. 113 and Equation (56)	35
Difference between heat capacity of saturated liquid and ideal gas, Figs. 114–115 and Equation (62), in range $T_r = 0.55 - 0.96$	25

Although these errors appear high, it must be remembered that they are estimated maximum values and that for many substances and conditions much better accuracy can be expected. Furthermore, several of the relations represent correction terms which are relatively small in themselves, and large errors in the corrections introduce but a small error in the property sought. In addition, if consistent use is made of the same series of generalized relations, many of the errors will tend to cancel in some applications such as energy balances.

It may be concluded that these generalizations are extremely useful and a great improvement over the assumption of ideal behavior or the use of relations based on simple corrections such as those of van der Waals. However, they must be used with care and judgment, with full realization of the errors which may result.

Thermodynamic Properties from Experimental Data. Thermodynamic properties derived from reliable experimental data are greatly to be preferred over estimations by any generalized method. The methods for the use of such data are similar to those employed in deriving the generalized relations and employ the same thermodynamic relations. However, the derivations are carried out in terms of absolute rather than reduced conditions and are restricted to the particular substance under consideration.

The thermodynamic properties of a substance may be completely evaluated from the following data covering the range of temperatures and pressures of interest.

1. *PVT* measurements of both the liquid and vapor state.
2. Heat capacities of the vapor, either measured at low pressure or calculated for the ideal state from spectroscopic data.
3. A value of absolute entropy.
4. Vapor pressure measurements.

From the vapor pressure and *pVT* data heats of vaporization are calculated from the Clapeyron equation. The variations of the enthalpy, entropy, and heat capacity of the vapors with pressure are calculated from Equations (10), (20), and (32). Enthalpies of the saturated liquid may be obtained by subtracting heats of vaporization from the enthalpies of the saturated vapor or may be developed from liquid-heat-capacity data calculated from Equation (62). If the compressed liquid region is of interest, calculations similar to those developed on a generalized data may be carried out from the experimental pVT data.

Although the foregoing procedure theoretically defines the complete thermodynamic behavior of a substance, it is generally desirable to verify the results by calculations from various types of supplementary data to detect possible inconsistencies in the experimental measurements. This general procedure is demonstrated by Benning and McHarness and coworkers[17–21] in evaluating the thermodynamic properties of fluorochloromethanes and ethanes recently developed as refrigerants. Vapor pressures were determined from 0.1 atm to the critical point.[18] PVT measurements were made on the vapors over the range of interest,[19] and the densities of the saturated liquids and vapors were determined up to the critical point.[20] Heat capacities of the vapors were measured[21] at atmospheric pressure. To verify the results of these data, measurements were also made of the heat capacity of the saturated liquid and the ratio C_p/C_v for the vapors by velocity of sound measurements.[21]

The liquid-heat-capacity data were used in combination with the vapor-heat-capacity measurements to calculate heats of vaporization by means of Equation (60). These results were compared with those obtained from the Clapeyron equation, using entirely different experimental data. Similarly, the measured C_p/C_v ratios were compared with ratios obtained from the measured heat capacities of the vapor at constant-pressure and constant-volume values calculated from the pVT data and Equation (s), Table XXIV, page 472.

For deriving precise thermodynamic data from experimental measurements it is desirable to use analytical methods rather than the less accurate graphical methods used in developing the approximate generalized relations. This requires that all data shall be expressed by accurate equations the constants of which are evaluated empirically. The following form was used by Benning and McHarness.

[17] A. F. Benning and R. C. McHarness, *Ind. Eng. Chem.*, **31**, 912 (1939).
[18] *Ind. Eng. Chem.*, **32**, 497 (1940).
[19] *Ind. Eng. Chem.*, **32**, 698 (1940).
[20] *Ind. Eng. Chem.*, **32**, 814 (1940).
[21] A. F. Benning, R. C. McHarness, W. H. Markwood, and W. J. Smith, *Ind. Eng. Chem.*, **32**, 975 (1940).

For vapor pressures,

$$\log p = A + \frac{B}{T} + C \log T + DT \tag{64}$$

This equation was found to deviate less than 0.3 per cent from the data. It is more convenient and more accurate than the generalized Equation (III-16), page 73, which contains only three constants but is difficult to evaluate from data.

The heat-capacity data were represented by Equation (VII-21), page 213. Similar equations were used for liquid densities over limited ranges.

The pVT data for the vapors were expressed by the Beattie-Bridgeman Equation (2). The constants were evaluated by Benning and McHarness[19] by the method of Buffington and Gilkey[22] which assumes that c is zero and thus considerably simplifies the operations. This simplified form is accurate where pT isometrics are linear.

THERMODYNAMIC CHARTS AND TABLES

For substances which are frequently used over wide ranges of temperature and pressure in both liquid and vapor phases, it is convenient to prepare tables giving values of the important properties over this entire range. Such tables are particularly indispensable for fluids used in power generation and refrigeration. It is also common practice to present such information graphically in the form of charts. Since all the properties of a pure component may be defined in terms of two independent properties, it is possible on a chart with two co-ordinate scales to construct contour lines for all related properties. For example, on a chart with temperature and entropy as co-ordinates, it is possible to construct additional contour lines for pressure, enthalpy, free energy, specific heat, and other properties. This procedure is limited by the confusion resulting from reading values from a multiplicity of intersecting lines. Tables have the advantage of permitting great accuracy in the reproduction of values, whereas charts are more compact and aid in visualizing the effects of changes.

The familiar steam tables have undergone extension and refinement with the result that thermodynamic properties are now known with a high degree of precision over wide ranges of temperatures and pressure for the saturated vapor and liquid, the superheated vapor, and the compressed liquid. Data for saturated vapors are generally tabulated in terms of the corresponding temperatures or pressures, whereas for the

[22] R. M. Buffington and W. K. Gilkey, *Ind. Eng. Chem.*, **23**, 254 (1931).

superheated vapors and compressed liquids tables of properties are required as functions of temperature at various constant pressures. Such tables are published as entire volumes devoted to a single substance.[23]

In thermodynamic calculations the five properties of pressure, volume, temperature, entropy, and enthalpy are of greatest interest. In the two-phase region of liquid and vapor, the additional variable of percentage vaporization or quality is introduced. Four types of charts, based on the following co-ordinate scales, are in common use for presenting these properties:

1. Pressure–volume.
2. Temperature–enthalpy.
3. Temperature–entropy
4. Enthalpy–entropy (Mollier chart).

Auxiliary lines may be drawn on each of these charts to show the other three properties or any other related property. Each chart has special advantages for certain applications, as is discussed in succeeding chapters, although each chart may be constructed to include all the data of the others. With known values of absolute entropies, it is also possible to construct additional charts and tables giving values of free energy and the total work function. Internal energy can be obtained from values of enthalpy, pressure, and volume.

A representative series of diagrams for ammonia is presented in Figs. 116–119 plotted from tables based on direct experimental data[24] up to 300 lb per sq in. and extended by generalizations above that pressure as indicated in broken lines. Since ammonia is extensively used as a refrigerant, values of the enthalpy and entropy are arbitrarily assigned zero values at the saturated-liquid state at $-40°F$. The absolute entropy of liquid ammonia at this state is 1.207 Btu/(lb)(°R).

The general shape and appearance of Figs. 116–119 are characteristic of such diagrams for substances of low molecular weight and low molal-heat capacity. The other common refrigerants and water are represented by diagrams of similar appearance.

Diagrams from Generalizations. Since few thermodynamic data are available for compounds of high molecular weight, it is generally necessary to resort to the generalized relations in order to develop diagrams for such substances. Such a series of diagrams for benzene is shown in

[23] J. H. Keenan and F. G. Keyes, "Thermodynamic Properties of Steam," John Wiley & Sons, New York (1936).

[24] *Natl. Bur. Standards U. S. Circ.* 142 (1923).

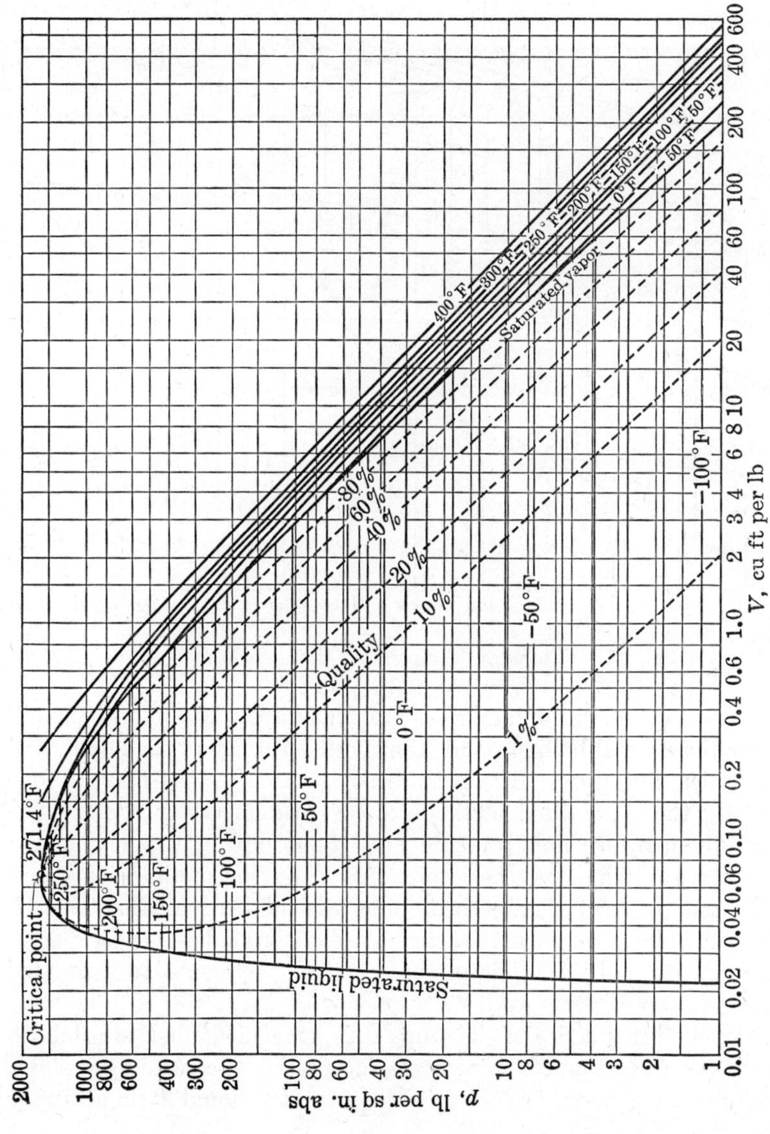

FIG. 116. Pressure–Volume Chart for Ammonia.

(Reproduced in " C.P.P. Charts ")

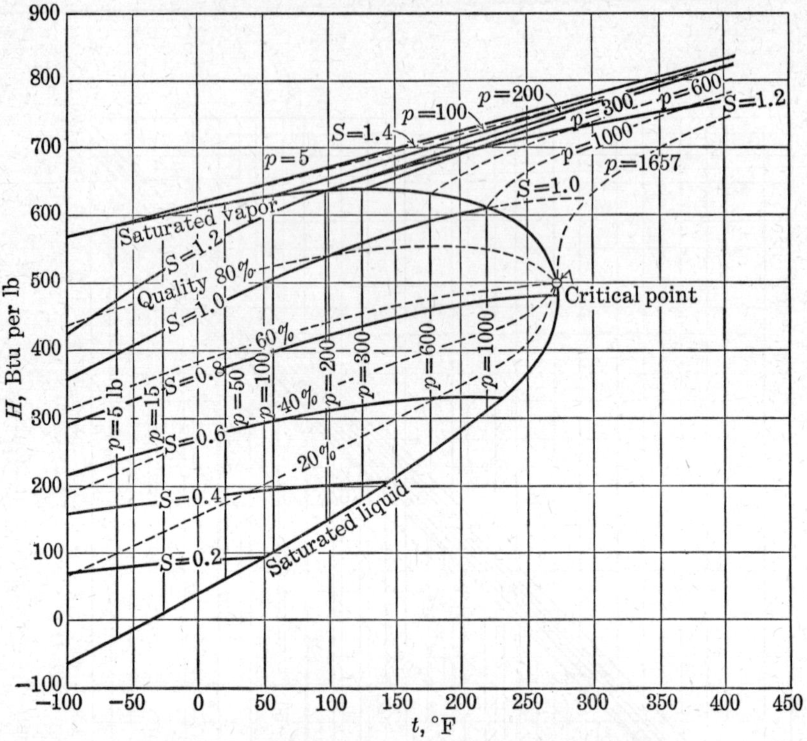

FIG. 117. Enthalpy–Temperature Chart for Ammonia
(Reproduced in "C.P.P. Charts")

Figs. 120–122. As a minimum requirement direct data must be available for the normal boiling point, the specific gravity of the liquid, and the chemical structure. The heat capacity of the vapor may be estimated by the generalized methods discussed in Chapter XVII. The critical temperature and pressure and the complete vapor-pressure–temperature relation may be approximated by the methods of Chapter III and the heat of vaporization by the methods of Chapter VII. In the case of benzene, these data were available from direct experimental measurements and accordingly were used in preference to estimated values.

In developing a series of diagrams it is convenient first to establish enthalpy as a function of temperature for a series of constant pressures. Enthalpies are generally referred to the saturated liquid at an arbitrary temperature, in this case 60°F. The enthalpy of the saturated vapor at the reference state is first calculated by correcting the normal heat of vaporization for temperature by means of Equation (VII-32), page 233. The enthalpy of the vapor at this temperature and zero pressure is

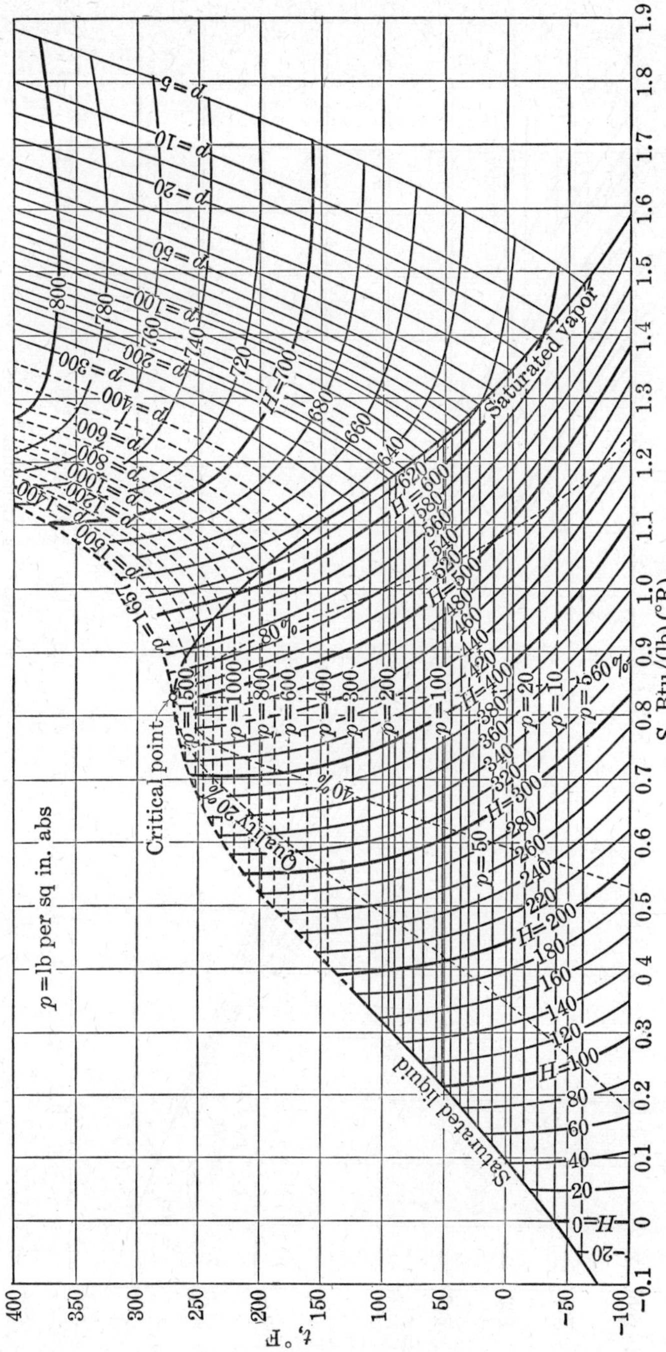

FIG. 118. Temperature–Entropy Chart for Ammonia.

(Reproduced in "C.P.P. Charts")

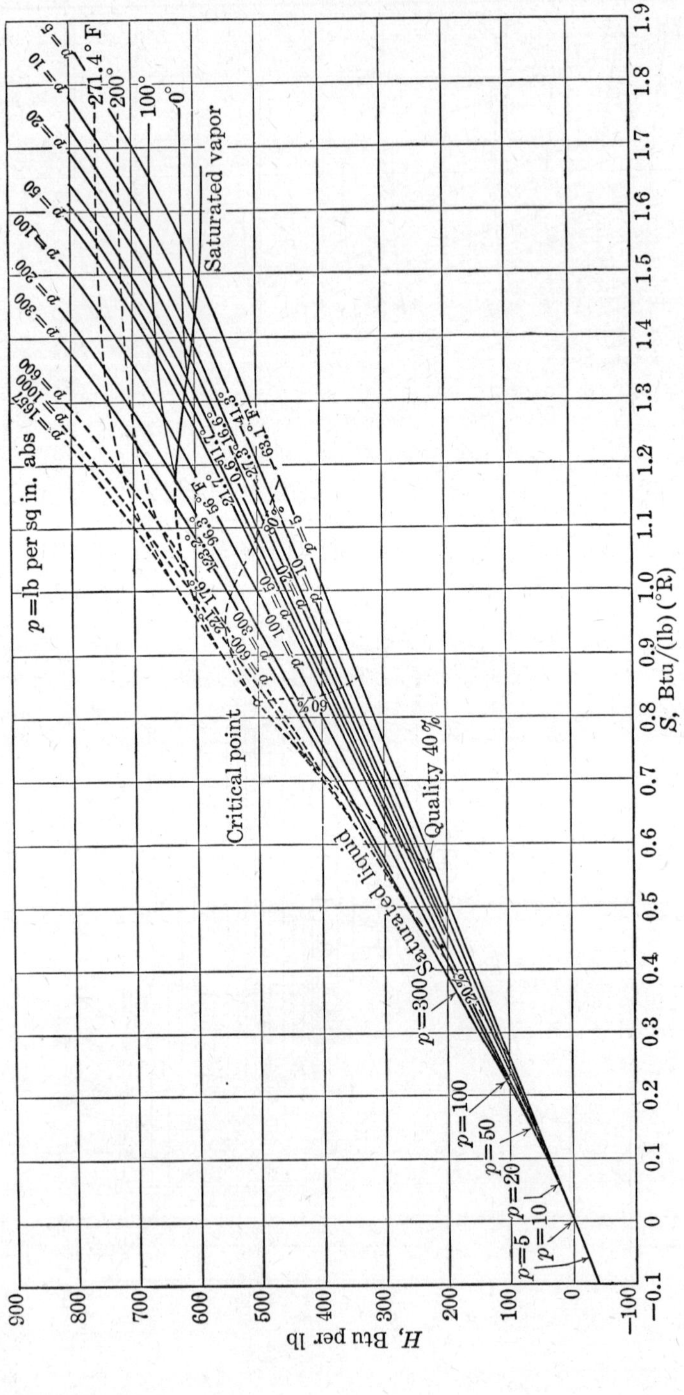

FIG. 119. Enthalpy–Entropy Chart for Ammonia.

(Reproduced in "C.P.P. Charts")

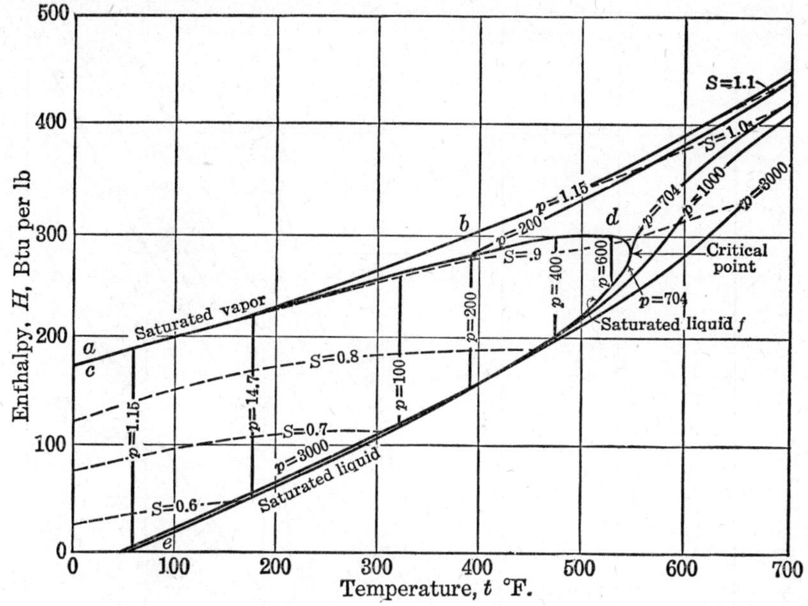

FIG. 120. Enthalpy–Temperature Chart for Benzene.

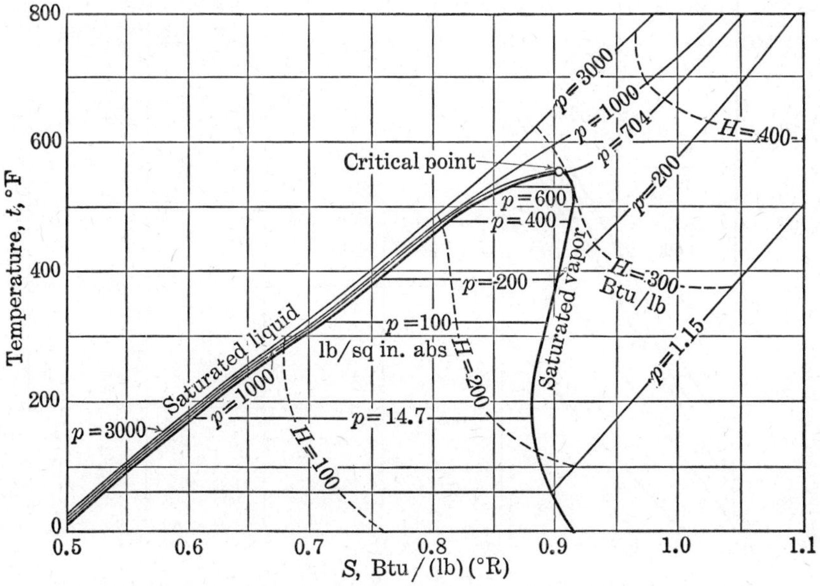

FIG. 121. Temperature–Entropy Chart for Benzene.

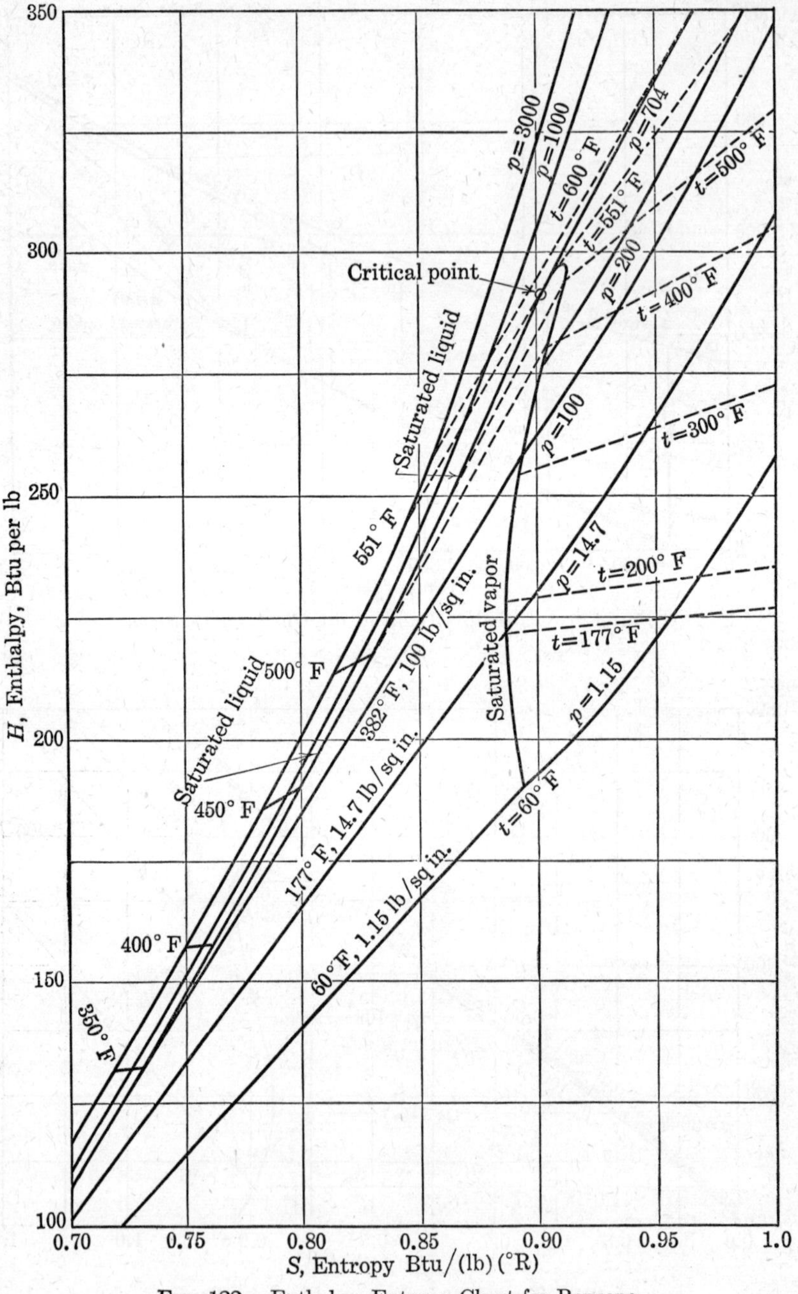

FIG. 122. Enthalpy–Entropy Chart for Benzene.

obtained by adding the correction from Fig. 105 corresponding to the reduced temperature and pressure of the saturated vapor. The line *ab*, Fig. 120, representing the enthalpy of the vapor at zero pressure H^* is then established from the heat capacity of the vapor, corresponding to zero pressure.

Lines representing the enthalpy of the superheated vapors at other pressures are determined by calculating series of values of $(H^* - H_p)_T$ from Fig. 106 for selected constant pressures and temperatures. These corrections are subtracted from line *ab*, establishing other constant pressure lines in the superheated-vapor region. The enthalpy of the saturated vapor is established in a similar manner. Values of $(H^* - H_p)_T$ are calculated at selected saturation temperatures and pressures and subtracted from the corresponding values of H^* from line *ab*, thus establishing line *cd*.

Line *ef* representing the enthalpy of the saturated liquid is established by subtracting heats of vaporization from corresponding values of the enthalpies of saturated vapor. The heats of vaporization are calculated from Equation (VII-32), page 233.

Constant-pressure lines in the compressed-liquid region are established from Fig. 110. It is convenient to establish the enthalpy of the liquid under the critical pressure over the entire temperature range and use this as a base to establish lines for other pressures.

This procedure may lead to discontinuities at the critical temperature for enthalpy lines at pressure higher than the critical pressure. Below the critical temperature these lines are established from Fig. 110 based on the liquid-expansion factor correlation, while at temperatures above the critical they are derived from Fig. 105 based on the gas-compressibility correlation. Discontinuities indicate inconsistencies in these two correlations and are arbitrarily smoothed, giving the greatest weight to the values calculated from the liquid correlation which appear to be more reliable in the critical region.

The temperature–entropy chart, Fig. 121, is established by a similar procedure. The use of absolute values of entropy instead of relative values greatly increases the utility of the charts. It has been established that the absolute entropy of liquid benzene at 60°F and its own vapor pressure is 0.530 Btu/(lb)(°R). The entropy of the saturated vapor at 60°F is obtained by adding the entropy of vaporization at 60°F to 0.530. The vapor pressure of benzene at this temperature is 1.15 lb per sq in. The entropy corresponding to ideal behavior at this temperature and pressure is obtained by adding the generalized entropy correction for deviation from ideal behavior, from Fig. 107. The entropy of superheated benzene vapor at 1.15 lb per sq in. for various temperatures, if ideal behavior is assumed, is obtained from the heat-capacity equation of ideal benzene vapor. The true entropy values at this pressure are then obtained by subtracting corrections read from Fig. 107. Entropies of the vapors at any given temperature and for other pressures are then determined by calculating the entropy of the ideal vapor at the existing pressure from that at 1.15 lb per sq in., and then subtracting the additional correction due to lack of ideal behavior, by use of Fig. 107. The entropy of the saturated liquid at any temperature is obtained by subtracting the entropy of vaporization from the entropy of the saturated vapor. Entropies of the compressed liquid are obtained by means of Fig. 111.

After the constant-pressure lines of Figs. 120 and 121 are established, they may be combined by cross-plotting to determine lines of constant entropy which are plotted on Fig. 120 and lines of constant enthalpy which are plotted in Fig. 121. Either of the charts then may be replotted in the form of Fig. 122.

Illustration 13. It is desired to calculate the enthalpy and entropy of benzene at 491°F and under the following conditions:

(*a*) The superheated vapor (dew point 60°F).
(*b*) The superheated vapor at 200 lb per sq in.
(*c*) The saturated vapor.
(*d*) The saturated liquid.
(*e*) The compressed liquid at the critical pressure.
(*f*) The compressed liquid at 3000 lb per sq in.

The following data are available:

Normal boiling point	177°F (637°R)(354°K)
Critical temperature	551°F (1011°R)(561°K)
Critical pressure	704 lb per sq in. (47.9 atm)
Liquid density	0.874 g per cc at 77°F
Molecular weight	78.04

Molal heat capacity in Btu per lb-mole per degree Rankine at zero pressure:

$$c_p^* = -5.70 + (54.14)(10^{-3})T - (13.41)(10^{-6})T^2$$

where T is in degrees Rankine.

Vapor pressure:

$$\log_{10} p = 7.3659 - \frac{2.8096}{T_r} - e^{-20(T_r - 0.188)^2} \text{ mm Hg}$$

Absolute entropy of saturated liquid at 60°F = 0.53 Btu/(lb)(°R). The reference state of zero enthalpy is the saturated liquid at 60°F.

Vapor pressure at 60°F:

$$T_r = \frac{520}{1011} = 0.514$$

$$\log_{10} p = 7.3659 - \frac{2.8096}{0.514} - e^{-20(514-0.188)^2} = 1.775$$

$$p = 59.2 \text{ mm or } 1.15 \text{ lb per sq in.}$$

Similarly, at 491°F, the vapor pressure is 470 lb per sq in.

Heat of vaporization at 60°F is calculated from Equation (VII-29), page 231. From Equation (VII-30), page 232,

$$B = \frac{\ln p_c/p_b}{\dfrac{1}{T_b - 43} - \dfrac{1}{T_c - 43}} = \frac{\ln 47.9}{\dfrac{1}{354 - 43} - \dfrac{1}{561 - 43}} = 3020$$

$$T_r = \frac{520}{1011} = 0.514; \qquad (T_r)_b = \frac{637}{1011} = 0.631$$

$$\left(\frac{T_b}{T_b - 43}\right)^2 = \left(\frac{354}{311}\right)^2 = 1.295$$

Substitution in Equation (VII-29), gives

$$\lambda = (0.95)(1.987)(3020)(1.295)\left(\frac{0.486}{0.369}\right)^{0.38} = 8230 \text{ cal per g-mole}$$

or

$$\frac{8230}{78.04}(1.8) = 190 \text{ Btu per lb}$$

Similarly, at 491°F, the heat of vaporization is calculated to be 85 Btu per lb. Enthalpy of saturated vapor at 60°F = 190 Btu per lb

$$\text{Entropy of saturated vapor at 60°F} = 0.53 + \frac{190}{520} = 0.896 \text{ Btu/(lb)(°R)}$$

At the low pressure of 1.15 lb per sq in., ideal behavior may be assumed, and these values may be taken as representing the ideal state.

(*a*) *At 491°F and 1.15 (lb)/(sq in.), if we assume ideal behavior:*

$$H^* = 190 + \frac{1}{78.04}\int_{520}^{951} [-5.70 + (54.14)(10^{-3}T) - (13.41)(10^{-6})T^2]\, dT$$

$$= 338 \text{ Btu per lb}$$

$$S^* = 0.896 + \frac{1}{78.04}\int_{520}^{951} \left[\frac{-5.70}{T} + (54.14)(10^{-3}) - (13.41)(10^{-6})T\right] dT$$

$$= 1.095$$

(*b*) *At 491°F, 200 lb per sq in.:*

$$T_r = \frac{951}{1011} = 0.94; \qquad p_r = \frac{200}{704} = 0.284$$

From Fig. 105, $\dfrac{\text{H}^* - \text{H}}{T_c} = 0.91; \qquad \text{H}^* - \text{H} = 920 \text{ or } \dfrac{920}{78.04} = 11.8$ Btu per lb.

$$H = 338 - 11.8 = 326 \text{ Btu per lb}$$

From Fig. 107, $s^* - s = 0.74$
or

$$S^* - S = 0.74/78.04 = 0.010 \text{ Btu/(lb)(°R)}$$

From Equation (XI-142),

$$S^* = 1.095 - \frac{1.987}{78.04} \ln \frac{200}{1.15} = 1.095 - 0.131 = 0.964$$

$$S = 0.964 - 0.010 = 0.954 \text{ Btu/(lb)(°R)}$$

(c) *Saturated Vapor at 491°F:*

$$\text{Pressure} = 470 \text{ lb per sq in.}$$

$$T_r = 0.94; \qquad p_r = 0.67$$

From Fig. 105,

$$(H^* - H) = (3.10)(1011) = 3130 \text{ Btu per lb-mole}$$

$$\text{or } 40.1 \text{ Btu per lb}$$

$$H = 338 - 40 = 298 \text{ Btu per lb}$$

From Fig. 107,

$$s^* - s = 2.80 \text{ Btu/(lb-mole)(R°) or } 0.036 \text{ Btu/(lb)(°R)}$$

$$S^* = 1.095 - \frac{1.987}{78.04} \ln \frac{470}{1.15} = 1.095 - 0.152 = 0.943$$

$$S = 0.943 - 0.036 = 0.907$$

(d) *Saturated liquid at 491°F:*

$$\lambda = 85 \text{ Btu per lb}$$

$$\Delta S_\lambda = \frac{85}{951} = 0.089$$

$$H = 298 - 85 = 213$$
$$S = 0.907 - 0.089 = 0.818$$

(e) *Compressed liquid at 491°F* and its critical pressure.

From Fig. 110,

$$\frac{\rho}{p_c\,\omega}\,(H_c - H) = -6.0$$

At 77°F, 1.0 atm, $T_r = 0.53$; $p_r = 0.0209$, $\rho_1 = 0.874$ g per cc

From Fig. 109,

$$\omega_1 = 0.13$$

For benzene:

$$\frac{\rho_1}{\omega_1} = 6.71 \text{ g per cc}$$

$$\frac{\rho_1}{p_c\omega_1} = \frac{(41.31)}{(47.9)}\ 6.71 = 5.8 \text{ g per cal}$$

Hence,

$$H_c - H = \frac{-6}{5.8} = -1.04 \text{ cal per g} \quad \text{or} \quad -1.9 \text{ Btu per lb}$$

$$H_c = 213 - 1.9 = 211 \text{ Btu per lb}$$

From Fig. 111,

$$\frac{\rho_1 T_c}{\omega_1 p_c}(S_c - S) = -12$$

$$\frac{\rho_1 T_c}{\omega_1 p_c} = \frac{41.3(6.71)(561)}{47.9} = 3240 \text{ (°K)(g)/cal}$$

$$S_c - S = \frac{-12}{3240} = -0.0037$$

$$S_c = 0.818 - 0.004 = 0.814 \text{ Btu/(lb)(°R)}$$

(*f*) *Compressed liquid at* 3000 *lb per sq in.:*

$$T_r = 0.94; \qquad p_r = 4.26$$

From Fig. 110,

$$\frac{\rho_1}{p_c \omega_1}(H_c - H) = 12.5$$

$$H_c - H = \frac{12.5}{5\ 8} = 2.16 \text{ cal per g} \quad \text{or} \quad 3.9 \text{ Btu per lb}$$

$$H = H_c - 3.9 = 211 - 3.9 = 207 \text{ Btu per lb}$$

From Fig. 111,

$$\frac{\rho_1 T_c}{\omega_1 p_c}(S_c - S) = 48$$

$$S_c - S = \frac{48}{3240} = 0.0148$$

$$S = S_c - 0.0148 = 0.814 - 0.015 = 0.799 \text{ Btu/(lb)(°R)}$$

The completed enthalpy–temperature, temperature–entropy and enthalpy–entropy charts for benzene, constructed from points calculated by the methods shown in this illustration, are shown in Figs. 120, 121, and 122.

Internal Energy, Free Energy, and Total Work Function. As was pointed out in Chapter XI, absolute values of these three properties U, G, and A are never known. However, relative values referred to any selected reference state may be calculated by the methods of the preceding sections.

The change in internal energy accompanying any change in conditions is derived from the corresponding enthalpy change. Thus, from the definition of enthalpy,

$$\Delta U = \Delta H - \Delta(pV) \tag{65}$$

Applying Equation (65) to determining internal energy relative to a reference state of zero enthalpy, designated by a zero subscript, gives

$$U = H - (pV - p_oV_o) \tag{66}$$

Similarly, by definition,

$$\Delta G = \Delta H - \Delta(TS) \tag{67}$$

or, expressing free energy relative to the reference state, gives

$$G = H - (TS - T_oS_o) \tag{68}$$

As was pointed out in Chapter XI, it is evident that a knowledge of absolute values of entropy is necessary for the determination of free-energy changes in all except isothermal operations.

Combination of Equation (65) with the definition of the total work function gives

$$\Delta A = \Delta H - \Delta(pV) - \Delta(TS) \tag{69}$$

or, expressing A relative to the state of zero enthalpy,

$$A = H - (pV - p_oV_o) - (TS - T_oS_o) \tag{70}$$

Illustration 14. The absolute entropy of liquid water at 25°C (77°F) is 16.75 entropy units per mole. Calculate the internal energy, free energy, and total work function, in Btu per lb of saturated water vapor at an absolute pressure of 125 lb per sq in., relative to the saturated liquid at 32°F. Use the enthalpy and entropy data of the steam tables, and assume the heat capacity of liquid water between 32°F and 77°F to be constant at 1.0 Btu per lb per °F, and neglect the effect of pressure on the entropy of the saturated liquid in this range.

Solution: The absolute entropy of the reference state is calculated from Equation (17):

$$S_o = \frac{16.75}{18} - 1.0 \ln \left(\frac{460 + 77}{460 + 32}\right) = 0.9306 - 0.0875 = 0.8431 \text{ Btu/(lb)(°R)}$$

From the steam tables, at the reference state,

$$p_o = 0.08854 \text{ lb per sq in.}$$

$$V_o = 0.01602 \text{ cu ft per lb}$$

For the saturated vapor at 125 lb per sq in. and 344.33°F,

$$V = 3.587 \text{ cu ft per lb}$$

$$H = 1191.1 \text{ Btu per lb}$$

$$S = 0.8431 + 1.5844 = 2.4275 \text{ Btu/(lb)(°R)}$$

Equation (66), converting pV from foot-pounds to Btu gives

$$U = 1191.1 - [(125)(3.587) - (0.08854)(0.01602)]\frac{144}{778}$$
$$= 1191.1 - 83.0 = 1108.1 \text{ Btu per lb}$$

From Equation (68),

$$G = 1191.1 - (804.33)(2.4275) + (492)(0.8431)$$
$$= 1191.1 - 1537.7 = -346.6 \text{ Btu per lb}$$

From Equation (70) and the preceding calculations of $\Delta(pV)$ and $\Delta(TS)$,

$$A = 1191.1 - 1537.7 - 83.0 = -429.6 \text{ Btu per lb}$$

As is evident from Illustration 14, the relative free energy and maximum work may decrease with increase in temperature. Complete tables or charts of the functions could be established either from the generalized correlations of properties or from experimental data. However, such data are of limited value, and the principle application of free energy is in the prediction of equilibrium in chemical reactions or solutions.

PROBLEMS

1. For the production of liquid oxygen it is desired that the gas be compressed to a pressure of 100 atm at a temperature of −90°C. Oxygen at a pressure of 14.5 lb per sq in. and a temperature of 22°C is compressed to these conditions. Calculate the volume of compressed gas resulting from 100 cu ft of the original: (*a*) From van der Waals' equation, (*b*) From the compressibility chart.

2. Using the Beattie–Bridgman equation, calculate the pressure exerted by 30 liters of air, measured at 20°C under a pressure of 740 mm of Hg, when compressed into a volume of 200 cc at a temperature of 0°C. It may be assumed that at the lower pressure the ideal-gas law is applicable.

3. For ethylene gas the value of pV at 20°C and 100 atm is 0.3600 referred to a value of unity at 0°C and 1 atm. The density of ethylene gas at standard conditions is 1.2604 g per liter. Calculate the volume of 1 g-mole at 20°C and 100 atm.

4. Calculate the volume occupied by 1 g-mole of carbon dioxide gas at its critical state when the product $pV = 0.3131$ referred to 1.00 at 0°C and 1 atm.

5. From the data of the International Critical Tables, calculate the compressibility factors of ethylene at a temperature of 20°C and at pressures ranging from 0 to 500 atm. Plot these compressibility factors against pressure in atmospheres and also against molal volumes in cubic centimeters per gram-mole.

6. From the data of problem 5, calculate the volume occupied by 50 g of ethylene at a temperature of 20°C and a pressure of 5000 lb per sq in.

7. Using the data of Figs. 100 and 102, calculate the density in pounds per cubic foot of nitrogen at a pressure of 900 lb per sq in. and a temperature of −40°C.

8. Using the data of Figs. 100–102, calculate the pressure necessary to compress 100 liters of nitrogen at a pressure of 745 mm of Hg and 23°C to occupy a volume of 2.0 liters at a temperature of 40°C.

9. Using the data of Figs. 100–102, calculate the temperature to which 1.2 lb of nitrogen must be cooled in order that it may exert a pressure of 50 atm when confined in a volume of 0.27 cu ft.

10. In a multistage compressor, carbon dioxide is compressed from a volume of 2 cu ft at a pressure of 100 lb per sq in. and a temperature of 22°C, to a volume of 0.4 cu ft at a temperature of 30°C. Calculate the pressure necessary, using Fig. 103.

11. Calculate the volume occupied by 10 lb of chlorine when compressed to a pressure of 125 lb per sq in. at a temperature of 30°C, using Fig. 103.

12. Methyl chloride for use in domestic refrigerators is sold in small cylinders having volumes of 0.15 cu ft. Calculate the weight of methyl chloride gas contained in a cylinder if the absolute pressure is 68 lb per sq in. and the temperature 20°C, using Fig. 103. (t_c = 143.1°C; p_c = 65.8 atm).

13. For high-pressure distribution in long pipe lines it is proposed to compress natural gas (methane) to an absolute pressure of 500 lb per sq in. Calculate its density in pounds per cubic foot at this pressure and a temperature of 80°F, using Fig. 103.

14. Sulfur dioxide is compressed from a pressure of 40 lb per sq in. and a temperature of 150°F to a pressure of 190 lb per sq in. at a temperature of 160°F. Calculate the ratio of the initial to the final volume, using Fig. 103.

15. The absolute entropy of carbon dioxide in the ideal state at 25°C and 1 atm is 51.08 entropy units per mole. For a temperature of 40°C and a pressure of 50 atm, calculate the absolute entropy, the enthalpy in calories per gram-mole referred to the ideal gas at 0°C, and the molal heat capacity, using Figs. 106–108.

16. Liquid sulfur dioxide has a density of 92.42 lb per cu ft at a temperature of 0°F and an absolute pressure of 10.35 lb per sq in. Calculate the density of the liquid at 150°F and 1000 lb per sq in. from Fig. 109.

17. Saturated liquid sulfur dioxide exerts a vapor pressure of 136.5 lb per sq in. at a temperature of 130°F and has an enthalpy of 181.24 Btu per lb and an entropy of 0.32472 unit per lb referred to the saturated liquid at −40°F. Calculate the enthalpy and entropy of the compressed liquid at 130°F and 3000 lb per sq in., using the data of problem 16.

18. Sulfur dioxide exists as a saturated liquid at 130°F and 136.5 lb per sq in. At these conditions calculate the molal heat capacity of the saturated liquid and the heat capacity of the liquid at constant pressure, using the heat capacity of the ideal gas (Table V, page 214) and the generalized correlations. Also calculate the heat capacity of the liquid at 130°F and 3000 lb per sq in.

19. Construct temperature–enthalpy, temperature–entropy, and enthalpy–entropy charts for methyl amine in Btu per lb referred to the saturated liquid at 0°F. Evaluate the properties of the saturated vapor and liquids at reduced temperatures of 0.7, 0.8, 0.85, 0.9, 0.95, 0.98, and 1.0. Establish constant-pressure lines at absolute pressures of 14.7, 100, 500, 1000, and 2000 lb per sq in. by calculations at the afore-mentioned temperatures and also at T_r = 1.1 and 1.2.

The properties of methyl amine are as follows:

Boiling point, −6.5°C — Critical temperature, 156.9°C
Critical pressure, 73.6 atm — Density at −11°C, 0.699 g per cc
Molal heat capacity of ideal vapor $= 4.02 + 30.72 \times 10^{-3}\,T - 8.70 \times 10^{-6}\,T^2$ where T = degrees Kelvin.
Heat of vaporization at −6.5°C, 368 Btu per lb

The vapor pressure may be obtained from Equation (III-16), page 73, which is most conveniently used by plotting a curve relating vapor pressure on a logarithmic scale to temperature on a uniform or reciprocal scale.

20. For methane gas at 122°F and 1000 lb per sq in. abs, calculate the values of

U, H, S, C_p, G, and A, per pound mole relative to the actual gas at 32°F and 1 atm pressure by the generalized method and also by the Benedict-Webb-Rubin equation of state. The absolute molal entropy of methane gas in the ideal state at 32°F and 1 atm pressure is 44.5 Btu/(lb-mole)(°R). The other necessary data are contained in Table XXI, page 336, and XXVII, page 483.

21. Benning and McHarness[17] determined the following constants for dichlorofluoromethane $CHCl_2F$ (molecular weight 102.9): In Equation (2),

$A_o = 20.54$ $a = -0.179$
$B_o = 0.286$ $b = 0.497$
$c = 0$ $R = 0.08206$
p = atmospheres T = degrees Kelvin
v = liters per gram-mole

In Equation (64),

$A = 38.2974$ $B = -2367.41$
$C = -13.0295$ $D = 0.0071731$
p = atmospheres T = degrees Kelvin

Heat capacity of liquid at 1 atm,

$$C_p = 0.2471 + 0.000189t$$

C_p = calories/(gram)(°C) t = degrees centigrade

Heat capacity of vapor at 1 atm,

$$c_p = 13.65 + 0.0249t° \text{ cal/(g-mole)(°C)}$$

Density of saturated liquid (−40°C to +70°C),

$\rho = 1.4256 - 2.316\,(10^{-3})t - (2.6)(10^{-6})t^2$ g per cc, where t is in degrees centigrade. Calculate by rigorous methods

(*a*) The heat of vaporization at −40°F.

(*b*) The enthalpy and entropy, referred to the saturated liquid at −40°F, of the saturated vapor and the saturated liquid at +120°F.

CHAPTER XIII

EXPANSION AND COMPRESSION OF FLUIDS

The most useful form of energy into which it is generally desirable to convert other forms is the mechanical work of motion as represented by a rotating shaft or a moving piston. *Mechanical work* is defined as the energy which is transferred by the effect of a force acting through a distance and is equal to the product of that force times the distance of action. Thus, when a fluid which is confined under pressure undergoes a change in volume, work is done as the result of the force of pressure moving through the distance corresponding to the volume change. Similarly, a flowing fluid may perform work through changes in volume, elevation, or external kinetic energy.

As pointed out in Chapter VII, work is a form of energy which is incapable of storage as such but is in transition from one form of stored energy to another. Mechanical energy in this transitory form is termed *shaft work*, as distinguished from the electrical and other forms of work which may be included in the work term of a complete energy balance. Shaft work is capable of transmission as such and may pass from one part of a system to another through either solid media such as shafts, pistons, gears, and belts or through fluid media such as hydraulic or pneumatic couplings and drives. However, no transmission device is perfect, and some work is always lost by degradation to a lower form of energy, made manifest as heat resulting from friction.

Differential Energy Balance. In Chapter VII a general energy balance was established for a complex chemical process involving several interconnected systems. By applying this energy equation to an infinitesimal change in the energy of a single system of constant mass m and constant composition, the following differential form results,

$$d(pV) + d\left(\frac{mu^2}{2g_c}\right) + d(mZ) + d(mE_\sigma) + dU = d'q - d'w \qquad (1)$$

where the energy terms represent in order, flow work, external kinetic energy, potential energy, surface energy, internal energy, heat added from the surroundings, and work done upon the surroundings. The work term $d'w$, in general, includes all possible forms of work, such as shaft, electrical, magnetic, and radiant. Equation (1) may also be written as

$$dE = d'q - d'w \tag{2}$$

where E is the total energy of the system including all the terms represented by the left side of Equation (1).

In the absence of electrical, magnetic, and radiant forms of work, $d'w$ represents mechanical or shaft work only. If changes in potential and external kinetic energy are also negligible, the shaft work results only from changes in flow work or from work of expansion. The evaluation of shaft work in such systems requires separate consideration of the special cases of nonflow and flow processes, in both isothermal and isentropic systems for reversible and irreversible processes.

Shaft Work in Nonflow Processes. In a nonflow process, as defined in Chapter VII, flow is absent in the initial and final states of the process, and changes in surface, kinetic, and potential energies are generally negligible. If only work of expansion is performed, Equation (1) then becomes

$$d'q = d'w_s + dU \tag{3}$$

where $d'w_s$ represents only mechanical work accomplished as the result of a volume change. In a reversible process $d'q = T\,dS$ and $dU = T\,dS - p\,dV$. Hence, for such a process Equation (3) becomes

$$d'w_s = p\,dV$$

or

$$w_s \text{ (nonflow)} = \int_1^2 p\,dV \tag{4}$$

Equation (4) is restricted to mechanically reversible nonflow processes. Where the expansion is not mechanically reversible $\int p\,dV$ is equal to the sum of the useful shaft work plus that lost in friction. In applying Equation (4) to actual expansion or compression operations care must be taken that the pressure used in the equation is that actually exerted on the face of the piston. The type of expansion and the existence of fluid turbulence and friction during expansion do not affect the validity of Equation (4) but do affect the amount of work performed in expanding a fluid between specified initial and final states.

Shaft Work in a Flow Process. In applying Equation (1) to a reversible-flow process, $d(pV) = p\,dV + V\,dp$; $dU = T\,dS - p\,dV$, and $d'q = T\,dS$. Where forms of work other than shaft work are negligible, Equation (1) becomes

$$d'w_s = -V\,dp - d\left(\frac{mu^2}{2g_c}\right) - d(mZ) - d(mE_\sigma) \tag{5}$$

Where changes in kinetic, potential, and surface energies are negligible, Equation (5) reduces to $d'w_s = -V\,dp$ or

$$w_s\ (\text{flow}) = -\int_1^2 V\,dp \tag{6}$$

Equation (6) represents the shaft work performed by any reversible-flow process with negligible kinetic-, potential-, and surface-energy changes and with no electrical, radiant, or magnetic work.

Equations (4) and (6) may be used for calculating the work performed by the volume or pressure changes of a fluid from its state properties, only provided these changes take place reversibly.

Shaft Work from the Energy Functions. For reversible processes under the conditions defined for Equations (4) and (6) the shaft work performed may be expressed as functions of the four energy functions. Thus, from the indicated equations of Table XXIV, combined with (4) and (6),

$$\text{(j)}\qquad \left(\frac{\partial A}{\partial V}\right)_T = -p \quad \text{or} \quad -(\Delta A)_T = \int_1^2 (p\,dV)_T = (w_s)_T\ (\textit{nonflow}) \tag{7}$$

$$\text{(k)}\qquad \left(\frac{\partial G}{\partial p}\right)_T = +V \quad \text{or} \quad -(\Delta G)_T = -\int_1^2 (V\,dp)_T = (w_s)_T\ (\textit{flow}) \tag{8}$$

$$\text{(j)}\qquad \left(\frac{\partial U}{\partial V}\right)_S = -p \quad \text{or} \quad -(\Delta U)_S = \int_1^2 (p\,dV)_S = (w_s)_S\ (\textit{nonflow}) \tag{9}$$

$$\text{(k)}\qquad \left(\frac{\partial H}{\partial p}\right)_S = V \quad \text{or} \quad -(\Delta H)_S = -\int_1^2 (V\,dp)_S = (w_s)_S\ (\textit{flow}) \tag{10}$$

From Equations (7–10) it is evident that:

(*a*) Under isothermal nonflow conditions, the work of reversible expansion is equal to the decrease in the total work function.

(*b*) Under isothermal flow conditions, the work of reversible expansion is equal to the decrease in free energy.

(*c*) Under isentropic nonflow conditions, the work of reversible expansion is equal to the decrease in internal energy.

(*d*) Under isentropic flow conditions, the work of reversible expansion is equal to the decrease in enthalpy.

From Equations (7–10) it is evident that the work of reversible expansion of a fluid under isothermal or isentropic conditions may be determined either by a pressure–volume integration or as the change of the appropriate energy function. The pV integration must follow the actual reversible path of the change, whereas the thermodynamic energy functions are state properties which are independent of path and are completely defined by the reference properties. Either method of calculating work may be employed with thermodynamic data based on either experimental measurements or generalized correlations. The choice of the method is determined by the type of data available. However, in general, it is more convenient to evaluate the thermodynamic energy functions at the terminal states of the system than to carry out integrations from equations of state.

Isothermal Nonflow Expansion. For the reversible isothermal expansion of any fluid under nonflow conditions, from Equation (7) and the definitions of the thermodynamic energy functions

$$(w_s)_T \text{ (nonflow)} = \int_1^2 p\, dV = -\Delta A = T\,\Delta S - \Delta U$$
$$= T\,\Delta S - \Delta H + \Delta(pV) \tag{11}$$

Equation (11) is restricted to reversible isothermal nonflow operations involving only shaft work and it neglects changes in kinetic, potential, and surface energies.

For ideal gases, $p\,dV$ is evaluated by differentiating the ideal-gas law at constant temperature. Thus, for 1 mole,

$$p\,dv = -v\,dp = -RT\frac{dp}{p} \tag{12)*}$$

Substitution of (12) in (11) and integration give

$$(w_s^*)_T \text{ (nonflow)} = -RT \ln \frac{p_2}{p_1} \tag{13)*}$$

For actual gases, $pv = zRT$. Differentiation at constant temperature gives

$$p\,dv + v\,dp = RT\,dz \tag{14}$$

Hence

$$(dw_s)_T \text{ (nonflow)} = p\,dv = -v\,dp + RT\,dz = -RTz\frac{dp}{p} + RT\,dz \tag{15}$$

or

$$(w_s)_T \text{ (nonflow)} = -RT\int_{p_1}^{p_2} z\,d\ln p + RT(z_2 - z_1) \tag{16}$$

The integral of Equation (16) can be evaluated graphically by plotting values of z as a function of log p and determining the area under the curve between the given pressure limits. Values of z may be obtained from Fig. 103 or from experimental measurements. Where an accurate equation of state such as the Beattie–Bridgman is available, the integration may be carried out analytically.

The foregoing graphical integration can be avoided by evaluating the work from the entropy and enthalpy terms of Equation (11). If the generalized correlations are employed, Equation (11) may be rearranged to give, for 1 mole of fluid,

$$\begin{aligned}(w_s)_T\,(\text{nonflow}) = T[(\text{s}_2^* - \text{s}_1^*) - (\text{s}_2^* - \text{s}_2) + (\text{s}_1^* - \text{s}_1)] \\ - [(\text{H}_2^* - \text{H}_1^*) - (\text{H}_2^* - \text{H}_2) + (\text{H}_1^* - \text{H}_1)] \\ + RT(z_2 - z_1)\end{aligned} \tag{17}$$

or, combining with Equation (XI–142), page 475,

$$\begin{aligned}(w_s)_T\,(\text{nonflow}) = T\left[R \ln \frac{p_1}{p_2} - (\text{s}_2^* - \text{s}_2) + (\text{s}_1^* - \text{s}_1)\right] \\ + T_c\left[\frac{(\text{H}_2^* - \text{H}_2)}{T_c} - \frac{(\text{H}_1^* - \text{H}_1)}{T_c}\right] \\ + RT(z_2 - z_1)\end{aligned} \tag{18}$$

The terms of Equation (18) may be obtained directly as functions of reduced temperature and pressure from Figs. 103, 106, and 107.

Isothermal Expansion, Flow Conditions. For the reversible isothermal expansion of any fluid under flow conditions, from Equations (8),

$$(w_s)_T\,(\text{flow}) = -\int_1^2 V\,dp = -\Delta G = -\Delta H + T\,\Delta S \tag{19}$$

Equation (19) is restricted to reversible isothermal flow conditions involving only shaft work and neglects changes in kinetic, potential, and surface energies.

For an ideal gas, per mole,

$$(w_s^*)_T\,(\text{flow}) = -RT \ln \frac{p_2}{p_1} \tag{20}^*$$

Thus for an ideal gas the work of reversible isothermal expansion is the same for flow as for nonflow conditions. This results from the fact that at constant temperature the change in flow work is zero for an ideal gas, or $d(pV) = 0$.

For an actual gas, per mole, $-v\,dp = -z(RT\,dp)/p$ or

$$(w_s)_T \text{ (flow)} = -RT \int_1^2 z\,d\ln p \tag{21}$$

Equation (21) is evaluated graphically by the method described for Equation (16) by using generalized compressibility factors or an accurate equation of state. The graphical or analytical integration can be avoided by evaluating the work from the entropy and enthalpy terms of Equation (19). If the generalized correlations of Figs. 106 and 107 are used, Equation (19) is rearranged as follows:

$$(w_s)_T \text{ (flow)} = T\left[R\ln\frac{p_1}{p_2} - (\text{s}_2^* - \text{s}_2) + (\text{s}_1^* - \text{s}_1)\right] - T_c\left[\frac{(\text{H}_1^* - \text{H}_1)}{T_c} - \frac{(\text{H}_2^* - \text{H}_2)}{T_c}\right] \tag{22}$$

Illustration 1. Calculate the work of reversible isothermal expansion when 1 lb-mole of ethylene gas initially at 564°R and 50 atm expands to 1 atm under nonflow and flow conditions. From Table XXV, $t_c = 9.7°\text{C}$; $p_c = 50.9$ atm

(*a*) Assuming *ideal behavior* for nonflow and flow conditions, from Equation (20),

$$(w_s^*)_T = -RT\ln\frac{p_2}{p_1} = +(1.987)(564)(2.303)\log 50 = 4{,}370 \text{ Btu}$$

(*b*) For *actual-behavior* nonflow conditions the value of $\int_1^2 z\,d\ln p$ in Equation (16) is obtained by plotting values of z against values of $\log p$ and integrating over the limits log 50 to log 1. The necessary data, read from Fig. 103 at a reduced temperature of 1.11 for reduced pressures from 0.0197 (1.0 atm) to 0.983 (50 atm) are tabulated as follows:

p_r	p (atm)	log p	z
0.0196	1.0	0.0	1.00
0.1	5.09	0.707	0.98
0.2	10.18	1.008	0.96
0.3	15.27	1.118	0.94
0.4	20.36	1.304	0.90
0.5	25.45	1.406	0.87
0.6	30.54	1.484	0.84
0.7	35.63	1.552	0.805
0.8	40.72	1.610	0.77
0.9	45.81	1.661	0.73
0.982	50.0	1.699	0.71

The area under this curve is determined to be −1.59. Then, substitution in Equation (16) gives

$$w_s = (1.987)(564)\,[(2.303)(1.59) + (1.00\text{–}0.71)] = 4440 \text{ Btu}$$

This problem can be solved without graphical integration by means of Equation (18).

From Fig. 106, at $T_{r2} = 1.11$; $p_{r2} = 0.0196$; $T_{r1} = 1.11$; $p_{r1} = 0.982$

$$(\text{H}_2^* - \text{H}_2) = (0.04)(509) = 20; \quad (\text{H}_1^* - \text{H}_1) = (2.30)(509) = 1170$$

From Fig. 107,

$$(\text{S}_2^* - \text{S}_2) = 0.03; \qquad (\text{S}_1^* - \text{S}_1) = 1.55$$

From Fig. 103, $z_2 = 1.00$; $z_1 = 0.71$

Substitution in Equation (18) gives

$$w_s = 564(1.987 \ln 50 - 0.03 + 1.55) - (1170 - 20) + (1.987)(564)(1.0 - 0.71)$$
$$= 4{,}430 \text{ Btu}$$

This is in agreement with the value obtained by the graphical method.

(c) For *actual-behavior flow conditions*, the work is expressed by Equation (21) which involves the same integral as evaluated in part *b*; hence,

$$w_s = (1.987)(564)(2.303)(1.59) = 4115 \text{ Btu}$$

Similarly, from Equation (22), the data developed in part *b*, give

$$w_s = 564(1.987 \ln 50 - 0.03 + 1.55) - (1170 - 20) = 4105 \text{ Btu}$$

Where thermodynamic charts are available for any particular system, the reversible work can be obtained directly. The applications of these charts for evaluating the reversible work term under isothermal, isentropic, flow and nonflow conditions are shown in Illustrations 2 and 3.

Illustration 2. From the data of Figs. 116–119, calculate the work performed in the reversible isothermal expansion of 1 lb of ammonia initially at 200 lb per sq in. and 150°F to 15 lb per sq in. under nonflow and flow conditions.

From Fig. 116,

$$V_1 = 1.740; \qquad V_2 = 25.5 \text{ cu ft per lb}$$
$$p_1V_1 = (200)(144)(1.740) = 50{,}100 \text{ ft-lb per lb}$$
$$p_2V_2 = (15)(144)(25.5) = 55{,}000 \text{ ft-lb per lb}$$
$$\Delta(pV) = 4{,}900 \text{ ft-lb per lb}$$

From Fig. 117,

$$H_1 = 671 \text{ Btu per lb}$$
$$H_2 = 697 \text{ Btu per lb}$$
$$\Delta H = 26 \text{ Btu per lb or } 20{,}200 \text{ ft-lb per lb}$$

From Fig. 118,

$$S_1 = 1.244 \text{ Btu}/(\text{lb})(°\text{R})$$
$$S_2 = 1.575 \text{ Btu}/(\text{lb})(°\text{R})$$
$$\Delta S = 0.331 \text{ Btu}/(\text{lb})(°\text{R}) = 257.5 \text{ ft-lb}/(\text{lb})(°\text{R})$$
$$T\Delta S = (610)(257.5) = 157{,}000 \text{ ft-lb per lb}$$

Under nonflow conditions, by Equation (11),

$$w_s = T\,\Delta S - \Delta H + \Delta(pV) = 157{,}000 - 20{,}200 + 4900 = 141{,}700 \text{ ft-lb per lb}$$

Under flow conditions, by Equation (19),

$$w_s = -\Delta H + T\,\Delta S = -20{,}200 + 157{,}000 = 136{,}800 \text{ ft-lb per lb}$$

Isentropic Nonflow Expansion. For reversible isentropic expansion under nonflow conditions, from Equation (9),

$$(w_s)_S \text{ (nonflow)} = \int_1^2 (p\,dV)_S = -(\Delta U)_S = [-\Delta H + \Delta(pV)]_S \quad (23)$$

Equation (23) applies to isentropic nonflow conditions where changes in kinetic, potential, and surface energies are negligible and where only mechanical work is involved. Calculation of the work is complicated by the fact that the final temperature or pressure after expansion is unknown and must be evaluated. *For an ideal gas*, from Equation (XI–137), page 475,

$$d\text{u}^* = c_v^*\,dT \quad (24)^*$$

The ideal-gas law may be written in differential form as $p\,dv + v\,dp = R\,dT$. Combining this expression with Equations (23) and (24), gives

$$p\,dv = -v\,dp + R\,dT = -c_v^* dT \quad (25)^*$$

Eliminating v from (25) and rearranging and combining with Equation (XI–134), page 474, gives

$$-\frac{R}{p}\,dp = -(c_v^* + R)\frac{dT}{T} = -c_p^*\frac{dT}{T} \quad (26)^*$$

Where moderate temperature changes are involved, it is generally satisfactory to assume a constant mean value of $(c_p^*)_m$ at the arithmetic mean temperature. Then Equation (26) is integrated to give

$$R \ln\left(\frac{p_2}{p_1}\right) = (c_p^*)_m \ln\left(\frac{T_2}{T_1}\right) \quad (27)^*$$

or

$$T_2 = T_1\left(\frac{p_2}{p_1}\right)^{\frac{R}{(c_p^*)_m}} \quad (28)^*$$

From Equations (23), (28), and (24),

$$(w_s^*)_S \text{ (nonflow)} = (c_v^*)'_m(T_1 - T_2) = [(c_p^*)'_m - R]T_1\left[1 - \left(\frac{p_2}{p_1}\right)^{\frac{R}{(c_p^*)_m}}\right] \quad (29)^*$$

It should be noted that $(c_p^*)_m$ and $(c_p^*)'_m$ are different averages which result from differential equations involving $d \ln T$ and dT, respectively.

The corresponding volume changes are obtained by combining the ideal-gas law with Equation (28). Thus,

$$V_2 = V_1 \left(\frac{p_2}{p_1}\right)^{\frac{R}{(c_p^*)_m} - 1} \quad \text{or} \quad \left(\frac{V_2}{V_1}\right)^{\frac{(c_p^*)_m}{(c_p^*)_m - R}} = \frac{p_1}{p_2} \tag{30*}$$

Equations (28), (29), and (30) are frequently written in terms of κ, the ratio of the heat capacities at constant pressure and constant volume. For an ideal gas,

$$\kappa = \frac{c_p^*}{c_p^* - R} = \frac{c_v^* + R}{c_v^*} \tag{31*}$$

Combination of (31) with (28), (29), and (30), respectively, gives

$$T_2 = T_1 \left(\frac{p_2}{p_1}\right)^{\frac{\kappa - 1}{\kappa}} \tag{32*}$$

$$(w_s^*)_S \text{ (nonflow)} = \frac{RT_1}{\kappa - 1}\left[1 - \left(\frac{p_2}{p_1}\right)^{\frac{\kappa - 1}{\kappa}}\right] \tag{33*}$$

$$\left(\frac{V_2}{V_1}\right)^{\kappa} = \left(\frac{p_1}{p_2}\right) \tag{34*}$$

Equations (27) to (34) are restricted to expansions of ideal gases involving small changes in heat capacity. Where the heat-capacity terms are not satisfactorily represented by constant mean values, empirical equations relating heat capacity to temperature should be used in the integrations of Equations (25) and (26), as shown in Equation (37).

For *actual gases* work is evaluated more readily from the terminal state properties using Equation (23) than from the pV integration. Since the expansion is isentropic, $\Delta s = 0$, or

$$\Delta s = s_2 - s_1 = (s_2^* - s_1^*) - (s_2^* - s_2) + (s_1^* - s_1) = 0 \tag{35}$$

From Equation (XI–142), page 475, for 1 mole

$$s_2^* - s_1^* = \int_{T_1}^{T_2} c_p^* \, d \ln T - R \ln \frac{p_2}{p_1} \tag{36*}$$

where c_p^*, the molal heat capacity for ideal behavior, is a function of temperature only.

Combination of (35) and (36) gives

$$\int_{T_1}^{T_2} c_p^* \, d \ln T - R \ln \frac{p_2}{p_1} - (s_2^* - s_2) + (s_1^* - s_1) = 0 \tag{37}$$

The integral of Equation (37) is evaluated analytically from the empirical heat-capacity equation for the ideal gas. The equation is then solved for the final temperature corresponding to a specified final pressure and initial state. This is best done by assuming a series of final temperatures and plotting the corresponding values of the left-hand side of the equation. For ideal gases the two $(s^* - s)$ terms become zero.

Once the initial and final states are defined the reversible work is calculated directly from Equation (23) from the terminal enthalpy and pV values.

If condensation occurs during the expansion, Equation (37) is applicable only until conditions of saturation are reached. A final temperature calculation from Equation (37) which is lower than the boiling point at the final pressure is proof of condensation and establishes the correct final temperature as the boiling point under the final pressure.

Where a chart of the thermodynamic properties of the system is available, the final temperature is obtained directly by the following:

$$s_2 = x_2 s_{2s} + (1 - x_2)\left(s_{2s} - \frac{\lambda_2}{T_2}\right) = s_{2s} - (1 - x_2)\frac{\lambda_2}{T_2} \tag{38}$$

or

$$\Delta s = s_2 - s_1 = (s_{2s}^* - s_1^*) - (s_{2s}^* - s_{2s}) + (s_1^* - s_1) - (1 - x_2)\frac{\lambda_2}{T_2} = 0 \tag{39}$$

where s_{2s} is the entropy per mole of the saturated vapor and λ_2 is the molal heat of vaporization at temperature T_2 which is calculated by the methods of Chapter VII. Equation (39) may be solved for quality x_2 by use of Equation (36), Fig. 107, and the temperature of condensation corresponding to the final pressure p_2.

Once the final quality is evaluated, the work is calculated from the enthalpy and pV changes of Equation (23). The enthalpies and volumes of the saturated liquid and vapor are separately evaluated by the methods previously demonstrated. If the generalized correlations are used, the complete expression for the shaft work per mole is as follows:

$$(w_s)_S\,(\text{nonflow}) = \int_{T_2}^{T_1} c_p^*\,dT - T_c\left[\frac{H_1^* - H_1}{T_c} - \frac{(H_2^* - H_2)}{T_c}\right] - R[z_1T_1 - x_2z_2T_2] + (1 - x_2)(\lambda_2 + p_2v_{L2}) \tag{40}$$

v_{L2} = molal volume of the liquid phase at the final conditions.

Ordinarily the volume of the liquid phase at the final conditions is negligible and the term p_2v_{L2} may be omitted.

Isentropic Expansion, Flow Conditions. *For an ideal gas,* since heat capacity is independent of pressure, Equation (10) may be written

$$(w_s^*)_S \text{ (flow)} = -\int_1^2 V\,dp = -\Delta H = -\int_1^2 C_p^*\,dT \qquad (41)^*$$

Equation (41) applies to the work in reversible isentropic expansion under flow conditions and neglects changes in external kinetic potential, and surface energies. Equation (41) combined with the ideal gas law for 1 mole gives

$$-\int_1^2 R\,d\ln p = -\int_1^2 c_p^*\,d\ln T \qquad (42)^*$$

If a constant mean value of heat capacity is assumed, Equation (42) integrates to give Equation (28), showing that for ideal gases the temperature changes in isentropic expansion are the same for flow and nonflow processes.

Combination of Equations (28) and (42) for 1 mole gives

$$(w_s^*)_S \text{ (flow)} = (c_p^*)'_m(T_1 - T_2) = (c_p^*)'_m T_1\left[1 - \left(\frac{p_2}{p_1}\right)^{\frac{R}{(c_p^*)_m}}\right] \qquad (43)^*$$

It may be noted that Equation (43) differs from (29) in the appearance of $(c_p^*)'_m$ instead of $(c_v^*)'_m$ as the multiplying coefficient.

The temperature and volume changes of isentropic flow expansion are expressed in terms of the heat-capacity ratio by Equations (32) and (34). By a similar development

$$(w_s^*)_S \text{ (flow)} = \frac{RT_1\kappa}{\kappa - 1}\left[1 - \left(\frac{p_2}{p_1}\right)^{\frac{\kappa-1}{\kappa}}\right] \qquad (44)^*$$

For an actual gas under *flow conditions* the procedure is similar to that followed in the isentropic nonflow case. Equations (35–39) are all directly applicable to the flow case since they express the condition of constant entropy. Thus, it may be concluded that for any fluid the temperature changes in isentropic expansion are the same for flow and nonflow processes.

If the final temperature is known, values of w_s follow from Equation (10). Thus, if the generalized correlations are used, the work per mole is expressed by

$$(w_s)_S \text{ (flow)} = \int_{T_2}^{T_1} c^*_p\,dT - T_c\left[\frac{H_1^* - H_1}{T_c} - \frac{H_2^* - H_2}{T_c}\right] + (1 - x_2)\lambda_2 \qquad (45)$$

When Equations (23) and (41) are compared, it is evident that the work in isentropic flow expansion is greater than that of a nonflow operation by the amount of $-\Delta(pV)$.

Illustration 3. Calculate the work of isentropic expansion of 1 lb-mole of ethylene gas when expanded from an initial pressure of 50 atm and 564°R to a final pressure of 1 atm under nonflow and flow conditions.

For nonflow conditions assuming ideal behavior, an equation for the heat capacity of ethylene, at zero pressure in the temperature range below 200°F is given in Table XXI, page 336.

$$c_p^* = 7.95 + 8.13 \times 10^{-11}(T)^{3.85}$$

Combining this equation with Equation (26) and integrating gives

$$\int_{564}^{T_2} \left[\frac{7.95}{T} + 8.13(10^{-11})(T)^{2.85}\right] dT = R \ln \frac{1}{50} = 7.95 \ln \frac{T_2}{564} + 2.11(10^{-11})[(T_2)^{3.85} - (564)^{3.85}]$$

This equation is solved graphically for T_2 giving $T_2 = 234$°R. The reversible work of expansion is obtained by combination of Equations (23) and (25) with the heat-capacity equation assuming no condensation:

$$(w_s^*)_S \text{(nonflow)} = -\int_{564}^{234} c_v \, dT = \int_{234}^{564} [7.95 + 8.13(10^{-11})(T)^{3.85} - R] \, dT = 2334 \text{ Btu per lb-mole}$$

$$(w_s^*)_S\text{(flow)} = \int_{234}^{564} [7.95 + 8.13(10^{-11})(T)^{3.85}] \, dT = 2990 \text{ Btu per lb-mole}$$

In actual behavior, condensation occurs at $T_2 = 304$°R, the boiling point at pressure of 1 atm. The properties at the initial and final state are as follows:

	1	2
Temperature, °R	564	304
Reduced temperature, T_r	1.11	0.598
Pressure, p, atm	50	1.0
Reduced pressure, p_r	0.982	0.0196
$(H^* - H)/T_c$ (Fig. 106)	2.3	0.39
$(S^* - S)$ (Fig. 107)	1.55	0.64
z (Fig. 103)	0.71	0.98

The molal heat of vaporization at 1 atm is estimated from the available experimental data as 6200 Btu per lb-mole. From Equation (36),

$$S_2^* - S_1^* = 7.95 \ln \frac{304}{564} + 2.11(10^{-11})[(304)^{3.85} - (564)^{3.85}] - 1.99 \ln \frac{1}{50} = -4.91 - 0.76 + 7.76 = 2.09$$

Substitution in Equation (39) gives

$$x = 1 - \frac{304}{6200}[2.09 - 0.64 + 1.55] = 0.85$$

The terms in Equation (40) may be evaluated individually, the volume of the liquid being neglected. Thus,

$$\text{H}_1^* - \text{H}_2^* = \int_{304}^{564} [(7.95 + 8.13(10^{-11})T^{3.85})]\,dT = 2417 \text{ Btu per lb-mole}$$

$$R(z_1T_1 - x_2z_2T_2) = 1.987[(0.71)(564) - (0.85)(0.98)(304)]$$

$$= 290 \text{ Btu per lb-mole}$$

$$T_c\left(\frac{\text{H}_1^* - \text{H}_1}{T_c} - \frac{\text{H}_2^* - \text{H}_2}{T_c}\right) = 508(2.3 - 0.39) = 972$$

$$(1 - x)\lambda = (0.15)(6200) = 930$$

Substituting in Equation (40)

$$(w_s)_S \text{ (nonflow)} = 2417 - 972 - 290 + 930 = 2085 \text{ Btu per lb-mole}$$

Substitution in Equation (45) gives

$$(w_s)_S \text{ (flow)} = 2417 - 972 + 930 = 2375$$

Illustration 4. Calculate from Figs. 116-119 the work of isentropic expansion of 1 lb of ammonia from an initial absolute pressure of 200 lb per sq in. and 150°F to a final absolute pressure of 15 lb per sq in. under nonflow and flow conditions.

From Fig. 118, following a vertical line from the initial conditions to the final pressure gives

$$t_2 = -27°\text{F}; \qquad x_2 = \frac{1.243 - 0.032}{1.394 - 0.032} = 89.0 \text{ per cent}$$

Similarly, from Fig. 119 it is found that

$$H_1 = 671 \text{ Btu per lb}$$

$$H_2 \text{ (saturated gas)} = 602.4 \text{ Btu per lb}$$

$$H_2 \text{ (saturated liquid)} = 13.6 \text{ Btu per lb}$$

$$H_2 \text{ (mixture)} = x(602.4) + (1 - x)(13.6) = 537.6 \text{ Btu per lb}$$

From Fig. 116, at the final temperature, pressure, and quality:

$$V_2 \text{ (saturated gas)} = 17.6 \text{ cu ft per lb}$$

$$xp_2V_2 = (0.89)(15)(144)(17.6) = 33{,}800 \text{ ft-lb per lb}$$

The volume of the liquid phase may be neglected.

For the initial conditions, $V_1 = 1.74$ (Fig. 116)

$$p_1V_1 = (200)(144)(1.74) = 50{,}100 \text{ ft-lb per lb}$$

Using the data of Illustration 2 for the initial state, we get

$$\Delta H = 538 - 671 = -133 \text{ Btu per lb or } -103{,}300 \text{ ft-lb per lb}$$

$$\Delta(pV) = 33{,}800 - 50{,}100 = -16{,}300 \text{ ft-lb per lb}$$

For nonflow conditions, from Equation (23),

$$w_s = 103{,}300 - 16{,}300 = 87{,}000 \text{ ft-lb per lb}$$

For flow conditions, from Equation (41),

$$w_s = 103{,}300 \text{ ft-lb per lb}$$

FREE EXPANSION

The unrestrained expansion of a gas is known as free expansion. Under conditions of no restraint no work is done, and under adiabatic conditions no heat is added. Free expansion under flow conditions is commonly known as throttling or as the Joule–Thomson effect referred to in Chapter XII.

The Joule–Thomson effect is measured experimentally by expanding the gas slowly and in steady flow through a well insulated porous plug; in this way potential work is lost, and no heat is allowed to enter or leave the system through the walls. The fluid flows reversibly into and out of the process, but the expansion step is completely irreversible. Since no heat is added or lost, the process takes place under conditions of constant enthalpy, according to Equation (VII–11), page 207. However a change in internal energy results if any change in flow energy pV occurs.

For an ideal gas, since enthalpy is independent of pressure, no change in temperature occurs in free expansion under either nonflow or flow conditions, and values of ΔH, $\Delta(pV)$, and ΔU are zero. However, an increase in entropy results which is equal to $-R \ln p_2/p_1$ by Equation (XI–142), and the values of ΔA and ΔG are each equal to $-T\,\Delta S$.

For nonideal gases, the temperature change in Joule–Thomson expansion is of great experimental and practical value. Such data are used for establishing the deviations from ideal behavior of real gases, particularly the effects of pressure on enthalpy and entropy. Free expansion under flow conditions is made use of industrially in the cooling and liquefaction of gases.

Free expansion under nonflow conditions, known as the Maxwell effect, is of little practical value. Because of the small heat capacity of a gas compared with that of the vessel, adiabatic conditions cannot be approached, and experimental measurements are of uncertain accuracy.

In the foregoing discussion it was assumed that kinetic energy changes were negligible as is generally the case. Where this is not true, the general energy-balance equation including kinetic-energy changes must be employed. Changes in both internal energy and temperature may result from the free expansion of an ideal gas at high velocities in either a flow or nonflow system.

Joule–Thomson Expansion. The Joule–Thomson coefficient is given by Equation (XII–35), page 501. Temperature changes in free expansion where only small changes are involved and where the heat capacity may be taken as constant may be calculated from Equation (XII–37). Where large temperature changes are involved, these calculations are more conveniently carried out through differences in state properties.

A Joule–Thomson expansion is represented on a thermodynamic chart by a line of constant enthalpy. By following such a line from the initial state to the final pressure, the final temperature and state are determined.

The state changes accompanying Joule–Thomson expansion also may be calculated from the generalized correlations of thermodynamic properties. Since the expansion is isenthalpic, where changes in kinetic energy, potential head, and surface energy are negligible,

$$\Delta H = \int_{T_1}^{T_2} c_p^* \, dT - T_c \left[\frac{H_2^* - H_2}{T_c} - \frac{H_1^* - H_1}{T_c} \right] = 0 \tag{46}$$

Equation (46) may be solved by assuming values of T_2 until the conditions of the equation are satisfied by graphical or trial methods.

If condensation occurs during the expansion, the final temperature will be the boiling point at the final pressure. Equation (46) must be modified to include the enthalpies of the liquid and vapor phases and the quality x_2. Thus,

$$\Delta H = \int_{T_1}^{T_2} c_p^* \, dT - T_c \left[\frac{H_{2s}^* - H_{2s}}{T_c} - \frac{H_1^* - H_1}{T_c} \right] - (1 - x_2)\lambda_2 = 0 \tag{47}$$

where λ_2 is the heat of vaporization at the final conditions and H_{2s} is the enthalpy per mole of the saturated vapor at state 2. Equation (47) may be solved for x_2 since the final temperature T_2 is known.

After the final temperature T_2 is evaluated, the changes in the other state properties follow. Thus,

$$\Delta(pv) = x_2 z_2 R T_2 + (1 - x_2) p_2 v_{L2} - z_1 R T_1 \tag{48}$$

where v_{L2} is the molal volume of the liquid phase.

$$\Delta S = \int_{T_1}^{T_2} c_p^* \, d \ln T - R \ln \frac{p_2}{p_1} - (S_{2s}^* - S_{2s}) + (S_1^* - S_1) - \frac{\lambda_2}{T_2}(1 - x_2) \tag{49}$$

$$\Delta U = \Delta H - \Delta(pv) = z_1 R T_1 - x_2 z_2 R T_2 - (1 - x_2) p_2 v_{L2} \tag{50}$$

$$\Delta A = z_1 R T_1 - x_2 z_2 R T_2 - (1 - x_2) p_2 v_{L2} - T_2 S_{2s} + \lambda_2 (1 - x_2) + T_1 S_1 \tag{51}$$

$$\Delta G = -T_2 S_{2s} + \lambda_2 (1 - x_2) + T_1 S_1 \tag{52}$$

If continuous free expansion is carried out at high velocities, the changes in kinetic energy are not negligible. In this case, from the general energy-balance Equation (VII-1), page 205, it is evident that if potential and surface energy changes are negligible a decrease in enthalpy occurs which is equal to the increase in kinetic energy. Thus, from an energy balance where no condensation occurs:

$$\Delta H = \frac{-m(u_2^2 - u_1^2)}{2g_c} = (H_2^* - H_1^*) - T_c\left(\frac{H_2^* - H_2}{T_c}\right) + T_c\left(\frac{H_1^* - H_1}{T_c}\right) \quad (53)$$

The final temperature T_2 may be evaluated from Equation (53), and the other state properties may be calculated from Equations (48–52).

Maxwell Expansion. As previously pointed out, adiabatic free expansion under nonflow conditions is characterized by constant internal energy. Thus,

$$\Delta U = 0 = \Delta H - \Delta(pV) \quad (54)$$

Equation (54) may be expressed in terms of generalized corrections and ideal behavior to give, in the absence of condensation,

$$\Delta U = \int_{T_1}^{T_2} c_p^* \, dT - T_c\left(\frac{H_2^* - H_2}{T_c}\right) + T_c\left(\frac{H_1^* - H_1}{T_c}\right) - z_2RT_2 + z_1RT_1 = 0 \quad (55)$$

The final temperature T_2 is determined by trial and error or by graphical solution of Equation (55). In the case of condensation, Equation (55) is modified by subtracting the term $(1 - x_2)(\lambda_2 + p_2v_{L2})$ and replacing z_2RT_2 by $x_2z_2RT_2$.

Thermodynamic charts of the type of Figs. 116–119 do not permit direct solution of problems in Maxwell expansion. A trial-and-error procedure is necessary in which a final state is assumed, and the assumption is tested by means of Equation (54).

From a knowledge of T_2, the changes in the other state properties may be evaluated from equations similar to (48) to (52).

Illustration 5. Estimate the changes in state properties in the free expansion of ethylene gas from 564°R and 50 atm to 1 atm under flow and nonflow conditions. The absolute entropy of ethylene at 77°F and 1 atm is 52.48 Btu/(lb-mole)(°R). The initial absolute entropy of the ethylene gas is 43.72 Btu/(lb-mole)(°R).

$$T_1 = 564°\text{R}; \quad T_{r1} = 1.11$$
$$p_1 = 50 \text{ atm}; \quad p_{r1} = 0.982; \quad z_1 = 0.71$$

From Fig. 106, $H_1^* - H_1 = 1170$
From Fig. 107, $S_1^* - S_1 = 1.55$

(a) *Flow conditions* (*Joule–Thomson effect*). The final temperature is evaluated by graphical solution of Equation (46). A trial value of $T_2 = 460°$R is assumed. Then

$$T_{r2} = 0.905; \quad p_{r2} = 0.0196$$

From Fig. 106, $$H_2^* - H_2 = \frac{0.0196}{0.03}(0.1)(509) = 33$$

The term 0.0196/0.03 represents the linear factor for pressures which are off the chart. At $p_r = 0.03$; $(H^* - H)/T_c = 0.10$. From the data of Illustration 3,

$$H_2^* - H_1^* = \int_{564}^{460} [7.95 + (8.13)10^{-11}T^{3.85}]dT = -1060$$

Substitution of these values into Equation (46) gives

$$\Delta H = -1060 - 33 + 1170 = 77 \text{ Btu per lb-mole}$$

By repeating the preceding procedure a value of $\Delta H = 0$ is obtained when $T_2 =$ 448°R. At this temperature there is no condensation and $x_2 = 1$. The change in entropy is evaluated from Equation (49).

At state 2,

$$T_{r2} = \frac{448}{508} = 0.882; \qquad p_{r2} = 0.0196; \qquad z_2 = 0.99$$

From Fig. 107, $$S_2^* - S_2 = \frac{0.0196}{0.03}(0.1) = 0.07$$

$$\int_{564}^{448} \left[\frac{7.95}{T} + (8.13)10^{-11}T^{2.85}\right] dT - R \ln \frac{1}{50} = 5.45$$

Substituting in Equation (49), yields

$$\Delta S = 5.45 - 0.07 + 1.55 = 6.93$$

Hence, $$S_2 = 6.93 + 43.72 = 50.65 \text{ Btu/(lb-mole)(°R)}$$

The change in internal energy is evaluated from Equation (50)

$$\Delta U = (0.71)(1.987)(564) - (0.99)(1.987)(448) = -88 \text{ Btu per lb-mole}$$
$$\Delta(TS) = 448(50.65) - (564)(43.72) = -1975 \text{ Btu per lb-mole}$$

$$\Delta A = \Delta U - \Delta(TS) = -88 + 1975 = 1887 \text{ Btu per lb-mole}$$
$$\Delta G = \Delta H - \Delta(TS) = 1975 \text{ Btu per lb-mole}$$
$$w = 0; \quad q = 0; \quad v_2 = (0.99)(0.729)(448) = 322 \text{ cu ft per lb-mole}$$

(*b*) *Nonflow conditions* (*Maxwell effect*). The final temperature is determined by graphical solution of Equation (55) by the same procedure followed in solving Equation (46) in part *a*. Thus, if 460°R is taken as a trial value of T_2, from part *a*,

$$H_2^* - H_1^* = 1060 \text{ and } -(H_2^* - H_2) + (H_1^* - H_1) = -33 + 1170$$

Values of z from Fig. 103 give

$$z_2RT_2 = (0.99)(1.987)(460) = 903$$
$$z_1RT_1 = (0.71)(1.987)(564) = 793$$

Substitution in Equation (55) gives $\Delta U = -1060 - 33 + 1170 - 903 + 793 = -33.$

Since $\Delta U > 0$, the value of 460°R is too high. At $\Delta U = 0$, the value of T_2 is found to be 458°R. Then

$$T_{r2} = 0.90; \qquad p_{r2} = 0.0196; \qquad z_2 = 0.99$$

TABLE XXIX

THERMODYNAMIC CHANGES IN THE EXPANSION OF ETHYLENE GAS

Initial conditions: 564°R and 50 atm, $s_1 = 43.72$; $s_1^* = 45.27$

Final conditions: T_2 °R and 1 atm

For the low temperature range below 660°R,

$T_c = 509$°R; $c_p^* = 7.95 + 8.13(10^{-11})T^{3.85}$; ($T$ = °R) $p_c = 50.9$ atm

x_2 = quality of vapor after expansion equals 1 except where indicated

Basis: 1 lb-mole; all energy terms in Btu.

Conditions of Expansion	T_2 °R	v_2 cu ft	Δs	q	w_s	Δ(pv)	Δu	Δh	Δa	Δg	Δ(Ts)
Assuming ideal behavior (no condensation) v_1 = 8.10 cu ft											
1. Isothermal, nonflow, reversible	564	410.9	7.75	4370	4370	0	0	0	−4370	−4370	4370
2. Isothermal, flow, reversible	564	410.9	7.75	4370	4370	0	0	0	−4370	−4370	4370
3. Isentropic, nonflow	234	170.5	0	0	2334	−656	−2334	−2990	12,616	11,960	−14950
4. Isentropic, flow	234	170.5	0	0	2990	−656	−2334	−2990	12,616	11,960	−14950
5. Irreversible, nonflow	564	410.9	7.75	0	0	0	0	0	−4370	−4370	4370
6. Irreversible, flow (Joule–Thomson)	564	410.9	7.75	0	0	0	0	0	−4370	−4370	4370
Actual behavior v_1 = 5.65 cu ft											
1. Isothermal, nonflow, reversible	564	410.9	9.30	5250	4386	336	864	1200	−4386	−4050	5250
2. Isothermal, flow, reversible	564	410.9	9.30	5250	4050	336	864	1200	−4386	−4050	5250
3. Isentropic, nonflow, reversible, $x_2 = 0.849$	304	185.0	0	0	2085	−290	−2085	−2375	9315	9025	−11400
4. Isentropic, flow, reversible, $x_2 = 0.849$	304	185.0	0	0	2375	−290	−2085	−2375	9315	9025	−11400
5. Irreversible, nonflow (Maxwell), $x_2 = 1.0$	458	329.0	7.13	0	0	110	0	110	1370	1480	−1370
6. Irreversible, flow (Joule–Thomson), $x_2 = 1.0$	448	322	6.93	0	0	88	−88	0	1887	1975	−1975

From Fig. 106, $\text{H}_2^* - \text{H}_2 = 33$

From Fig. 107, $\text{S}_2^* - \text{S}_2 = \left(\frac{0.0196}{0.032}\right)(0.10) = 0.06$

From Equation (49),

$$\Delta\text{S} = \int_{564}^{458} \frac{c_p^* dT}{T} - R \ln \frac{1}{50} - 0.06 + 1.55 = 7.13$$

$$\text{S}_2 = 7.13 + 43.72 = 50.85 \text{ Btu/(lb-mole)(°R)}$$

$$\Delta(pV) = z_2RT_2 - z_1RT_1 = (0.99)(1.987)(458) - (0.71)(1.987)(564) = 110 \text{ Btu per lb-mole}$$

$$\Delta\text{H} = \Delta\text{U} + \Delta(pv) = 0 + 110 = 110 \text{ Btu per lb-mole}$$

$$\Delta(T\text{S}) = T_2\text{S}_2 - T_1\text{S}_1 = (458)(50.85) - (564)(43.72) = -1370 \text{ Btu per lb-mole}$$

$$\Delta\text{A} = \Delta\text{U} - \Delta(T\text{S}) = 1370 \text{ Btu per lb-mole}$$

$$\Delta\text{G} = \Delta\text{H} - \Delta(T\text{S}) = 110 + 1370 = 1480 \text{ Btu per lb-mole}$$

$$v_2 = (0.99)(0.729)(458) = 329 \text{ cu ft per lb-mole}$$

In Table XXIX are summarized the changes in point and path properties when ethylene gas is expanded from 50 atm and 564°R to 1 atm under various conditions of restraint and no restraint when both ideal and actual behaviors are assumed. It may be observed that for both ideal and actual behavior the changes in point properties under reversible conditions are identical for both flow and nonflow conditions, but work and heat terms are dependent upon the path and conditions of flow. For all reversible conditions $q = \int T\,dS$, and each of the work terms is identified with one of the energy functions. In free expansion, the heat and work terms are zero, and the changes in point properties for actual behavior are different for flow and nonflow conditions. It is interesting to note that free expansion cools ethylene gas to 458°R without condensation, whereas by isentropic expansion the gas is cooled to 304°R with 15 per cent condensed.

CYCLIC PROCESSES

A series of operations so conducted that all changes are periodically repeated in the same order is a *cyclic process*. This may comprise a sequence of either nonflow or flow operations or a combination of the two types, either repeated on the same substance or involving a new mass of substance for each cycle. For example, a reciprocating air compressor produces a repeated series of nonflow operations in which the conditions at a selected point of any cycle duplicate those at the same point of any other cycle. However, a new mass of gas is involved in each cycle, and a flow result is produced by the sequence of nonflow cycles.

On the other hand, in a steam-power plant the change may be continually recirculated through a series of operations such as feed-water heating, evaporation, superheating, expansion, and condensation. Such a cycle is essentially composed of flow operations but may involve component nonflow steps such as the expansion of steam in a reciprocating engine.

Nonflow cyclic operations repeated on the same mass of substance have no present practical applications but are of great value in establishing the theoretical behavior of cyclic processes in general.

Reversible Cycles. A cyclic process which is wholly reversible must of necessity be composed of individual steps each of which is in itself reversible. This requires that at no point in the cycle may there be degradation of higher forms of energy to heat through mechanical or fluid friction, nor may there be any other irreversible steps such as free expansion of a fluid or the transfer of heat under a finite temperature difference.

The concept of reversibility as discussed in Chapter XI requires that a reversible cycle may be operated in either a forward or a reverse direction and must return to any given initial state without change in any of the state properties of the system. Thus, it follows that *in any reversible cycle the algebraic sum of the changes in any state property for all steps must be equal to zero.* Thus, for any reversible cycle,

$$\Sigma\Delta S = \Sigma \int \frac{d'q}{T} = 0 \tag{56}$$

The Carnot Cycle. It has been previously stated that the maximum shaft work accomplished as a result of any change in state of a fluid is obtained when the change takes place reversibly without friction or turbulence. The Carnot engine has been conceived as a hypothetical device which represents this maximum achievement in a cyclic process and hence is used as a standard in evaluating the efficiency and performance of all actual mechanical cycles for transforming heat energy into mechanical energy.

In the Carnot engine, a gas is contained in a cylinder equipped with a frictionless piston connected to a frictionless receiver of mechanical energy. The cylinder walls and piston are impervious to heat, but the cylinder head is interchangeable alternately to an impervious or to a highly conducting plate. In the first stage of the cycle, that of isothermal expansion, heat is supplied to the gas through the conducting cylinder head from a source of heat maintained at a constant temperature T_1 and flows into the fluid as a result of an infinitesimal temperature drop. The gas is allowed to expand isothermally and reversibly with an influx of heat equal to q_1. The reversible work of expansion from state a to state b is $\int_a^b p\,dV$.

The first stage is followed by a reversible adiabatic or isentropic expansion which is accomplished theoretically by attaching the nonconducting head to the cylinder and permitting expansion to continue. Work of adiabatic expansion is performed at the expense of the internal energy of the fluid and in expanding from state b to state c is equal to $\int_b^c p\,dV$. In this stage the temperature decreases to $T_2 + dT$.

A receiver of heat is now provided by replacing the impervious cylinder head with a conducting plate in contact with the receiver at a constant temperature T_2. In the third stage, an isothermal compression is performed wherein work of compression $\int_c^d p dV$ is performed on the gas, and an amount of heat $-q_2$ flows into the receiver at T_2. In the fourth and last stage, the cylinder head is made nonconducting, and compression is continued under isentropic conditions from state d to the original state a with a reversible work of compression equal to $\int_d^a p\,dV$.

Since by completing the cycle the fluid is returned to its original state, its net internal energy change is zero. The same is true of all other point properties. Since no heat was received or lost during either adiabatic change, it follows from the law of conservation of energy that the net work done, w_{net}, must equal $q_1 + q_2$. This is equal to the difference between the amount of heat received from the reservoir at temperature T_1 and the amount rejected to the receiver at temperature T_2. Thus,

$$w_{\text{net}} = \int_a^b p\,dV + \int_b^c p\,dV + \int_c^d p\,dV + \int_d^a p\,dV$$

$$= \Sigma \int p\,dV = q_1 + q_2 \quad (57)$$

The change in entropy of the fluid during isothermal expansion is $\Delta S_1 = S_b - S_a = q_1/T_1$ and in isothermal compression is $\Delta S_2 = S_d - S_c = q_2/T_2$. Since the change in entropy over the entire cycle is zero, and the change during each reversible adiabatic stage is zero, it follows that $\Delta S_1 = -\Delta S_2$. Hence,

$$q_1 + q_2 = \Delta S_1(T_1 - T_2) \quad (58)$$

Since $q_1 + q_2 = w_{\text{net}}$ and $q_1 = T_1\,\Delta S_1$, it follows that

$$\frac{w_{\text{net}}}{q_1} = \frac{q_1 + q_2}{q_1} = \frac{T_1 - T_2}{T_1} \tag{59}$$

The ratio w_{net}/q_1 expressed by Equation (59) is termed the *thermodynamic efficiency* of the cycle. It will be recognized that an efficiency of 100 per cent can be realized only when the temperature of the receiver is at absolute zero.

The operation and efficiency of the Carnot cycle is shown graphically by straight lines on the temperature–entropy diagram, Fig. 123. During the first stage, the reversible isothermal expansion at temperature T_1 is represented by the horizontal line *ab*; the heat removed from the source is represented by the area $S_a abS_b$ or $T_1\,\Delta S$. During the second stage, the reversible adiabatic expansion is represented by the vertical line *bc* with no change in entropy and with no heat added. In the third stage, heat is rejected at T_2 along horizontal line *cd* and is represented by area $S_a\,dcS_b$ or $T_2\,\Delta S_1$. In the last stage the adiabatic compression moves along a vertical line *da* of constant entropy and with no heat rejected. The net work done w_{net} is represented by area *abcd* equal to $q_1 + q_2$. The thermodynamic efficiency of the cycle may be visualized as the ratio of area *abcd* to area $abcS_bS_ad$.

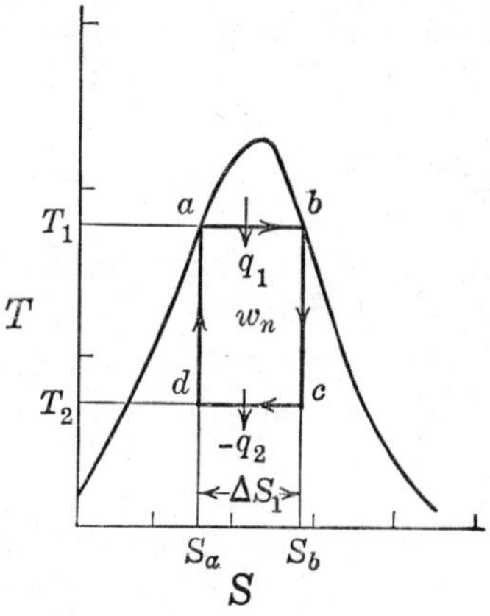

FIG. 123. The Carnot Cycle.

The Carnot cycle also may be operated by starting with a saturated liquid and vaporizing it during the first stage. This condition is indicated on Fig. 123 by the curved line which represents the temperature–entropy relationship of the saturated vapor and liquid of the working fluid. On such a vapor cycle two phases exist in the cylinder at all points except *a* where the system is completely liquefied and *b* where it is completely vaporized. Other methods of operation might involve varying degrees of liquefaction at the various points of the cycle, but in all cases the energy transformations are expressed by Equations (57–59).

The concept of the Carnot engine permits visualization of the significance of the *Carnot principle* and the second law of thermodynamics. The Carnot principle involves two basic conclusions which follow directly from Equations (56–59) and the second law of thermodynamics.

1. No self-acting engine which operates by the absorption of heat at one constant temperature and by the rejection of heat at another lower constant temperature can be more efficient than an engine operating on a reversible cycle with absorption and rejection of heat at only these same two temperature levels. If such an engine of higher efficiency existed, it could be used to drive a reversible Carnot engine operating as a heat pump. The net result of the two engines would then be the transfer of heat from a lower to a higher temperature, in contradiction of the second law.

2. The thermodynamic efficiencies of all reversible cycles which absorb heat at the same constant temperature and reject heat at the same constant lower temperature must be equal, regardless of the nature of the system or the fluid. This conclusion follows from Equation (59) which expresses efficiency, independent of system or fluid characteristics. Although this equation was derived by consideration of the Carnot engine, it is applicable to any other engine operating between two constant temperatures as required by Equation (59).

The Availability of Thermal Energy. The Carnot principle furnishes a basis for quantitatively expressing the availability of thermal energy in performing mechanical work. Thus, Equation (59) expresses the maximum fraction of the heat energy q_1 which can be converted into work by the most efficient machine possible when receiving heat from a temperature T_1 and rejecting heat at temperature T_2. The quantity $(q_1 + q_2)$ may thus be termed the available portion of q_1 with respect to temperature T_2. Similarly, q_2 is the unavailable portion with respect to this same temperature of rejection. An expression for the fractional unavailability of energy is obtained by rearranging Equation (59).

$$\text{Fraction unavailable} = \frac{-q_2}{q_1} = \frac{T_2}{T_1} \tag{60}$$

The temperature of rejection is ordinarily fixed by the temperature of the atmosphere or the available cooling water. As the temperature level of the energy source is reduced, its availability is decreased, and, when the temperature of the source becomes equal to the rejection temperature, heat energy becomes wholly unavailable, even in an ideal engine.

POWER-PLANT CYCLES

The production of shaft energy from the great store of chemical energy in fuels is accomplished by the highly irreversible process of combustion, whereby heat is released at high temperatures and transferred to a working fluid at a much lower temperature. In a steam-power plant part of the heating value of the fuel is transferred to water in a boiler; the steam generated expands against the piston of a steam engine, or through a turbine and is then condensed with rejection of heat at a lower temperature to a cold fluid flowing in a condenser; the condensate is pumped back into the boiler. A flow diagram of these operations is

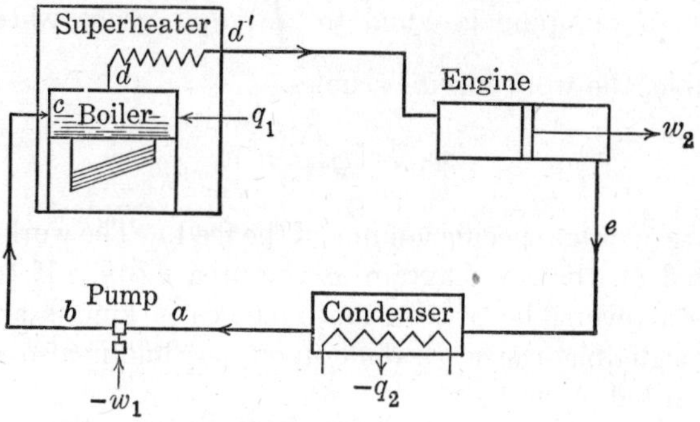

FIG. 124. Rankine-Cycle Power Plant.

shown in Fig. 124. When both expansion of steam and pumping of the liquid condensate occur reversibly, the *Rankine cycle* is followed. This cycle differs from the Carnot in that each stage is carried out in a different apparatus and particularly in that condensation is completed at the

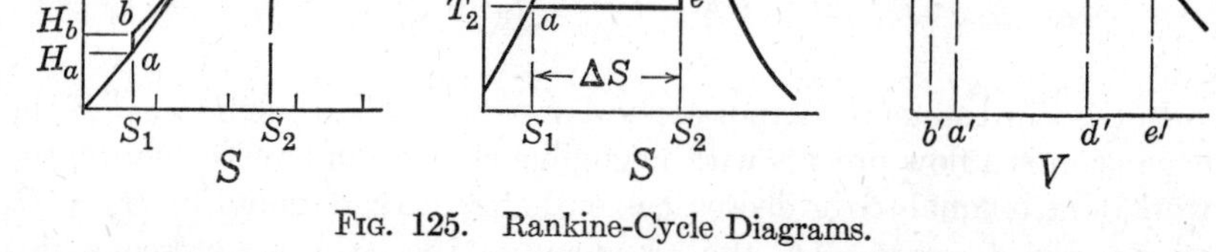

FIG. 125. Rankine-Cycle Diagrams.

low temperature and pressure of the condenser instead of by isentropic compression. The condensate is pumped into the boiler to mix with the hot liquid.

The changes taking place in the Rankine cycle are shown graphically in the *HS*, *TS*, and *pV* diagrams in Figure 125. In the first stage the feedwater is pumped from pressure p_2 to p_1 with an increase in enthalpy from H_a to H_b, as shown on *HS* diagram, and with a negligible rise in temperature, shown as a to b, on the TS diagram. The water enters the boiler subcooled, that is, at a temperature below that corresponding to its equilibrium pressure. The entropy remains constant during pumping, since the process is assumed to be reversible and adiabatic, as indicated by the vertical line ab on the HS diagram.

The work of pumping is equal to $\int V\,dp$. Since water is nearly incompressible, the work input becomes

$$-w_1 = (p_b - p_a)v_f \tag{61}$$

where v_f is the average specific volume of the feed. The work of pumping is represented on the pV diagram as the area $mban$. If the pumping operation is a reversible process, and changes in kinetic and potential energy are negligible, the work done upon the fluid is also equal to the increase in enthalpy, or

$$-w_1 = H_b - H_a \tag{62}$$

In the second stage the water is heated along line bcd, becomes saturated at point c and temperature T_1 and then vaporizes at constant temperature T_1 and pressure p_1 to form saturated vapor at d. The TS diagram shows the wide departure from isentropic conditions in heating the liquid from T_2 to T_1, as represented by the slope of the line bc and the area bcc'. The gain in enthalpy is represented by $H_d - H_b$ on the HS diagram. If kinetic-energy terms are neglected, the heat supplied is equal to $H_d - H_b$ or $\int_1^2 T\,dS$, which is represented on the TS diagram by the area $S_1abcdeS_2$. Thus,

$$q_1 = H_d - H_b = \int_{S_1}^{S_2} T\,dS \tag{63}$$

In the third stage the vapor expands isentropically from d to e. If the expansion is a flow process with negligible changes in kinetic energy, the work done is equal to the decrease in enthalpy as represented by $H_d - H_e$ on the HS diagram. On the pV diagram the work done during this stage of isentropic expansion is represented by the area $mden$.

In the condensation stage the enthalpy loss is represented by $H_e - H_a$ on the HS diagram and as area S_1aeS_2 on the TS diagram, both of which equal the heat $-q_2$ absorbed by the condenser.

The net work of the cycle represents the engine work minus the small work input to the pump. Since the process is cyclic with no change in any point property, it follows from (XII-2) that the net work done is equal to the heat absorbed less the heat rejected if kinetic energy changes are negligible. Thus,

$$w_{\text{net}} = -\int_{p_1}^{p_2} V\,dp - (p_1 - p_2)V_f = (H_d - H_e) - (H_b - H_a)$$
$$= q_1 + q_2 \quad (64)$$

The net work done during the cycle is represented by $(H_d - H_e) - (H_b - H_a)$ on the HS diagram, by the area *abde* on the pV diagram and by the area *abcde* on the TS diagram.

The ideal thermodynamic efficiency of the Rankine cycle is represented by

$$\frac{w_{\text{net}}}{q_1} = \frac{q_1 + q_2}{q_1} = \frac{(H_d - H_e) - (H_b - H_a)}{H_d - H_b} \quad (65)$$

or, if pump work is neglected,

$$\frac{w_{\text{net}}}{q_1} = \frac{H_d - H_e}{H_d - H_a} \quad (66)$$

The significance of the efficiency of the Rankine cycle is best visualized from a study of the TS diagram where it is represented by the ratio of the area *abcde* to area S_1bcdeS_2. Any change in conditions which will enlarge the former area with respect to the latter will improve the efficiency. It is evident that the efficiency of the Rankine cycle is lower than that of a Carnot cycle operating between the same extreme temperatures. This results from the fact that in the Rankine cycle a portion of the heat absorbed is delivered to the process at intermediate temperatures along the line *bc* representing the heating of the liquid to saturation at temperature T_1. This difference is shown in Fig. 126a where the efficiency of the Carnot cycle is represented as $\dfrac{\text{area } (A)}{\text{area } (A + C)}$ and of the Rankine cycle as $\dfrac{\text{areas } (A + B)}{\text{area } (A + B + D + C)}$.

The efficiency of the Rankine cycle can be increased by increasing the temperature and pressure of the saturated steam supply or by lowering the temperature at which condensation takes place. For the same temperature change it is evident from Fig. 126b that a decrease in the condensation temperature is more effective than an increase in the temperature of the saturated steam. Increased efficiency obtained by lowering the temperature of condensation is limited by the temperature of the cooling-water supply. For boilers operating at the same pressure, greater efficiency can be expected from this source in cold than in hot climates. Several degrees temperature difference must be allowed for the flow of heat from condensate to cooling water to permit economical designs of condensers at the expense of reduced efficiency of conversion of heat into work.

An increase in efficiency produced by increasing the temperature of saturated steam is limited by the convergence of the saturation lines as shown on the *TS* diagram. It is evident that the gain due to increased temperature and pressure diminishes as the temperature approaches the

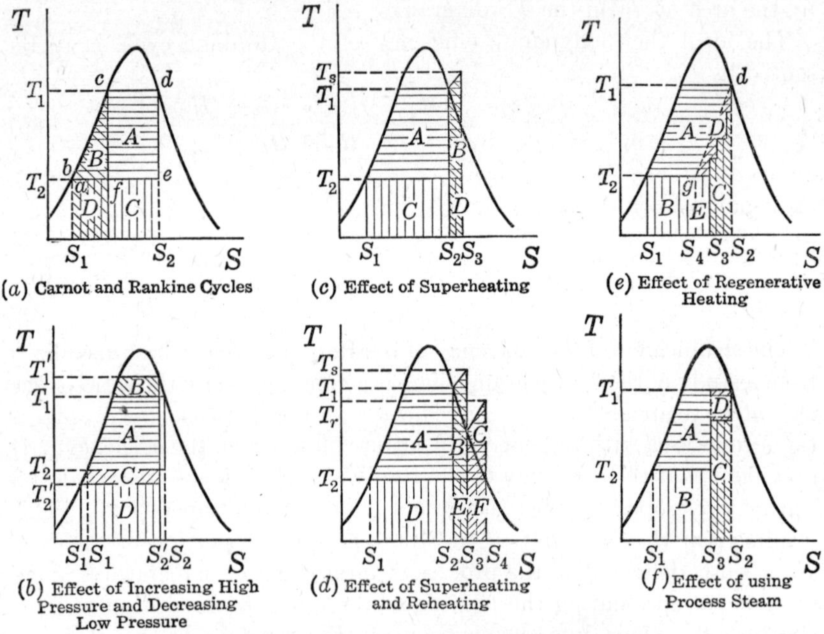

FIG. 126. Improvements in Cycle Efficiencies.

critical. The maximum useful temperature and pressure are also limited by the cost and strength of available materials of construction. To obtain the advantage of high temperature without the disadvantage of high pressures, some fluid other than water may be resorted to, such as mercury or diphenyl. A dual power plant employing both mercury and water has been used for increased efficiency of stationary power plants.

Improvements over the Rankine Cycle. Improvement in efficiency over the Rankine cycle is obtained by superheating the vapors before expansion as shown in Fig. 126c. This also increases the quality of steam after expansion and thus reduces erosion and friction due to condensate in the engine or turbine. The temperature of superheating is limited by the available materials of construction. Higher efficiencies are also obtained by *reheating* in combination with multistage turbines or engines. In this scheme the superheated steam at high pressure partially expands in the first stage and is then again superheated in an auxiliary heat exchanger in the furnace. The reheated steam is expanded in a second stage as shown in Fig. 126d. This reheating increases efficiency and eliminates condensation in the expansion stages of the engine or turbine. The advantage of reheating is partly offset by the cost of the auxiliary heat-exchange surface required which operates at a lower pressure than the primary steam supply.

Regenerative heating of the feedwater to the boiler is another means of increasing efficiency where multistage turbines or engines are used. In this scheme steam is bled from the turbine progressively as expansion proceeds, and this stream of fluid of steadily decreasing temperature is used to heat the accumulating condensate progressively and countercurrently as it is returned to the boiler. If this heat exchange could be accomplished with a minimum temperature drop, the efficiency of the Rankine cycle could be made to approach that of the Carnot cycle. This hypothetical continuous preheating scheme is designated by dotted line *dg* in Fig. 126e. The efficiency in the limiting case will approach the ratio of $A/A + B$ which is that of the Carnot cycle. Actually the results are limited by the finite steps in multistage expansion, as represented by the four stages in Fig. 126e. By regenerative heating more steam is actually supplied to the turbines per horsepower-hour developed, but less heat is rejected to the condenser. Again the advantage of regenerative heating is offset by the cost of auxiliary heat-exchange equipment, but the gain in efficiency is real, and distinct gains in economy of operation are achieved in large power plants.

Process Steam. Of particular interest in chemical manufacture is the use of process steam. Where multistage power production is available, process steam can most economically be supplied by generating steam at the highest pressure available and bleeding off the desired amount at the temperature required for the process, from the lowest stage permissible. Where a quantity of heat represented by area C on Fig. 126f is desired for process heating at temperature T', an amount of work energy equivalent to area D can be obtained by generating the steam first at the higher temperature T_1. This work can be achieved at little additional fuel cost and little increased boiler capacity. The work accompanying the generation of process steam in this manner is virtually obtained by the absorption of an equivalent amount of heat. Obviously, to generate process steam at temperature T' would permit no power generation, and all heat would be rejected to the process.

Heat Exchangers. The heat removed from any fluid in cooling loses least in entropy and hence retains highest availability if this transfer is made under the smallest temperature drop. The transfer of heat approaches reversibility as the temperature difference diminishes towards zero, whereas direct mixing of a hot stream with a cold stream is most inefficient and wasteful. Ideal heat transfer is approached by countercurrent exchangers. The actual temperature drop for the transfer of heat will be fixed by the cost of the exchanger.

A comparison of the advantage of countercurrent heat exchange with that of parallel flow or mixing can be illustrated by considering a given mass of a hotter fluid of a given heat capacity exchanging heat with the same mass of a colder fluid of the same heat capacity. If the hotter fluid cools countercurrently from temperature T_1 to T_2, and the colder fluid is heated from temperature T_2 to T_1, then the loss in entropy of the hot fluid is exactly compensated by the gain in entropy of the cold fluid, and there is no net loss in entropy. The heat exchange is reversible in this limiting case. In the case of parallel flow or mixing, the hotter fluid is cooled to temperature T_3, and the colder fluid is heated to temperature T_3. The hotter fluid loses entropy to the extent of $(C_p)_m \ln \dfrac{T_3}{T_1}$, and the colder fluid gains in entropy by the amount $(C_p)_m \ln \dfrac{T_3}{T_2}$. The net gain in entropy is the algebraic sum of the two changes or

$$\Delta S = (C_p)_m \ln \left(\frac{T_3^2}{T_1 T_2} \right) \tag{67}$$

It can be shown mathematically that for all possible values of T_3 in Equation (67), $\Delta S > 0$.

Engine Efficiency and Performance. The gross work performed by a Rankine cycle results from the flow and expansion of the fluid through the engine. As previously pointed out, the net work is slightly lower than this, because of the work required for pumping. Maximum gross work is obtained by ideal isentropic expansion as expressed by Equation (45). However, actual engines as a result of mechanical friction and deviations from isentropic conditions of the fluid deliver as shaft work an amount of energy less than that calculated from Equation (45). The ratio of the shaft work actually delivered to that ideally obtainable by isentropic expansion between the existing terminal conditions is termed the *engine efficiency*. Thus, from Equation (45),

$$\text{Engine efficiency} = \frac{\text{actually delivered } w_s}{H_1 - H_2} \tag{67a}$$

Engine efficiencies vary widely with size and design, ranging from as high as 80 per cent for large well-designed turbines and reciprocating engines down to 30 per cent or less for small units of inferior design.

Engine performance is commonly expressed by the *steam rate* or *water rate* which is defined as the pounds of steam used per horsepower-hour produced. One horsepower-hour equals 2545 Btu. It is evident that the steam rate is affected both by the operating conditions and the engine efficiency. The *ideal water rate* for any engine may be calculated from Equation (45). The actual steam rate is then the ideal rate divided by the engine efficiency.

Ideal steam rates vary from as low as 5 to 10 lb with superheated high-pressure steam and vacuum exhaust to more than 40 lb where high exhaust pressures are used for the production of process steam. With saturated steam at 100 to 200 lb per sq in. in exhausting at atmospheric pressure ideal steam rates vary from 18 to 24 lb.

The factors of mechanical design and operation which determine engine efficiency and performance are discussed in texts on mechanical engineering thermodynamics.[1–3]

Illustration 6. Calculate the ideal thermodynamic efficiency of a steam-power plant operating on the Rankine cycle in which dry saturated steam is generated at a gauge pressure of 150 lb per sq in. and expanded through an engine exhausting at atmospheric pressure. Compare this efficiency with that of a Carnot cycle operating between the same temperatures, and calculate the ideal water rate of the engine.

Solution: The following data are taken from the steam tables:[1]

Gauge pressure, lb per sq in.	0	150
Absolute pressure, lb per sq in.	14.7	164.7
Saturation temperature, t, °F	212	365.9
Enthalpy of saturated vapor, Btu per lb	1150.4	1195.6
Enthalpy of saturated liquid, Btu per lb	180.1	338.4
Entropy of saturated vapor, Btu/(lb)(°R)	1.757	1.562
Entropy of saturated liquid, Btu/(lb)(°R)	0.312	0.523
Specific volume of saturated liquid, cu ft per lb	0.0167	0.0182

From Equations (38) and (39), the nomenclature of Fig. 125 is used,

$$S_d - S_e = 0 = 1.562 - [1.757x + (1 - x)(0.312)] = 0$$
$$x = 0.865, \text{ the quality after expansion}$$
$$H_e = (1150.4)(0.865) + (180.1)(0.135) = 1019.5 \text{ Btu per lb}$$

From Equation (61),

$$\text{Work of pumping} = (H_b - H_a) = (164.7 - 14.7)(144)(0.0167)/778$$
$$= 0.465 \text{ Btu per lb}$$

[1] J. H. Keenan and F. G. Keyes, "Thermodynamic Properties of Steam," John Wiley & Sons, New York (1936).

[2] P. J. Kiefer and M. C. Stuart, "Principles of Engineering Thermodynamics," John Wiley & Sons, New York (1930).

[3] J. H. Keenan, "Thermodynamics," John Wiley & Sons, New York (1941).

From Equation (65),

$$\frac{w_{\text{net}}}{q_1} = \frac{1195.6 - 1019.5 - 0.465}{1195.6 - (180.1 + 0.465)} = 0.173 \text{ or the ideal thermodynamic efficiency is } 17.3 \text{ per cent}$$

From Equation (59) the corresponding Carnot cycle efficiency is

$$\frac{w_{\text{net}}}{q_1} = \frac{365.9 - 212}{365.9 + 460} = 0.186 \quad \text{or} \quad 18.6 \text{ per cent}$$

The ideal work obtained from the steam in the engine is given by Equation (10).

$$(w_s)_s(\text{flow}) = 1195.6 - 1019.5 = 176.1 \text{ Btu per lb}$$

$$\text{Ideal water rate of engine} = \frac{2545}{176.1} = 14.5 \text{ lb per hp-hr}$$

Thus, if the design of the engine were such as to result in an engine efficiency of 60 per cent the over-all thermodynamic efficiency would be 10.4 per cent, and the actual water rate of the engine 24.2 lb per hp-hr.

COMPRESSION OF GASES

Gases are compressed for transportation, storage, and liquefaction, or to maintain either high or subatmospheric pressures in chemical and metallurgical processes. Although compression is accompanied by an increase in temperature, this temperature rise is rarely the object of the operation, except in special applications such as the ignition of fuel in the Diesel engine or in vapor compression evaporation. Where a high temperature of the compressed gas is desired, the temperature rise in compression is useful but must be considered as incidental to the operation, since a temperature increase is produced more economically through direct heating than by the dissipation of valuable shaft work.

Mechanical gas compressors are of three general types:

1. Reciprocating compressors.
2. Positive-displacement rotary blowers.
3. Turbocompressors and fans.

The first two types operate with a sequence of intermittent operations including a nonflow process. Thus, a charge of gas is taken in at the suction pressure p_1 and compressed to the final pressure p_2 along *be* on the pV diagram of Fig. 127, if the operation is adiabatic. This step is nonflow in character but is immediately followed by the discharge of the compressed gas along line *ed*. Since all the compressed gas flows into the apparatus at the suction conditions and out of the apparatus at the discharge conditions, the net effect of the over-all operation is a flow process even though an intermediate nonflow step is involved. The over-all work requirements and changes in thermodynamic properties

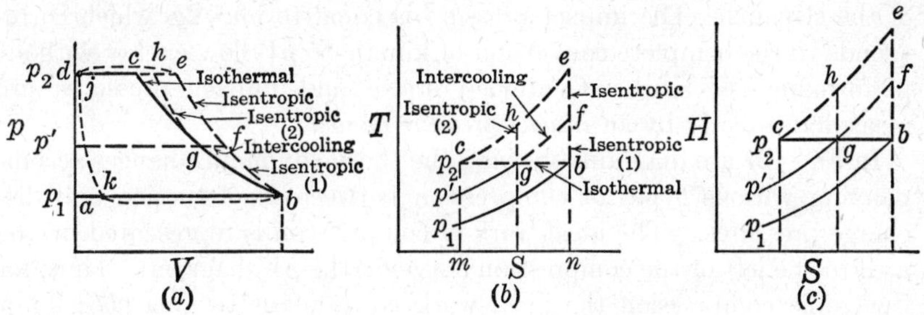

FIG. 127. Compression of Gases.

accordingly follow the equations developed for expansion or compression under flow conditions.

Turbocompressors and Fans. Equipment of this general type operates essentially in a continuous manner. Although within the machine high velocities and kinetic energies may be developed, the design is generally such that these intermediate kinetic energies are largely transformed to flow energy at the suction and discharge. Accordingly, from an over-all standpoint, kinetic energies are generally negligible as compared with the work of compression if the ratio of discharge to suction pressure is greater than 1.1. Under such conditions the theoretical work requirements may be calculated from Equation (22) or (45) if conditions of reversibility are approached.

Where small pressure changes are involved, changes in the kinetic energies of the suction and discharge streams may constitute an appreciable portion of the work requirement. These effects are taken into account in isentropic compression by combination of Equations (5) and (41), with changes in potential and surface energies being neglected.

$$-(w_s)_S = \int_1^2 V\,dp + \frac{m(u_2^2 - u_1^2)}{2g_c} \tag{68}$$

$$= (H_2 - H_1)_S + \frac{m(u_2^2 - u_1^2)}{2g_c}$$

Similarly, for isothermal compression, by combining Equations (5) and (19),

$$-(w_s)_T = (H_2 - H_1)_T - T(S_2 - S_1) + \frac{m(u_2^2 - u_1^2)}{2g_c} \tag{69}$$

Thus, the work input is increased by the difference between the kinetic energies of the discharge and suction streams.

The kinetic-energy terms in Equations (68) and (69) may be omitted if the effective pressures in the suction and discharge streams are both taken as the sums of the respective *static pressures* and *impact pressures* of the streams. The impact pressure is equal to $mu^2/2g_c$ which corresponds to the complete conversion of kinetic to pV flow energy without performance of work. Combined static and impact pressures are measured directly by an *impact-pressure gauge.*

In Fig. 127 are diagrams showing the thermodynamic changes accompanying various types of compression between fixed suction and discharge pressures. The ideal work of compression is represented by the area to the left of the compression curve on the pV diagram. Thus, for isentropic compression the ideal work corresponds to area *abed.* For isothermal compression the corresponding work requirement is represented by area *abcd.* It is evident that isothermal operation is highly advantageous as regards power consumption. In general, it is attempted to design compressors to approach ideal isothermal operation at the lowest feasible temperature. In this way power requirements are reduced by expenditures for cooling. The optimum design is that corresponding to minimum over-all costs of compression and cooling.

Single-Stage Compressors. Large-size compressors and fans ordinarily approximate adiabatic operation and would behave in accordance with Equation (68) if the operation were reversible. However, irreversibility results from fluid friction and turbulence as well as from mechanical friction in the equipment, both of which increase the work requirement. Fluid friction also increases the final temperature of the fluid above the value calculated from Equation (68). The departure of the operation from ideal reversibility is expressed by *an isentropic compression efficiency factor* η_S which is defined as the percentage ratio of the ideal work requirement, calculated from Equation (68) for the pressure increase obtained, to the total shaft work input to the compressor. Thus,

$$\eta_S = \frac{(H_2 - H_1)_S + \dfrac{m(u^3 - u_1^2)}{2g_c}}{w_S \text{ (total)}} (100) \tag{70}$$

where H_2 is the enthalpy of the discharged gas at the calculated temperature resulting from isentropic compression. Inasmuch as work of compression is a minimum in a reversible isothermal operation, another index of performance, termed *the isothermal compression efficiency,* η_T is also used.

$$\eta_T = \frac{\text{work of reversible isothermal compression}}{\text{total shaft-work input to the compressor}} (100)$$

$$= \frac{(G_2 - G_1)_T + \dfrac{m(u_2^2 - u_1^2)}{2g_c}}{w_S \text{ (total)}} (100) \tag{71}$$

The over-all efficiency includes compression efficiency and the effectiveness of the cooling system in approaching isothermal operation.

Compression efficiencies of commercial machines range from 40 to 65 per cent for fans and from 65 to 85 per cent for turbocompressors. The power required for a specified compression duty is determined by calculating the work per unit quantity of fluid from Equation (68) and multiplying by the rate of flow per unit time. If the work is in foot-pounds per pound-mole and the flow in pound-moles per minute, the result is divided by 33,000 to obtain the required horsepower.

Reciprocating Compressors. As previously pointed out, an ideal reciprocating compressor would operate on the cycle *abed* of the pV diagram of Fig. 127 if the compression were isentropic. In such a case, the work requirement would be expressed by Equation (68). Actual compressors of industrial sizes closely approximate adiabatic operation but deviate from reversibility as a result of fluid and mechanical friction. A further complication is introduced by the necessity of providing a certain *clearance volume* between the cylinder head and the piston at its extreme position. This clearance volume is represented by *dj* in Fig. 127. At the end of the stroke the clearance volume is filled with compressed gas which expands to k, reducing the effective suction volume to kb.

The *volumetric efficiency* η_v is defined as the percentage ratio of the volume of gas, measured at p_1, actually taken into the cylinder during the suction stroke to the volumetric displacement of the piston. The volumetric efficiency determines the capacity of a machine for a specified duty while the compression efficiency determines the total work input required per unit quantity of gas compressed. The total power required for a machine of given displacement operating between specified pressures is increased by increased volumetric efficiency and reduced by increased compression efficiency.

Compression efficiencies are reduced by mechanical friction and by the pressure drops through the suction and discharge valves. As a result of these pressure drops the actual suction pressure is somewhat below p_1 and the discharge pressure above p_2, as indicated by the dotted lines between *ab* and *je* on Fig. 127. The increased work resulting from the throttling through valves represents a major loss, and proper design of valve openings is of great importance. Compression efficiencies are increased by any cooling of the gas during compression which will cause the operation to approach isothermal conditions, as indicated in the pV diagram of Fig. 127. Cylinders of small compressors are frequently water-cooled, but water jackets are not highly effective on large machines because of the small ratio of surface to volume. In some designs cooling water is injected into the gas during compression.

Volumetric efficiency is reduced by pressure drop through the suction valves and by increased clearance volume. The problems of compressor design have been discussed in detail by York[4] who concludes that the gas retained in the clearance volume of large machines approximates isentropic expansion as the piston starts on the suction stroke. In this case the ideal work of compression per unit quantity of gas discharged is unaffected by the clearance volume which may be looked upon as containing an isolated quantity of gas alternately compressed and expanded in a reversible manner without any net consumption of power. Reduction in volumetric efficiency due to clearance can be calculated if isentropic expansion of the clearance volume on the intake stroke is assumed, but further reduction due to pressure drop through valves and parts can be determined only experimentally.

Variation of the clearance volume offers an effective means of controlling the capacity of a compressor driven at a constant speed. Machines are constructed with clearance "pockets" of variable volume which permit operating the machine at reduced capacities without waste of power. An increase of the clearance volume greatly reduces the volumetric efficiency with relatively little effect on the compression efficiency.

Volumetric efficiencies of industrial compressors range from 50 to 90 per cent, whereas compression efficiencies, based on the work requirement of isentropic compression, vary from 75 to 90 per cent.

Illustration 7. A single-stage compressor is to compress 800 cu ft per min of ammonia gas at 0°F and an absolute pressure of 15.0 lb per sq in. to 75 lb per sq in. Calculate the actual horsepower required to drive the compressor and the piston displacement in cubic feet per minute, assuming a compression efficiency of 76 per cent and a volumetric efficiency of 85 per cent.

From Fig. 119 and Equation (10), for isentropic compression at the special conditions,

$$-(w_s)_S = (H_2 - H_1)_S = 715 - 617 = 98 \text{ Btu per lb}$$

The ideal discharge temperature is 200°F.

At 0°F, 15 lb per sq in., the specific volume of ammonia gas from Fig. 116 is 18.92 cu ft per lb.

$$\text{Compression rate} = 800/18.92 = 42.3 \text{ lb per min}$$

$$\text{Theoretical horsepower} = \frac{(42.3)(98)(778)}{(33,000)} = 97.8 \text{ hp}$$

$$\text{Actual horsepower} = \frac{97.8}{0.76} = 129 \text{ hp}$$

$$\text{Piston displacement} = \frac{800}{0.85} = 940 \text{ cu ft per min}$$

Multistage Compression. Because of mechanical difficulties associated with high ratios of discharge to suction pressure a single-stage recip-

[4] R. York, Jr., *Ind. Eng. Chem.*, **34**, 355 (1942).

rocating compressor is ordinarily not designed for a compression ratio greater than 4 or 5 to 1. Turbocompressors are limited to much lower ratios per stage. In order to develop high overall compression ratios it is necessary to use multistage compression in which the one stage discharges to the suction of the next.

An important advantage of multistage operation is that it permits *intercooling* of the gas between stages and thus approaches the same suction temperature for each stage. This results in a reduction of the work of compression and approaches isothermal conditions where several stages of small compression ratios are employed. A two-stage operation with intercooling is indicated in Fig. 127. Isentropic compression in the first stage along *bf* is followed by cooling at constant pressure p' along *fg*. The final pressure p_2 is reached by isentropic compression in the second stage along *gh* with a final temperature intermediate between that of *e* resulting from single-stage isentropic compression and the isothermal value of *c*.

It is evident that the work required for the two-stage operation is less than that of the single stage by the area *fghe*. If more stages with intercooling were employed, the work would be still further reduced, approaching the ideal isothermal case. For multistage compression, the *equivalent compression efficiency* is defined as the ratio of the work for multistage isentropic compression with complete intercooling between stages divided by the actual shaft-work input.

The total work in a multistage compression is the sum of the work in the individual stages. Thus, from Equation (44), if a constant value of κ and a constant suction temperature T_1 are assumed and kinetic energy change is neglected, the total work per mole for two-stage compression of an ideal gas becomes

$$-w_s^* = \left(\frac{RT_1\kappa}{\kappa - 1}\right)\left\{\left[\left(\frac{p'}{p_1}\right)^{\frac{\kappa-1}{\kappa}} - 1\right] + \left[\left(\frac{p_2}{p'}\right)^{\frac{\kappa-1}{\kappa}} - 1\right]\right\} \tag{72*}$$

If the intermediate pressure is p', the work required will be a minimum value when dw_s^*/dp' is zero. The optimum intermediate pressure is determined by equating dw_s^*/dp' from Equation (72) to zero. Thus,

$$\frac{dw_s^*}{dp'} = \left(\frac{RT\kappa}{\kappa - 1}\right)\left[\left(\frac{1}{p_1}\right)^{\frac{\kappa-1}{\kappa}}\left(\frac{\kappa - 1}{\kappa}\right)(p')^{\frac{-1}{\kappa}} + (p_2)^{\frac{\kappa-1}{\kappa}}\left(\frac{1-\kappa}{\kappa}\right)(p')^{\frac{1-2\kappa}{\kappa}}\right] = 0$$

$$p'^{\frac{2(\kappa-1)}{\kappa}} = (p_1p_2)^{\frac{\kappa-1}{\kappa}}$$

$$p' = \sqrt{p_1p_2} \tag{73*}$$

Rearrangement of (73), gives

$$\frac{p'}{p_1} = \frac{p_2}{p'} = \sqrt{\frac{p_2}{p_1}} \qquad (74)^*$$

Thus, for best operation, each stage should have the same compression ratio, equal to the square root of the over-all compression ratio. Where s stages are employed it may be demonstrated similarly that optimum results are obtained when the compression ratio in each stage is equal to

$$\frac{p''}{p'} = \left(\frac{p_2}{p_1}\right)^{\frac{1}{s}} \qquad (75)^*$$

If (75) is combined with an equation of the form of (72), it follows that the work per stage is constant, and the over-all work per mole is given by

$$-w_s^* = \left(\frac{sRT_1\kappa}{\kappa - 1}\right)\left[\left(\frac{p_2}{p_1}\right)^{\frac{\kappa-1}{s\kappa}} - 1\right] \qquad (76)^*$$

By means of Equation (76) the number of stages for optimum economy of operation may be estimated by consideration of the increased equipment and maintenance costs which accompany an increased number of stages with a resultant reduction in power requirement.

For compression of nonideal gases the problem of calculating optimum intermediate pressures involves trial-and-error methods. A simplified method was proposed by York[4] which may be used where economy is of great importance. However, for many cases the over-all work requirement is not greatly affected by small variations of the intermediate pressures from the optimum values, and Equation (75) may be used for establishing the intermediate pressures. Once these pressures are established, the work and volumetric capacity for each stage are calculated from Equation (45) and the pV data. Equation (75) may lead to serious departure from optimum pressure conditions when vapor mixtures are dealt with if large amounts of condensate are separated in the intercoolers. Where this condition is encountered, the intermediate pressures may be adjusted to keep the work in different stages approximately equal. If only two stages are involved, a series of intermediate pressures may be assumed and the corresponding over-all work requirements calculated; the optimum intermediate pressure corresponds to the minimum of a curve relating work to intermediate pressure.

In the isentropic compression of gases of low molecular weight and low molal heat capacity there is no tendency for condensation in the compressor. This is evident from the *TS* diagram for ammonia, Fig. 118. Adiabatic compression of the saturated vapor results in superheating. However, gases of high molal heat capacity behave quite differently, as may be seen from the *TS* diagram for benzene in Fig. 121. If saturated benzene vapor at 100°F is isentropically compressed, it is first superheated, and then at a pressure of 655 lb per sq in. it becomes saturated, and condensation begins. Formation of condensate in compressors is frequently hazardous, and great care must be exercised in designing compression systems for vapors of this type. Formation of condensate in the intercoolers of multistage compressors is common when vapor mixtures are being compressed, and facilities for its separation from the uncondensed gas passing to the next stage are readily provided.

REFRIGERATION

The purpose of refrigeration is to produce a region of temperature below that of the atmosphere. The over-all result is the extraction of heat from the low-temperature region and its rejection at the higher temperature of the surroundings or of the available cooling medium. Thus, in effect a refrigeration machine is a heat pump performing the reverse function of an engine. Refrigeration is of constantly increasing importance in the preservation of foods, in air conditioning, and in chemical engineering operations where low temperatures are required for the control of reactions, the condensation of vapors, or the crystallization of solids.

Two general types of refrigeration cycles are in common use. Both depend upon the attainment of a cold region by the vaporization of a refrigerant fluid from an evaporator at a low temperature and pressure. In *compression refrigeration*, shown diagrammatically in Fig. 129, mechanical energy is used to compress the vapor leaving the evaporator so that the heat absorbed at the low temperature of the evaporator may be rejected at the temperature level of the condenser. A refrigerant fluid is employed to absorb heat at low pressure and temperature and is then compressed to a higher pressure and temperature where it gives off heat, usually with condensation. Expansion to the lower pressure and temperature completes the cycle. This latter is usually free expansion; isentropic expansion has rarely proved feasible.

In *absorption refrigeration*, shown diagrammatically in Fig. 131, heat which is absorbed at a low temperature and pressure is rejected at an intermediate temperature and high pressure after its temperature level has been increased by addition of heat from a high-temperature source such as the combustion of a gas. In this scheme no mechanical energy is required except in some systems for pumping the liquid absorbent, and the refrigerant is compressed from the low pressure at which it absorbs heat to the high pressure at which it rejects heat through intermediate absorption or adsorption at a low-pressure level. The refrigerant is released from the absorbent or adsorbent at a high pressure through the addition of heat from the high-temperature source. This method has the advantage of minimizing or eliminating moving mechanical parts and may effectively utilize low-pressure exhaust steam or fuel gas as the source of heat.

Compression Refrigeration. A Carnot engine operating in reverse represents an ideal reversible refrigeration cycle. The *TS* diagram of

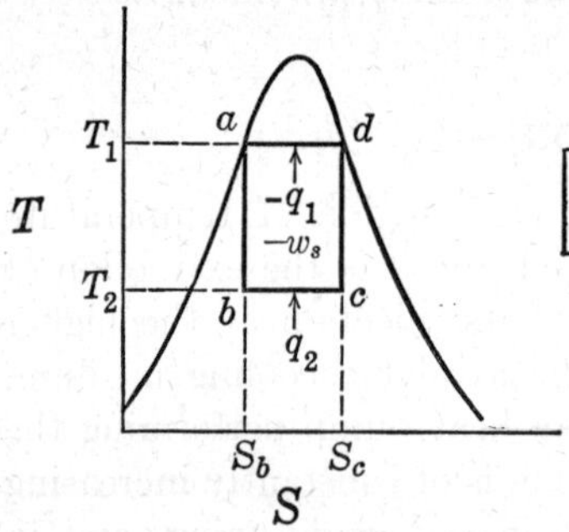

FIG. 128. Reverse Carnot Cycle.

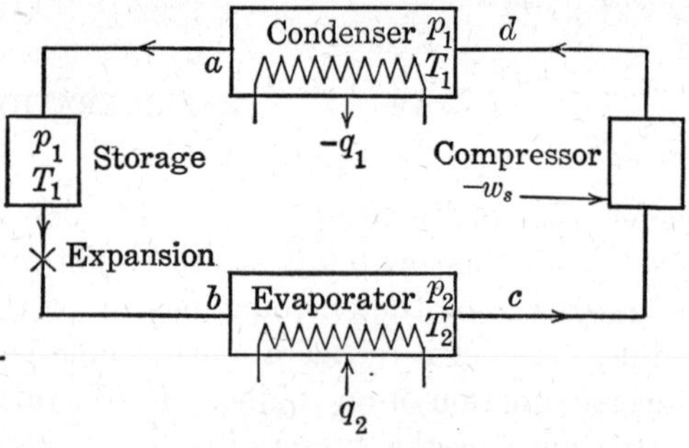

FIG. 129. Vapor Compression Refrigeration.

such a cycle operating with saturated vapor is shown in Fig. 128. A quantity of heat q_2 is absorbed in increasing the quality of the vapor at temperature T_2 along *bc*. Isentropic compression along *cd* raises the temperature to T_1 where a quantity of heat $-q_1$ is rejected through condensation of the vapor along *da*. The liquid is isentropically expanded and partially vaporized along *ab* to temperature T_2 thus completing the cycle.

As was developed in consideration of the Carnot engine, page 557, the net work done by the system is the difference between the heats absorbed at the higher and lower-temperature levels, and Equations (58) and (59) are directly applicable to the reverse cycle operating as a heat pump. However, in this case both w_s and $S_b - S_c$ are negative. It follows that the amount of heat rejected at the high temperature is greater than that absorbed at the low temperature by the amount of the work done on the system.

For comparing the performance of refrigeration machines a *coefficient of performance* is used as an index. This is defined as the ratio of the heat absorbed in the evaporator to the work done on the fluid. Thus, from Equation (59) the coefficient of performance

$$\frac{q_2}{-w_s} = \frac{T_2 \Delta S_2}{\Delta S_2(T_1 - T_2)} = \frac{T_2}{T_1 - T_2} \tag{77}$$

where $\Delta S_2 = S_c - S_b$. According to the Carnot principle, the coefficient of performance of any reversible refrigeration machine operating between temperatures T_1 and T_2 is expressed by Equation (77), regardless of the nature of the refrigerant or the steps of the cycle.

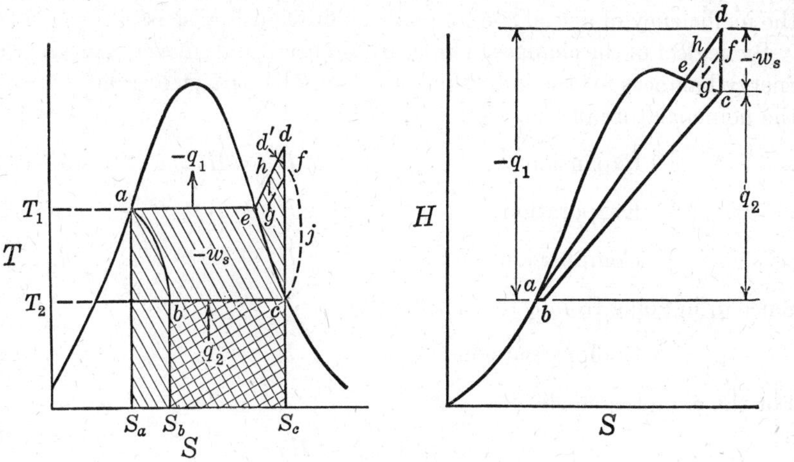

FIG. 130. Refrigeration Cycle.

The capacity of a refrigeration machine is ordinarily expressed in tons of ice-making capacity per 24 hr. *One ton of refrigeration capacity is capable of absorbing in 24 hr the heat evolved in freezing 1 ton of water at 32°F.* On this basis, 1 ton refrigeration capacity = 200 Btu per min = 4.71 hp. To define completely the characteristics of a refrigeration operation, it is necessary to specify the coefficient of performance, the refrigeration capacity, and the two temperature levels of heat absorption and rejection.

Actual compression refrigeration cycles differ from the ideal Carnot heat pump in that each step is carried out in a separate piece of equipment with the refrigerant fluid circulating from one to another as indicated in the flow diagram of Fig. 129. The corresponding *TS* and *HS* diagrams are shown in Fig. 130. The liquid refrigerant leaves the storage tank at temperature T_1 and pressure p_1 and is expanded through a valve along *ab*. This expansion is substantially isenthalpic free expansion and is hence highly irreversible and accompanied by a large increase in entropy, $S_b - S_a$. Partial vaporization and reduction in temperature to T_2 occurs in this expansion. Heat is absorbed at temperature T_2 along *bc* until vaporization is completed. The saturated vapors are then isentropically compressed to p_1 along *cd*. The superheated vapor at *d* is cooled at constant pressure, becoming saturated at *e* and condensing along *ea*.

This cycle differs from the ideal Carnot cycle in the irreversible nature of the expansion and the fact that saturated vapor leaving the evaporator is compressed to a superheated state. Further deviations result from the inefficiency of actual compressors, as previously discussed.

By neglect of the changes in kinetic, potential, and surface energies, the energy balances for the individual steps may be written as follows; with the nomenclature of Fig. 130:

$$\text{Expansion} \qquad H_a = H_b \tag{78}$$

$$\text{Evaporation} \qquad H_b + q_2 = H_c \tag{79}$$

$$\text{Compression} \qquad H_c - w_s = H_d \tag{80}$$

Since q_1 in going from d to a is negative,

$$\text{Cooling and condensation} \qquad H_d = H_a - q_1 \tag{81}$$

For the complete cycle,

$$-w_s = -(q_1 + q_2) = H_d - H_c \tag{82}$$

The ideal coefficient of performance then becomes, from (78), (79), and (82),

$$\frac{q_2}{-w_s} = \frac{H_c - H_a}{H_d - H_c} \tag{83}$$

where $-w_s$ represents the net work done on the fluid and not the total work input to the compressor.

The coefficient of performance may be visualized from the *TS* diagram of Fig. 130 in which the double-crosshatched area represents q_2 and the single-crosshatched area represents $-w_s$. The coefficient of performance is the ratio of these two areas. It is evident that the coefficient of performance would be greatly improved if the free expansion along *ab* were replaced with isentropic expansion in an engine, the work from which would be transmitted to the compressor. This scheme is rarely used because of mechanical complications.

In developing Equation (83) it was assumed that the compression step is isentropic; actually some heat is generally absorbed in the piping leading to the compressor, and during compression heat is at first gained by conduction through the walls of the compressor. As the temperature is increased in the compression stroke, the temperature difference between the gas and its surroundings is reduced. As a result the compression follows some curved path such as the broken line *cjd'*. The final temperature reached may be either higher or lower than that resulting from isentropic compression. Similarly, the work of compression may be either increased or decreased by such deviations from adiabatic conditions.

The coefficient of performance of a vapor compression cycle may be improved by two-stage compression with intercooling. However, the selection of the optimum intermediate pressure is different from that employed in ordinary compression in which the suction temperatures of all stages are the same.

In order that intercooling may be carried out with the same higher-temperature cooling medium that is used to maintain temperature T_1, it is necessary that compression be carried out adiabatically to a temperature greater than T_1. The gas is then cooled to approximately T_1, and compression is continued in another stage. A two-stage operation of this type is indicated by lines *cfghe* of Fig. 130. In this way work is saved corresponding to the area *fghd* on the *TS* diagram. A modification of this scheme is to cool during the second stage of compression by interjecting liquid refrigerant from the reservoir.

Refrigeration cycles are also operated with a gas such as air as the refrigerant without condensation at any stage. A gas refrigeration cycle comprises isentropic compression, isobaric cooling, isentropic expansion, and isobaric absorption of heat corresponding to the cycle of Fig. 127b. The power generated in the isentropic expansion is utilized to aid in driving the compressor. Despite its approach to Carnot efficiency, this type of refrigeration has been outmoded because of its mechanical complexity and the relative difficulty of heat transfer to a gas.

Many different fluids are used in vapor-compression cycles. Ammonia, sulfur dioxide, carbon dioxide, methyl chloride, and the chlorofluoroparaffins, the so-called Freons, are the most popular. Selection of the refrigerant is governed by the working temperatures, the limitation regarding the bulk of the compressor, the hazards of very high pressures, and many other considerations such as toxicity, inflammability, and corrosiveness.

Illustration 8. A compression refrigeration plant using propane is to be designed for a capacity of 50 tons of refrigeration per day at an evaporator temperature of 30°F. Cooling water is available at 90°F, and the condenser is to be designed for a condensation temperature of 100°F. It is assumed that the compression is adiabatic with a compression efficiency for the compressor of 85 per cent and a volumetric efficiency of 75 per cent.

Calculate:

(a) The pounds of propane circulated per hour.

(b) The heat removed in the condenser, Btu per hour.

(c) The power required to drive the compressor.

(d) Volumetric displacement of compressor, cubic feet per minute.

(e) The coefficient of performance.

In the absence of tables or charts on the thermodynamic properties of the refrigerant, this problem can be solved by the following generalized procedure.

Basis of calculations: 1 lb-mole of propane circulated.

Let H_a = enthalpy of liquid propane at 100°F
H_b = enthalpy of liquid propane at 30°F
H_c = enthalpy of saturated propane vapor at 30°F
H_d = enthalpy of propane vapor leaving compressor
H_e = enthalpy of saturated propane vapor at 100°F

Corresponding subscripts are used for entropies. For the reference state, use liquid propane at 30°F; $S_b = 0$; $H_b = 0$. Heat capacity of propane vapor, low-temperature range (page 336):

$$c_p^* = 7.95 + 0.00397T^{1.25} \text{ Btu/(lb-mole)(°R)}$$
$$p_c = 43.0 \text{ atm}; \qquad T_c = 666°\text{R}$$

Conditions in	*Condenser*	*Evaporator*
From Table IIa	p = 13.69 atm	4.83 atm
	T = 560°R	490°R
	p_r = 0.318	0.113
	T_r = 0.840	0.735
From Fig. 103:	z = 0.78	0.91
From Fig. 107:	$S^* - S$ = 1.75 Btu/(lb-mole)(°R)	1.10
From Fig. 106:	$(H^* - H)/T_c$ = 1.8	0.88
From Table XIII:	λ = 6040 Btu per lb-mole	7250

For isentropic compression from Equation (39):

$$S_d - S_c = (S_d^* - S_c^*) - (S_d^* - S_d) + (S_c^* - S_c) \tag{a}$$

$$S_d^* - S_c^* = \int_{490}^{T_d} \left[\frac{7.95 + 0.00397(T)^{1.25}}{T}\right] dT - R \ln \frac{13.69}{4.83}$$

$S_d^* - S_d$ depends upon the temperature T_d of propane vapor leaving condenser and $S_c^* - S_c = 1.10$.

By trial-and-error solution of Equation (a), $T_d = 566°R$

Hence, $T_{rd} = 0.85$ $\qquad p_{rd} = 0.318$
$S_d^* - S_d = 1.62$ $\qquad (H_d^* - H_d)/T_c = 1.7$

From Equation (45),

$$-(w_s)_S \text{ (flow)} = H_d - H_c = \int_{490}^{566} (7.95 + 0.00397T^{1.25})\, dT - 666(1.7 - 0.88)$$
$$= 822 \text{ Btu per lb-mole}$$

Since $H_e = 7250$, $H_d = 822 + 7250 = 8072$ Btu per lb-mole

$$H_e - H_d = (H_e^* - H_d^*) - T_c \left[\frac{(H_e^* - H_e)}{T_c} - \frac{(H_d^* - H_d)}{T_c} \right]$$

$$= \int_{566}^{560} (7.95 + 0.00397T^{1.25})\, dT - 666(1.8 - 1.7) = -180$$

$$H_e = 8072 - 180 = 7892$$
$$H_a = 7892 - 6040 = 1852$$

Heat absorbed in evaporator = (50)(60)(200) = 600,000 Btu per hour

(*a*) Propane circulation rate $= \dfrac{600{,}000}{7250 - 1852} = 111$ lb-moles per hr or 4890 lb per hr

(*b*) Heat transfer duty of condenser = (111)(8072 − 1852) = 691,000 Btu per hr
Ideal work of compression = 111(822) = 91,300 Btu per hr

(*c*) Power to drive compressor $= \dfrac{91{,}300}{(0.85)(2545)} = 42.2$ hp at 85 per cent compression efficiency

$$\text{Volumetric intake} = (111)(359)\frac{490}{(492)}\frac{1}{4.83}(0.91) = 7650 \text{ cu ft per hr}$$

(*d*) Piston displacement of compressor at 75 per cent volumetric efficiency =

$$\frac{7650}{(60)(0.75)} = 170 \text{ cu ft per min}$$

(*e*) Coefficient of performance $= \dfrac{(600{,}000)(0.85)}{(91{,}300)} = 5.58$

An ammonia compression refrigeration system operating between the same temperature levels and with the same compression efficiency requires 46.8 hp and gives a coefficient of performance of 5.04. This compression indicates that use of propane as the fluid results in slightly improved performance as compared to ammonia, although the difference indicated is almost within the possible error of the generalized correlations. Propane has the advantage of a relatively high heat capacity which results in little superheating in the compression step.

Absorption Refrigeration. The most widely applied absorption refrigeration cycle uses ammonia as the refrigerant and water as the absorbent. A flow diagram of such a process is shown in Fig. 131. This flow diagram indicates a type of operation which may be used for large-scale process refrigeration. Many other modifications of the basic ammonia absorption cycle are employed to meet particular specific situations. Liquid ammonia from the accumulator is expanded freely through an expansion valve, where it is partly vaporized and then passes to the evaporator. The ammonia vapors leave the evaporator to an absorber where they are absorbed by weak ammonia liquor. In order to reduce the amount of weak liquor circulated per unit mass of ammonia absorbed, it is necessary to cool the absorbent close to the temperature of the available cooling water and to provide intermediate cooling surface in the absorber.

The strong ammonia liquor on its way from the absorber to the stripper is passed through a heat exchanger where it is heated countercurrently by the hot weak liquor on its way from the stripper to the absorber. The weak liquor after cooling in the heat exchanger enters the top of the absorber. The strong liquor is fed to the stripping column where heat is added by means of a reboiler to evaporate the ammonia. The vapors

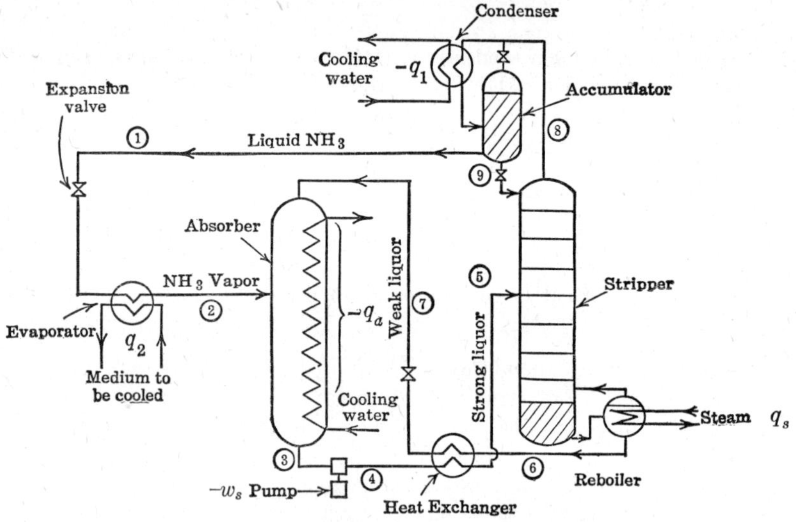

Fig. 131. Ammonia Absorption Refrigeration.

from the top of the stripping column are condensed, and a portion of the liquid is returned as reflux to the top of the stripping column which contains bubble decks or packing to effect fractional distillation. It is evident that the stripper must be operated at a pressure high enough to produce condensation of the ammonia at the temperature attainable with the available cooling water.

Absorption refrigeration cycles are also in common use in which a simple still, with or without a partial condenser or dephlegmator, replaces the stripping column. In other modifications even the pump used for pumping liquid to the still is eliminated. This is accomplished by mixing with the refrigerant an inert gas of low density such as hydrogen which permits the entire system to be kept at a constant total pressure. However, different partial pressures of the refrigerant are maintained for condensation and for evaporation. Circulation is maintained by thermal convection induced by the still, and absorption proceeds by the partial-pressure gradient of the refrigerant from vapor to liquid phase. Descriptions of such units which are used largely in household refrigeration are given by Keenan.[3]

The problems of design and analysis of absorption refrigeration systems involve a series of energy balances embracing the entire cycle and each of its parts. Thus, for the over-all process, with reference to Fig. 131, if changes in kinetic, potential, and surface energies are neglected,

$$q_2 + q_s + q_1 + q_a - w_s = 0 \tag{84}$$

where q_2 = heat absorbed in evaporation
$-q_1$ = heat removed in the condenser
q_s = heat added to system in the stripper reboiler
$-q_a$ = heat removed in the absorber
$-w_s$ = work done upon the system by the pump

Energy balances for individual steps involve only enthalpy and heat terms except where the pump is included. Thus, for example, for the stripper,

$$m_5H_5 + m_9H_9 + q_s = m_8H_8 + m_6H_6 \tag{85}$$

where m_5 is the mass of the stream 5 in Fig. 131, and H_5 is the corresponding enthalpy per unit mass. The other symbols have similar significance.

Since mixtures of varying concentrations are involved throughout the cycle, energy balances are most readily evaluated by means of an enthalpy-concentration chart, such as described in Chapter VIII. Fig. 132 is such a chart for the enthalpies of the ammonia–water system at various pressures and temperatures and for all compositions and phases. Enthalpies are based upon 1 lb of the fluid and are referred to liquid ammonia at −40°F and to liquid water at 32°F, each at its respective vapor pressure at these temperatures. This chart is a modification of one prepared by Bosnjakovic[5] by the methods described in Chapters VII and XIV.

The lowest group of lines on the chart represents the enthalpies of various solid phases of ice and ammonia with the indicated eutectic and congruent points. The next lowest group of curves represents the enthalpies of saturated solutions over the entire range of composition for various temperatures and pressure. The third group from the bottom is

[5] "Technische Thermodynamik II," Theo. Steinkopf, Dresden and Leipzig (1935).

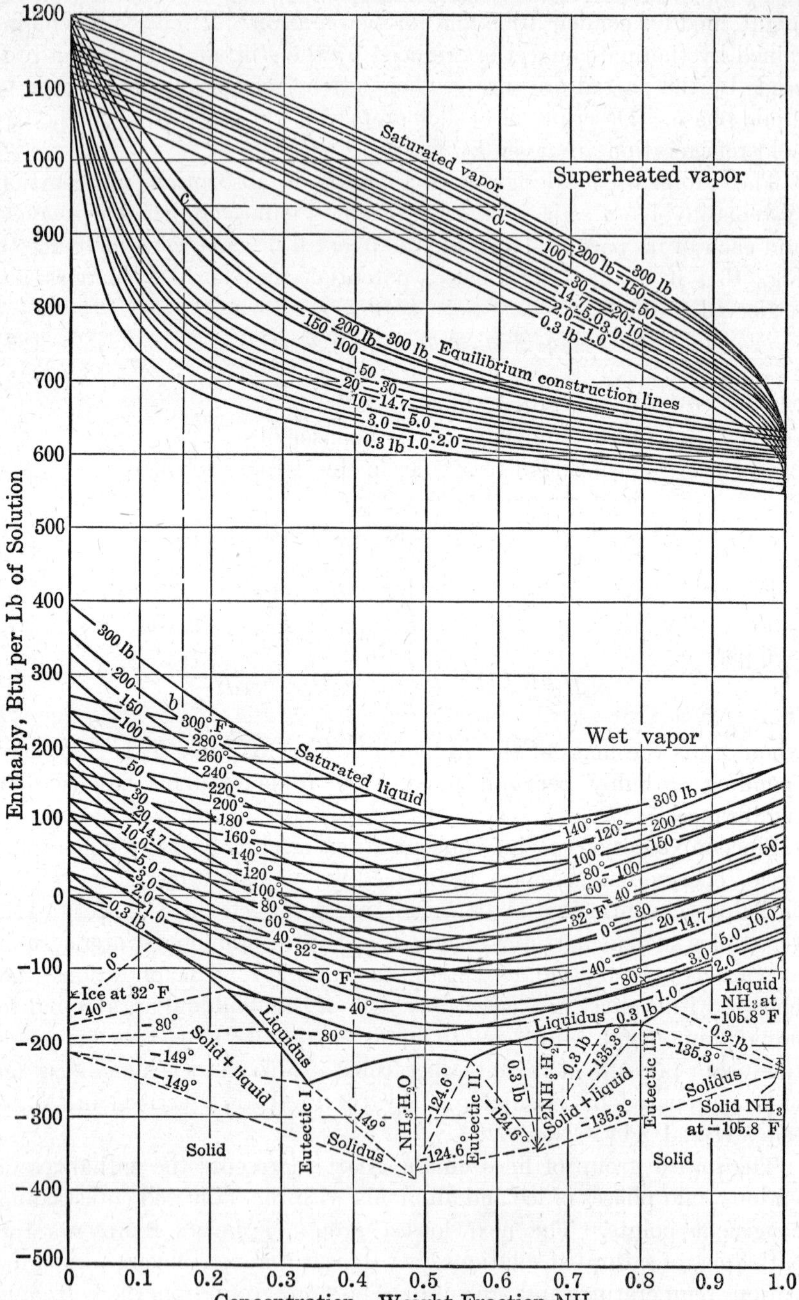

FIG. 132. Enthalpy Concentration of the Ammonia – Water System. (Reference: Water at 32°F, liquid ammonia at – 40°F).
(Reproduced in "C.P.P. Charts")

construction lines for obtaining vapor–liquid equilibrium relations. The next higher group of lines represents the enthalpies of saturated vapors covering the entire range of composition at various temperatures and pressures.

Conditions of vapor–liquid equilibria may be established by selection of points of equal temperature and pressure in the saturated liquid and vapor curves. Thus, point *b* represents saturated liquid containing 16 per cent NH_3 at 300°F and 200 lb per sq in. By following a vertical line from *b* to *c* and a horizontal line from *c* to *d* the composition of the vapor in equilibrium with the liquid is found to be 60 per cent ammonia, on the corresponding pressure line. The enthalpies required in dealing with problems in absorption refrigeration may be read directly from Fig. 132.

Illustration 9. An ammonia-absorption refrigeration plant of the type shown in Fig. 131 is to be designed for the same duty as the propane-compression plant of Illustration 8. The capacity is to be 50 tons per day with an evaporator temperature of 30°F and cooling water such that condensation and cooling can be carried out at a temperature of 100°F. In order that the evaporator pressure may be high and that no difficulty due to formation of ice in the low-temperature sections of the plant will ensue, it is desired that the ammonia from the stripper shall contain only 0.5 per cent by weight of water. It may be assumed that the absorber produces contacting of such effectiveness that the ammonia concentration of the strong liquor is 95 per cent of the equilibrium value. It also may be assumed that vaporization is completed in the evaporator under conditions such that the total vapor formed is at all times in equilibrium with the residual liquid.

The design study is to be based on the following arbitrary specifications:

(1) Temperature of weak liquor leaving the heat exchanger, 120°F.

(2) Reflux ratio of stripper [weight of stream (9) /weight of stream (1)] = 0.50.

(3) Ammonia content of weak liquor, 20 per cent by weight.

(4) Temperature of strong liquor leaving absorber, 100°F.

These specifications might be varied over wide ranges with resulting changes in the design of the plant for the given duty. These changes would affect both the capital and direct operating costs. A series of design studies in which such specifications are varied is necessary in order to arrive at an optimum design which results in minimum over-all operating costs.

In this design study it is permissible to neglect:

(1) Heat exchange between the equipment and its surroundings.

(2) Variations in pressure in both the high- and low-pressure sections of the plant.

(3) Variation in temperature of the ammonia–water mixture in the evaporator. The evaporator pressure may be conservatively established as corresponding to complete vaporization at the specified temperature.

On this basis calculate a complete material and energy balance of the plant together with the temperatures and pressure at all points. Also calculate:

(*a*) The heat-transfer duties of the condenser, absorber cooler, heat exchanger, and reboiler, in Btu per hour.

(*b*) The hydraulic horsepower of the pump.

(*c*) The pressure and quantity of dry saturated steam required for heating the reboiler with a 10°F temperature difference.

Solution: The material and energy balances are established simultaneously from the data of Fig 132, starting with the evaporator. These calculations are summarized in Table A where the streams are identified by the numbers of Fig 131. The results in this table are reported in more significant figures than are justified by the accuracy of the data in order that the numerical accuracy of the calculations may be readily verified by over-all balances.

(1) The pressure of stream (1) is read from Fig. 132 as 210 lb per sq in.; the equilibrium pressure of liquid containing 99.5 per cent NH_3 at 100°F.

(2) The pressure of stream (2) is the equilibrium pressure of saturated vapor containing 99.5 per cent NH_3 at 30°F, which is read from Fig. 132 by a trial-and-error procedure. A trial value of the equilibrium pressure is assumed, and a horizontal line is followed from the point representing saturated vapor at this pressure and 99.5 per cent NH_3 over to the corresponding construction line. A vertical line is followed downward from this point to the 30°F line of the saturated liquid curves, and the corresponding equilibrium pressure is read. This second approximation pressure is used to repeat the operation until consistent pressures are obtained. In this manner it is found that the evaporator pressure is approximately 45 lb per sq in. and that the liquid in equilibrium with the vapor leaving the evaporator contains 77 per cent NH_3.

TABLE A

Material and Energy Balances of Ammonia Absorption Refrigeration Plant

Stream	NH_3 *Per Cent by Wgt*	*t°F*	*Abs Pressure, lb per sq in.*	*H Btu per lb*	*m lb per hr*	*mH Btu per hr*
1	99.5	100	210	152	1229.5	186,880
2	99.5	30	45	640	1229.5	786,880
3	40.0	100	45	−5	4887.3	−24,440
4	40.0	100+	210	−4.4	4887.3	−21,660
5	40.0	217	210	128.8	4887.3	629,432
6	20.0	290	210	220	3657.8	804,720
7	20.0	120	45	42	3657.8	153,628
8	99.5	110	210	650	1844.3	1,198,800
9	99.5	100	210	152	614.8	93,450

The rate of stream (2) is fixed by the heat-transfer duty of the evaporator and the differences in specific enthalpies of streams (1) and (2). Thus,

$$m_1 = (50)(200)(60)/(640 - 152) = 1229.5 \text{ lb per hr}$$

(3) The ammonia content of the strong liquor is 95 per cent of the equilibrium concentration at 100°F and 45 lb per sq in., which is read from Fig 131. Thus, NH_3 in stream (3) $= 0.95 \times 42 = 40$ per cent. It may be assumed that the enthalpy of a liquid which is not completely saturated with NH_3 is the same as that of a saturated solution of the same NH_3 content and at the same temperature. On this basis $H_3 = -5$ Btu per lb.

The rate of stream (3) is determined by the over-all material balance and the ammonia balance of the absorber. Since from the over-all material balance $m_7 = m_3 - 1229.5$,

$$0.40m_3 = (m_3 - 1229.5)0.20 + (1229.5)(0.995)$$
$$m_3 = 4{,}887.3 \qquad m_7 = 3{,}657.8$$

(4) The work done by the pump is calculated from Equation (61). The data of the International Critical Tables indicate a specific gravity of approximately 0.86 for a solution at these conditions.

$$-w_s = \frac{(210 - 45)(144)(4887.3)}{(62.4)(0.86)} = 2{,}160{,}000 \text{ ft-lb per hr}$$
$$= 2{,}780 \text{ Btu per hr}$$

$$\text{Power input to fluid} = \frac{2{,}160{,}000}{(60)(33{,}000)} = 1.09 \text{ hydraulic hp}$$

In order to determine the power required for the pump drive, the hydraulic horsepower must be divided by the fractional efficiency of the pumps.

$$\text{Enthalpy increase of the fluid in the pump } \frac{2780}{4887.3} = 0.57 \text{ Btu per lb}$$

(5) The temperature of the feed to the stripper is determined by an energy balance around the heat exchanger. The quantity of stream (7) is determined and its enthalpy may be evaluated from the specified temperature of 120°F. The temperature of stream (6) is the equilibrium temperature of a solution containing 20 per cent NH_3 at a pressure of 210 lb per sq in., which is read from Fig. 132 as 290°F with an enthalpy of 220 Btu per lb. The enthalpy of stream (5) is then calculated from the specific enthalpies of Table A and the energy-balance equation $m_5H_5 = m_6H_6 + m_4H_4 - m_7H_7$ or

$$m_5H_5 = 804{,}720 - 21{,}660 - 153{,}628 = 629{,}432 \text{ Btu per hr}$$
$$H_5 = 629{,}432/4887.3 = 128.8 \text{ Btu per lb}$$

This specific enthalpy corresponds to a temperature of approximately 217°F on the saturated liquid lines of Fig. 132 which is above the saturation temperature of 202°F for a 40 per cent solution at 210 lb per sq in. Thus, the feed to the stripper is partially vaporized by the heat exchanger. The heat-transfer duty of the exchanger is 804,720 − 153,628 = 651,092 Btu per hr.

(6) The heat input to the reboiler is determined by an over-all energy balance around the stripping column. The rate of stream 9 is fixed by the specified reflux ratio and $m_8 = m_9 + m_1$. The corresponding enthalpies are read from Fig. 132 as listed in Table A. The temperature of stream (8) is the temperature of the saturated liquid which is in equilibrium with vapors containing 99.5 per cent NH_3 at 210 lb per sq in. From Fig. 132 it is seen that this dew-point liquid has a composition of approximately 85 per cent NH_3 and a temperature of 110°F. From Table A,

$$q_s = m_8H_8 + m_6H_6 - m_5H_5 - m_9H_9 = 1{,}198{,}800 + 804{,}720 - 629{,}432 - 93{,}450$$
$$= 1{,}280{,}638 \text{ Btu per hr}$$

The heat removed by the condenser is expressed by

$$-q_1 = m_8H_8 - m_1H_1 - m_9H_9 = 1{,}198{,}800 - 186{,}880 - 93{,}450 = 918{,}470 \text{ Btu per hr}$$

The heat removed in the absorber is expressed by

$$-q_a = m_7H_7 + m_2H_2 - m_3H_3 = 153{,}628 + 786{,}880 - (-24{,}440) = 964{,}948 \text{ Btu per hr}$$

These results may be substituted in Equation (84) as a verification of the calculation. Thus,

$$600{,}000 + 1{,}280{,}638 - 918{,}470 - 964{,}948 + 2780 = 0$$

(7) The steam used for heating the reboiler must be at a temperature of 300°F which corresponds to an absolute pressure of 67 lb per sq in. or a gauge pressure of 52 lb per sq in. If it is assumed that the condensate is withdrawn at 300°F, heat to the reboiler per pound of steam is the heat of vaporization which from the steam table is 910 Btu per lb.

$$\text{Reboiler steam rate} = \frac{1{,}280{,}638}{910} = 1407 \text{ lb per hr.}$$

If the results of Illustrations 8 and 9 are compared, it is seen that with the particular specifications employed identical refrigeration duties are performed by a 44.3-hp mechanical drive in the one case and by 1,407 lb per hr of 52-lb steam in the other. If the heating steam were used to operate the drive of the compression machine an engine with a steam rate of 34 lb per hp-hr would be required. This corresponds to an engine efficiency of 65 per cent if the exhaust is at atmospheric pressure. The heat to cooling water in the absorption cycle is $(1.96)10^6$ Btu per hr as compared to $(0.696)10^6$ in the compression machine. This difference in cooling-water requirement is greatly reduced if a steam drive with a condenser is used for the compressor.

It may be concluded that the relative desirability of compression and absorption refrigeration cycles is largely dependent upon particular local conditions. The absorption operation is favored by the availability of low-cost exhaust steam for heating the reboiler. The choice of a refrigeration cycle thus involves consideration of the entire power-steam balance of the plant. As previously mentioned, many modifications of the absorption cycle are possible in order to adapt it to local conditions.

Other binary systems have been used for absorption refrigeration, but the ammonia–water is the most common. The chief difficulty in the development of other binary systems is in the selection of an absorbent of high absorption capacity and ease of separation, and one that is nontoxic, noncorrosive, and not decomposed by prolonged use at the stripper temperatures.

Refrigeration cycles based upon the adsorption of refrigerant vapors upon solid adsorbents have been used to a limited extent and may develop to greater importance. Such schemes operate on an intermittent cycle, adsorbing vapors at low temperature and pressure in one chamber, while another chamber is evolving previously adsorbed vapors at a higher temperature and pressure. Adsorbents of high capacity such as silica gel, activated alumina, or charcoal may be used with a variety of fluids. The disadvantages of these methods are the intermittent nature of the operation and the difficulty of transferring heat to a solid material. The thermodynamic problems of adsorption refrigeration may be handled by the same general principles as demonstrated here for the absorption systems. It is desirable to develop an enthalpy-composition chart for the adsorbed system to facilitate calculations. Such a chart is shown in Fig. 62, page 300, for water vapor on silica gel.

Low-Temperature Liquefaction of Gases. It is frequently desired to liquefy gases such as air which are normally far above their critical temperatures. The object may be to separate components of a gaseous mixture by subsequent fractional distillation or to produce very low-temperature refrigeration, below the practical working ranges of the common refrigerants.

In order to obtain liquefaction of such gases, combined cooling and compression must be produced. For example, air must be cooled to $-140.7°C$ before it can be liquefied even at high pressures. When working with gases of very low critical temperatures, it may be necessary

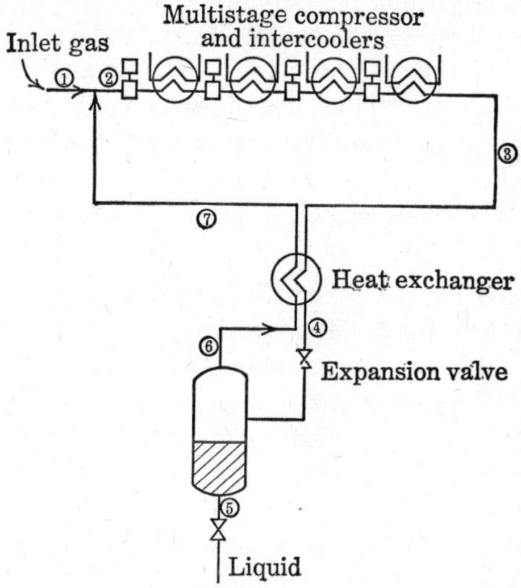

FIG. 133. Low-Temperature Liquefaction of Gases.

to precool by means of conventional refrigeration. The lower temperatures required for liquefaction are then produced by the self-cooling cycle shown diagrammatically in Fig. 133.

The gas to be liquefied enters the system at temperature T_1 and pressure p_1 and is combined with recycled gas which drops the temperature slightly to T_2. The gas is then compressed to p_3 with an increase in temperature. Multistage compression is generally employed with rejection of heat $-q_1$ in the intercoolers. The compressed gas is cooled at the interstages with water or a cooling fluid circulated by a conventional refrigeration system. The compressed gas is passed to a heat exchanger where it is cooled by a return flow of the same gas after its expansion. The cold compressed gas is then expanded to a lower pressure with a corresponding drop in temperature. This expansion may be carried out isentropically, and the power developed may be used in the compressor drive or for other purposes. This scheme is termed the Claude process. More commonly, however, the simple Linde process involving free expansion through a throttling valve is used, even though the temperature drop is much less, and no work is recovered. The cold expanded mixture of liquid and gas passes to a separator where the liquid is collected and the uncondensed gas is returned to the heat exchanger and thence to the compressor inlet, completing the cycle.

The design or performance of a gas-liquefaction plant may be developed without difficulty from principles already demonstrated. With reference to the stream numbers of Fig. 133, it is desirable to base the design on specified values of p_1, p_3, p_5, T_1, T_3, and T_7. Energy balances involving x, the fraction of the compressed stream which is liquefied, may then be established. Thus, if free expansion is used and changes of kinetic and potential energy and the exchange of heat with the surroundings are neglected,

$$H_3 = xH_5 + (1 - x)H_7 \tag{86}$$

Since H_5 is fixed by the saturation conditions at p_5, Equation (86) may be solved directly for x. Evaluation of x permits establishment of a complete material balance to correspond to the specified rate of production of liquid. The intermediate temperatures and the heat-transfer and compression duties are then readily calculated. If absorption of heat from the surroundings is appreciable, these items should be included in the energy balance.

It is evident that the Linde process is applicable only to gases which cool in free expansion. With reference to Fig. 105, it is seen that free expansion cooling is not obtained at reduced temperatures above 6. Hence, for example, in order to liquefy hydrogen, it is necessary to precool to 100°K before the Linde process may be employed.

VAPOR-RECOMPRESSION EVAPORATION

The evaporation of water from nonvolatile solutes or suspensions is generally accomplished by multiple-effect evaporators, such as illustrated in Fig. 29, page 180. In this scheme several pounds of water may be evaporated by the use of 1 lb of heating steam, depending upon the number of effects used. A somewhat similar result can be produced in more compact and simple equipment by the use of the vapor-recompression cycle, as shown diagrammatically in Fig. 134.

A dilute solution at a temperature T_1 is charged to the evaporator containing the boiling concentrated solution at temperature T_2 and pressure

p_2. As a result of the elevation of the boiling point of the solution over that of pure water, the vapors evolved are superheated. These vapors are isentropically compressed to p_4, the pressure of the fresh steam supply, and combined with this supply to the evaporator. Condensate is withdrawn substantially at the saturation temperature of the heating steam. Concentrated solution is continuously withdrawn from the bottom of the evaporator to maintain a constant liquid level and concentration.

By this method it is possible to evaporate several pounds of water per pound of steam supply through the addition of a relatively small amount

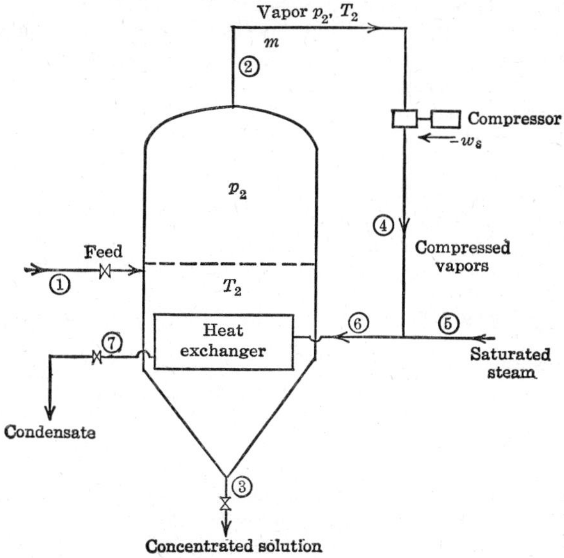

FIG. 134. Vapor-Recompression Evaporation.

of energy as shaft work in the compressor. However, the cost per unit of energy as shaft work is much higher than as heat. The following equation expresses the energy balance of an evaporator in which m_1 is the feed rate in pounds per hour, m_2 is the rate of evaporation, and m_5 is the rate of supply of fresh steam.

$$m_1H_1 + m_5H_5 = (m_2 + m_5)H_7 + (m_1 - m_2)H_3 + w_s \qquad (87)$$

where the H values represent enthalpies per pound corresponding to the designations on Fig. 134. For specified values of m_1, m_2, p_2 and p_4, the work done on the vapors in the compressor is calculated by the methods previously demonstrated. The heating steam requirement m_5

is then obtained directly from Equation (87) and the ratio of evaporation to steam m_2/m_5 may be calculated.

This scheme of evaporation is of interest where power costs are low and fuel costs high. Such a situation might exist in the vicinity of large hydroelectric-power developments.

PROBLEMS

1. Calculate from the changes in the energy functions, the work done, the heat added, and the changes in T, v, S, U, H, A, and G in the reversible expansion to 1 atm by each of the following methods of 1 lb-mole of chlorine originally at an absolute pressure of 500 lb per sq in. and 250°F. The initial absolute entropy of chlorine is 48.76 Btu/(lb-mole)(°R). Carry out these calculations (I) assuming ideal behavior, (II) taking into account deviations from ideal behavior by the generalized correlation. Express work in foot-pounds and the energy functions in Btu.

(*a*) Nonflow isothermal.
(*b*) Flow isothermal.
(*c*) Nonflow isentropic.
(*d*) Flow isentropic.
(*e*) Maxwell expansion (irreversible).
(*f*) Joule–Thomson expansion (irreversible).

The physical properties of chlorine are as follows:

$$c_p^* = 8.28 + 0.00031T \text{ Btu/(lb-mole)(°R)} \qquad [T = \text{°R}]^6$$

$$\text{Normal boiling point} = 430\text{°R}$$
$$\lambda = 8780 \text{ Btu per lb-mole at } 430\text{°R}$$
$$T_c = 750\text{°R}; \qquad p_c = 76.1 \text{ atm}$$
$$\text{Molecular weight} = 70.9; \qquad \rho_L = 1.56 \text{ g per cc at } 430\text{°R}$$

2. Repeat the calculations of parts *a*II and *b*II of problem 1 by graphical integration of the pV changes based on the generalized compressibility-factor correlation.

3. Propane at an absolute pressure of 200 lb per sq in. and 120°F is expanded through a valve to an absolute pressure of 39 lb per sq in. Using the generalized thermodynamic correlations and the vapor-pressure equation from Table IIa, page 73, calculate the temperature on the downstream side of the valve and the quality or degrees of superheat of the vapor above saturation before and after expansion.

4. The expansion referred to in problem 3 might also be carried out isentropically in an engine. Calculate the temperature and degrees of superheat or quality after expansion in this manner and the ideal work performed per pound-mole.

5. A Rankine-cycle power plant generates dry saturated steam at a gauge pressure of 400 lb per sq in. which is isentropically expanded through engines to an exhaust gauge pressure of 10 lb per sq in. for use in process heating. It may be assumed that the condensate is returned to the boilers at its saturation temperature under the exhaust pressure by an isentropic pump.

(*a*) By means of the data of the steam tables, determine the number of pounds of dry saturated low-pressure steam available for process heating per 100 lb of high-pressure steam generated.

[6] K. K. Kelley, *U. S. Bur. Mines. Bull.* 371 (1934).

(b) Calculate the ideal thermodynamic efficiency of the cycle. Compare this result to the Carnot efficiency.

(c) Calculate the ideal water rate of the engine.

6. Determine the effect on the results of problem 5 of superheating the steam 100°F.

7. A pump is driven by a steam engine which delivers 20 hp and operates on dry saturated steam at 400 lb gauge pressure and exhausts at an absolute pressure of 2.0 lb per sq in. The engine efficiency is 70 per cent. Using data from the steam tables, calculate the pounds per hour of steam required and the heat-transfer duty of the condenser in Btu per hr. Compare the ideal steam rate of the engine with that of problem 5c.

8. A single-stage compressor is required to compress 450 cu ft per min of CO_2 measured at 60°F and 14.5 lb per sq in. abs from an absolute pressure of 5 lb per sq in. and 80°F to an absolute pressure of 20 lb per sq in. Assuming isentropic compression, calculate the horsepower required to drive the compressor, the required piston displacement in cubic feet per minute and the discharge temperature, assuming a volumetric efficiency of 77 per cent and a compression efficiency of 83 per cent.

9. The compression duty of problem 8 might also be handled by a two-stage machine with intercooling to 80°F. Calculate the horsepower required to drive the compressor in this operation and the discharge temperature, assuming the efficiencies given in problem 8.

10. Anhydrous HCl ($t_b = -85$°C; $t_c = 51.4$°C; $p_c = 81.6$ atm) at an absolute pressure of 30 lb per sq in. and 80°F is to be compressed to 450 lb per sq in. abs at a rate of 100 cu ft per min measured at 60°F and 14.5 lb per sq in. A two-stage machine is to be used with intercooling to 80°F. The compression efficiency based on isentropic compression in each stage is 80 per cent and the volumetric efficiency 65 per cent. The heat capacity of HCl in the ideal state is given in Table V. The intermediate suction pressure may be calculated from Equation (73).

Assuming isentropic compression in each stage, calculate:

(a) The horsepower required.

(b) The required piston displacement, cubic feet per minute each stage.

(c) The discharge temperature.

11. A compressor having a 75 per cent mechanical efficiency is to compress 1.5 lb-moles of ammonia per minute from an absolute pressure of 15 lb per sq in. and 100°F to 200 lb per sq in. abs. The operation is to be in two stages, each of which may be assumed to be isentropic with intercooling to 115°F. By use of Figs. 116–119 calculate the power required to drive the compressor when designed for the optimum intermediate pressure. Calculate the discharge temperature from each stage and the heat-transfer duty of the intercooler, and compare the optimum intermediate pressure with that calculated from Equation (73).

12. (a). The refrigeration duty of Illustration 8 is to be performed by a plant using ammonia as the refrigerant, and with the same temperature levels. Assuming the same compressor efficiencies, calculate the power input, piston displacement, coefficient of performance, and heat input to the condenser for the ammonia machine. Compare these values and also the operating temperatures and pressures with those of the propane system.

(b) Repeat the design for an evaporator temperature of 0°F.

13. Design an ammonia–water absorption refrigeration plant for the conditions specified in Illustration 9 except that the evaporator temperature is to be maintained at 0°F and the reflux ratio increased to 0.9 because of the more dilute solutions handled. Compare the heat-transfer duties, circulation rates, and steam requirement with those of Illustration 9.

14. It is desired to design a plant for the liquefaction at atmospheric pressure of 2000 lb per day of ethylene. The ethylene is available at a pressure of 1 atm and 90°F. Develop a process design for the plant on the basis of compression to 50 atm in a four-stage compressor with intercooling to 90°F. The compressed ethylene is cooled to 90°F, and the recycle gas is heated to 70°F in the heat exchanger. Gain or loss of heat by the equipment may be neglected and isentropic compression assumed in each stage of compression. Equation (75) may be used to establish the intermediate pressures in multistage compression. In liquefaction the gas is expanded irreversibly through a free expansion valve with self-cooling as shown in Fig. 133.

Calculate:

(*a*) The fraction of the compressor discharge which is liquefied.

(*b*) The temperatures at all points of the cycle.

(*c*) The horsepower input to the gas in each stage of the compressor.

(*d*) The heat-transfer duties of the cooler and heat exchanger in Btu per hour.

(*e*) Repeat calculations for isentropic expansion during liquefaction.

15. It is desired to concentrate continuously 1000 lb per hr of a 10 per cent calcium chloride solution in a vapor-recompression evaporator to 40 per cent concentration. The dilute solution enters at 100°F. The evaporator is operated at an absolute pressure of 3.0 lb per sq in. Dry saturated heating steam is supplied at an absolute pressure of 20 lb per sq in.

Calculate:

(*a*) The steam required from outside the cycle in pounds per hour.

(*b*) The pounds of water evaporated per pound of outside steam.

(*c*) The horsepower input to the compressor at 75 per cent compression efficiency.

In solving this problem the enthalpy-concentration chart for the calcium-chloride–water system, Fig. 54, page 281, may be used.

CHAPTER XIV

THERMODYNAMICS OF SOLUTIONS

The thermodynamic principles which are developed in previous chapters are applied in Chapters XII and XIII to problems involving essentially pure substances. The general relations of Chapter XI are in no way restricted to pure substances and may be validly applied to solutions and mixtures of all types. Such applications are complicated by the introduction of composition as an added variable.

COMPRESSIBILITY OF GASEOUS MIXTURES

The laws of additive pressures and of additive volumes discussed in Chapter II, page 37, when considered separately, do not necessarily imply ideal gases. In general, these laws apply with much greater accuracy than the ideal-gas law and are reasonably satisfactory even where pressures are moderately high. As previously pointed out, either Dalton's or Amagat's law may hold for actual gases, but together they hold rigidly only for ideal gases. Which of these laws gives the better approximation depends upon the conditions and the nature of the gaseous mixture. For example, for mixtures of argon and ethylene Dalton's law of additive pressures is the more nearly accurate; for mixtures of nitrogen and hydrogen the law of additive volumes holds better.

Mean Compressibility Factors. A simple empirical equation of state for a mixture of gases may be written in a form similar to Equation (XII-3), thus,

$$pV = z_m n_t RT \tag{1}$$

where

p = total pressure of mixture

V = total volume of mixture

z_m = mean compressibility factor of mixture

n_t = total number of moles of gases in mixture

Equation (1) permits calculation of the pressure, temperature, and volume relationships of gaseous mixtures of known compositions if the mean compressibility factors are known. The mean compressibility factor is a function of the pressure, temperature, and composition of the mixture. By the same methods which were described for a pure gas, the compressibility factors of a mixture of fixed composition may be determined and expressed graphically as a function of pressure, temperature, or molal volume. In order to obtain complete information regarding a particular system, these measurements must be repeated with mixtures of different compositions. Because of the great amount of experimental work involved, only a few simple systems of binary gaseous mixtures have been investigated and these only over relatively narrow ranges of conditions.[1]

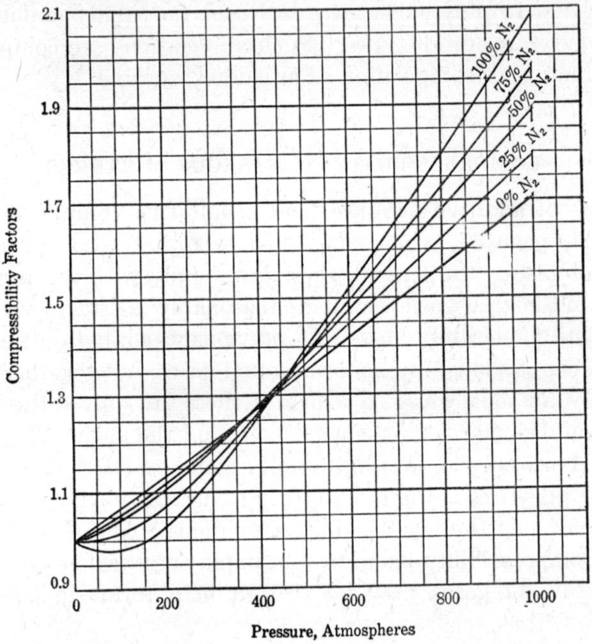

FIG. 135. Compressibility Factors of Mixtures of Hydrogen and Nitrogen at 0°C.

Bartlett[2] has carried out a series of careful determinations for mixtures of hydrogen and nitrogen at a temperature of 0°C. The results of this investigation are shown graphically in Figs. 135 and 136. In Fig. 135 is shown the variation of the mean compressibility factor with pressure for mixtures of various molal compositions. In Fig. 136 is shown the variation of the compressibility factor with the composition of the mixture at various pressures. If Amagat's law is assumed to apply, an ex-

[1] "International Critical Tables," III, 17.
[2] E. P. Bartlett, *J. Am. Chem. Soc.*, **49**, 687–701, 1955–1958 (1927).

pression for the mean compressibility factor follows:

$$z_m = \frac{z_A n_A + z_B n_B + z_C n_C}{n_t} \tag{2}$$

where

z_A = compressibility factor of component A at the temperature and total pressure of the mixture

This equation permits the approximate calculation of the mean compressibility factor of a mixture, at any conditions of temperature and pressure, from the compressibility factors of the component gases at the

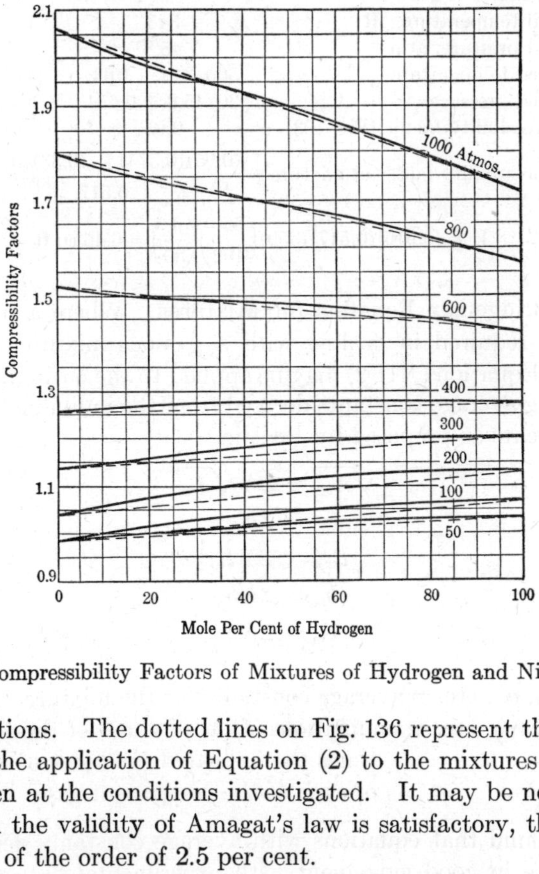

FIG. 136. Compressibility Factors of Mixtures of Hydrogen and Nitrogen at 0°C.

same conditions. The dotted lines on Fig. 136 represent the results obtained by the application of Equation (2) to the mixtures of hydrogen and nitrogen at the conditions investigated. It may be noted that for this system the validity of Amagat's law is satisfactory, the maximum error being of the order of 2.5 per cent.

Equations (1) and (2) may be used in conjunction with compressibility data on the pure components obtained either by direct measurement or from Fig. 103. This procedure yields fairly satisfactory results where the compressibility factor of none of the individual components is smaller than 0.5. Serious errors result if the conditions of the mixture are close to the critical points of any of the components. This method involves questionable extrapolations of Fig. 103 for components whose liquid-vapor pressure at the temperature of the mixture is less than the total pressure of the mixture.

Illustration 1. A gaseous mixture containing 6 lb of methane and 4 lb of ethylene is compressed to 30 atm at 100°F. Calculate the volume occupied by the compressed gas.

Basis: 10 lb of mixture:

	Methane	*Ethylene*
Pound-moles of gas	0.374	0.143
Critical temperature, °R	343	510
Critical pressure, atm	45.8	50.9
Reduced temperature	560/343 = 1.63	560/510 = 1.10
Reduced pressure	30/45.8 = 0.654	30/50.9 = 0.59
Compressibility factor (Fig. 103)	0.95	0.83

$$\text{Mean compressibility factor of mixture} = \frac{(0.95)(0.374) + (0.83)(0.143)}{0.517} = 0.916$$

$$V = (0.916)(0.517)(359)\left(\frac{560}{492}\right)\left(\frac{1}{30}\right) = 6.46 \text{ cu ft}$$

Beattie–Bridgeman Equation for Mixtures. Where a high order of precision is required in dealing with a gaseous mixture, the Beattie–Bridgeman Equation (XII-2) may be applied to the entire mixture using the following average constants derived from the individual constants of the components as proposed by Beattie.[3]

$$\sqrt{A_{om}} = \Sigma N \sqrt{A_o}$$

$$a_m = \Sigma N a$$

$$B_{om} = \Sigma N B_o$$

$$b_m = \Sigma N b$$

$$c_m = \Sigma N c$$

where

A_{om}, a_m, etc. = average constants for the mixture

$\Sigma N\sqrt{A_o}$, ΣNa, etc. = summations of the products of the mole fractions times the indicated functions of the constants of each of the component gases

Beattie found that equations with average constants derived in this manner were in good agreement with experimental data on the compressibility of mixtures. For the systems investigated the maximum error encountered was 0.55 per cent.

[3] J. A. Beattie, *J. Am. Chem. Soc.*, **51**, 19 (1929).

This method requires a knowledge of the Beattie–Bridgman constants of each component gas and is of limited utility for that reason.

Critical Phenomena of Gaseous Mixtures. The behavior of a mixture in the region of the critical point is best understood by reference to the pressure–temperature diagram of Fig. 137.

Curve AC' represents the vapor–pressure of a pure compound having the critical temperature and pressure corresponding to point C'. For a single-component system the area to the left of this curve represents the region of the liquid phase, and the area below it is the region of the vapor phase.

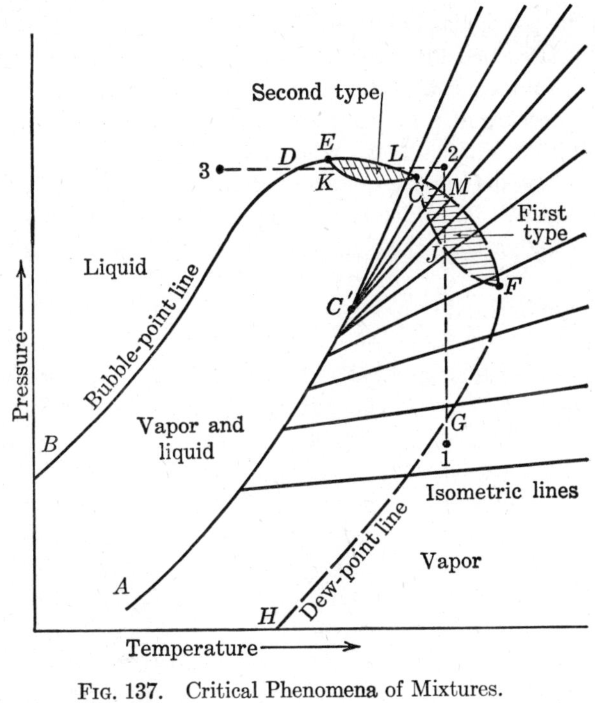

FIG. 137. Critical Phenomena of Mixtures.

The sloping straight lines on Fig. 137 are lines of constant volume for the single-component system and are termed *isochors* or *isometrics.* It has been found that for most substances the isometrics are approximately straight, represented by the equation,

$$p = mT - b \tag{3}$$

where m and b are constants dependent on the volume and the substance under consideration.

Curve $BDECFGH$ represents what is termed the *border curve* of a mixture having the same average volatility as that of the pure compound represented by line AC', but made up of two or more substances of different volatilities. The area enclosed by the border curve represents a two-phase region in which both liquid and vapor are present in equilibrium. Line BDE represents conditions of initial vaporization, to the left of which is the region of complete liquefaction termed the *bubble-point line.* Line HGF termed the *dew-point line* represents conditions of initial condensation below which is a region of complete vaporization.

Point C represents the critical point of the mixture. This critical point does not necessarily correspond to a maximum temperature at which the liquid phase can exist as in the case with a pure component, but rather the particular point on the border-line curve where the vapor and liquid phases become indistinguishable, or where the bubble-point and dew-point lines meet. In general, both the critical temperature and the critical pressure of a mixture are higher than those of a pure compound having the same average volatility.

It may be noted that the dew-point line passes through a maximum temperature at F. Thus, in the case of a mixture, liquid may exist at a temperature higher than the critical temperature. This maximum temperature F on the border curve is termed the *critical-condensation* temperature. Similarly, in the case of many mixtures the bubble-point line passes through a point of maximum pressure E, higher than the critical pressure. The areas $FJCM$ and $EKCL$ represent regions of *retrograde condensation.* If the mixture at the condition of point 1 is compressed at constant temperature, a more dense phase appears at point G on the dew-point line. As the pressure is further increased, the quantity of this more dense phase increases to a maximum at point J and then diminishes, disappearing entirely when point M is reached. This type of retrograde condensation occurring in area $FJCM$ is called the *first type.* If the liquid mixture at conditions of point 3 is heated at constant pressure, a less dense phase appears at point D on the bubble-point line, reaches a maximum at point K, and then diminishes, disappearing entirely at point L. The type of retrograde condensation which occurs in area $EKCL$ is called the *second type.*

The entire area outside the border curve is a region of homogeneous fluid in which no phase separations occur. Moving clockwise in the region about the border curve, the liquid phase merges imperceptibly into the vapor phase.

Retrograde condensation is of particular interest in petroleum production. When natural gas is withdrawn from high-pressure wells, liquid gasoline condenses upon release of pressure. The residual gas may then be recompressed and recycled to the underground oil-bearing formation where it again reaches equilibrium with the liquid oil. As a result low-boiling fractions of oil enter the gas phase and are subsequently recovered by the retrograde condensation accompanying the expansion step. A full discussion of this operation is given by Katz and Kurata.[4]

It may be noted from Fig. 137 that retrograde condensation of the first type occurs when a constant-temperature line twice crosses the dew-point line (broken line), and retrograde condensation of the second type

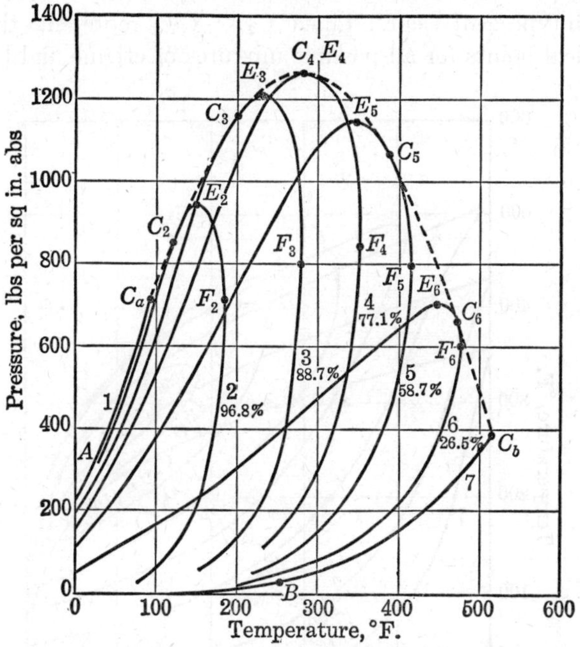

FIG. 138. Pressure–Temperature Diagrams for Mixtures of Ethane and n-Heptane at Various Compositions.

occurs when a constant-pressure line twice crosses the bubble-point line (solid line). Both types of retrograde condensation occur in a given mixture when the critical point C lies on the border curve between the maximum pressure E and the maximum temperature F. When the border curves of a binary system are obtained for the entire range of compositions, it is found that the critical temperatures occur at different points on the border-line curves relative to E and F.

These effects are illustrated in Fig. 138 showing the border curves of various mixtures of ethane and heptane as determined by Kay.[5]

[4] D. L. Katz and F. Kurata, *Ind. Eng. Chem.*, **32**, 817 (1940).

[5] W. B. Kay, *Ind. Eng. Chem.*, **30**, 459 (1938), reprinted with permission.

Line AC_a is the vapor–pressure curve of pure ethane, and point C_a is its critical point. Line BC_b has the same significance for pure heptane. The lines 2–6 are the border curves for five different mixtures containing the indicated mole percentages of ethane. The corresponding critical points of the mixtures are designated as C_2–C_6, respectively, the corresponding maximum pressure as E_2–E_6, and the maximum temperatures as F_2–F_6. It may be noted that for compositions 2 and 3 only retrograde condensation of the first type can occur, whereas for compositions 5 and 6 both types can occur. Line $C_aC_2 \cdots C_6C_b$ represents the envelope of all critical points for all possible mixtures of ethane and heptane.

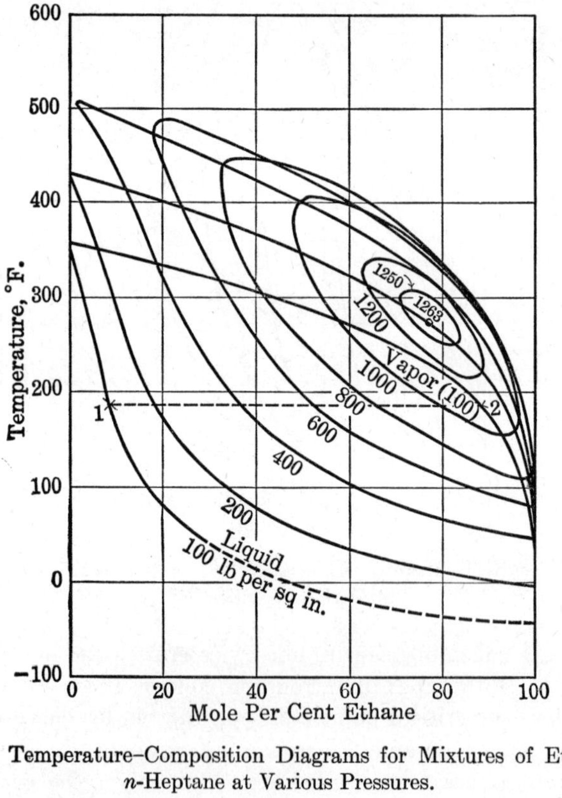

FIG. 139. Temperature–Composition Diagrams for Mixtures of Ethane and *n*-Heptane at Various Pressures.

If the envelope curve were of such shape that it approached point C_b with a positive slope, it would indicate the existence of mixtures having critical temperatures higher than that of the higher boiling component. These mixtures would have critical points lying in a clockwise direction beyond the corresponding points E and F, both of which would be on the bubble-point line. Such mixtures could exhibit only retrograde condensation of the second type. This situation is possible but unusual.

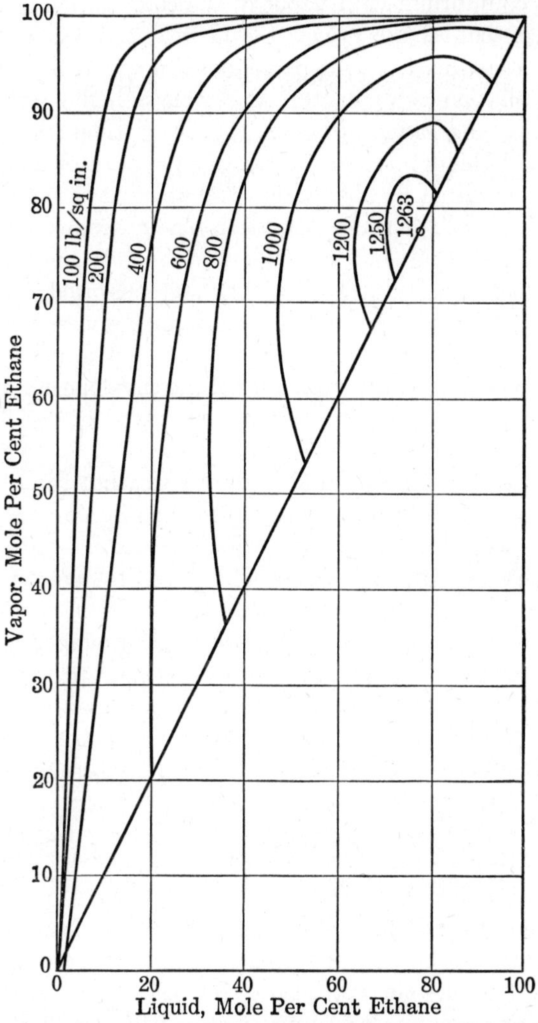

FIG. 140. Vapor–Liquid Mole Fraction Relations for Ethane–*n*-Heptane at Various Pressures.

Isobaric boiling-point curves and liquid–vapor composition curves of the ethane–heptane system[5] at different pressures are plotted in Figs. 139 and 140.

Each envelope on Fig. 139 corresponds to the indicated constant pressure. The upper curve relates temperature to vapor composition and the lower to liquid composition. Thus, at a pressure of 100 lb per sq in. and a temperature of 190°F liquid of the composition corresponding to point 1 is in equilibrium with vapor of the composition of point 2. It may be noted that as the pressure is increased the range of the two-phase region is reduced. This behavior is of great importance in the design of distillation equipment. At pressures higher than 396 lb per sq in. pure heptane cannot exist in the liquid phase in contact with ethane gas, and this represents the highest pressure at which pure heptane can be separated by distillation. At pressures between 396 and 712 lb per sq in., distillation can theoretically separate mixtures into pure ethane and liquid solutions of ethane in heptane. At pressures above 712 lb per sq in. pure ethane cannot be separated, and distillation can produce separation only into two mixtures, one rich and one poor in ethane. The maximum pressure at which a two-phase system can exist is 1263 lb per sq in., the maximum of the envelope curve of Fig. 138. This is the maximum pressure at which any type of distillation of the system can be conducted.

Pseudocritical Point. It was shown by Kay[6] that the pVT data of a gaseous mixture may be satisfactorily correlated by the generalized compressibility-factor curves of Fig. 103 if *pseudocritical* temperature and pressure are used for the calculation of reduced temperatures and pressures. For all except pure compounds or mixtures of compounds differing little in physical properties, the pseudocritical temperature and pressure are less than the true critical temperature and pressure.

It was found that the pseudocritical properties are approximately equal to the molal averages of the critical properties of the components. Thus, the pseudocritical temperature T_c' of a mixture is approximately the molal average of the critical temperatures of the components, and the pseudocritical pressure p_c' is approximately the molal average of the component critical pressures.

Kay's measurements indicated some differences between correct pseudocritical properties and the molal averages of the properties of the components, but an improved general method of estimating pseudocritical properties has not yet been developed. Kay[6] and later Smith and Watson[7] proposed more accurate methods for use with hydrocarbon mixtures.

Once the pseudocritical point is established, pVT calculations on a mixture are carried out exactly as for a pure compound, the average molecular weight of the mixture being used. Compressibility factors are determined from Fig. 103, the pseudoreduced temperature and pres-

[6] W. B. Kay, *Ind. Eng. Chem.*, **28**, 1014 (1936).
[7] R. L. Smith and K. M. Watson, *Ind. Eng. Chem.*, **29**, 1408 (1937).

sure being used. However, it must be emphasized that this method applies only to the vapor phase and breaks down at conditions within the border curve of the mixture. The pseudocritical point itself is located in this region.

Illustration 2. Calculate the volume occupied by 1 lb of a mixture of 59.9 mole per cent ethylene and 40.1 per cent argon at a temperature of 25°C and a pressure of 100 atm.

	Ethylene	*Argon*
Critical temperature, °C	9.7	−122
Critical pressure, atm	50.9	48
Molecular weight	28	39.9

Pseudocritical temperature $T'_c = (0.599)(283) + (0.401)(151) = 230°\text{K}$
Pseudocritical pressure $p'_c = (0.599)(50.9) + (0.401)(48) = 49.7$ atm
Average molecular weight $= (0.599)(28) + (0.401)(39.9) = 32.7$
Pseudoreduced temperature $T'_r = 298/230 =$ 1.3
Pseudoreduced pressure $p'_r = 100/49.7 =$ 2.01
Compressibility factor (Fig. 103) = ... 0.69

$$\text{Volume} = \frac{(359)(0.69)}{32.7}\left(\frac{298}{273}\right)\left(\frac{1}{100}\right) = \text{.................... } 0.083 \text{ cu ft}$$

The experimental measurements of Masson and Dolley indicated a compressibility factor of 0.712 as compared to the value of 0.69 derived from Fig. 103 and the pseudocritical point in the preceding illustration.

It is believed that use of the pseudocritical point together with reduced compressibility factors is the most satisfactory generalized method for handling pVT calculations on gaseous mixtures, particularly at conditions near the critical. In this region it is more reliable than either Dalton's or Amagat's laws.

COMPRESSIBILITY OF LIQUID SOLUTIONS

When two miscible liquids at the same temperature are mixed together, the resulting solution may exhibit ideal behavior, that is, possess properties which are additive functions of the properties of the pure components. Thus, the volume and enthalpy of the solution may be the exact sum of the volumes and enthalpies of the components. In this case the density of the solution is the sum of the products of the densities of the components multiplied by their respective volumetric fractions, and the specific heat of the solution is the sum of the products of the specific heats of the components multiplied by their respective weight fractions.

Nearly ideal solutions are formed by closely related homologs of a series of organic compounds such as heptane and octane or benzene and toluene at atmospheric temperatures. However, if the temperature is increased and the liquids are maintained at saturation, the deviations from ideal behavior increase. Finally, conditions are reached at which the more volatile component is above its critical temperature, and large deviations from additivity of volumes result, depending upon the pressure under which this component is confined. Thus, even adjacent homologs form ideal liquid solutions only at low temperatures corresponding to low vapor pressures for both the solution and the individual components.

Ideal Systems. An *ideal system* is defined[8] as a group of components which tend to form ideal liquid solutions at low temperatures where the saturated vapor of each component behaves as an ideal gas. A *nonideal system* does not show ideal behavior as a liquid in any temperature range. For example, chemically dissimilar materials such as alcohol and benzene never form ideal liquid solutions. Chemical homologs generally behave as approximately ideal systems but do not form ideal solutions at all conditions. Thus, solutions of liquid ethane and heptane at atmospheric temperatures depart widely from ideal behavior but at lower temperatures approach ideality. Many somewhat dissimilar materials but of low polarity may be treated as ideal systems with accuracy satisfactory for many purposes. The greatest deviations from the behavior of the ideal system are encountered with highly polar compounds such as water, ammonia, the lower alcohols, and ketones.

Density of Liquid Solutions. It was found by Gamson and Watson[8] that the empirical correlation of Fig. 109 may be applied to solutions in ideal systems through the use of pseudoreduced temperatures and pressures. If components A, B, C, $\cdots$ may be considered as an ideal system, at some low temperature T_0 they will form an ideal liquid solution in which the volumes of the components are additive. Thus,

$$V_0 = V_{A_0} + V_{B_0} + V_{C_0} + \cdots = \frac{m_A}{\rho_{A_0}} + \frac{m_B}{\rho_{B_0}} + \frac{m_C}{\rho_{C_0}} + \cdots \tag{4}$$

Combining (4) with Equation (XII-38), page 502, gives

$$V_0 = \frac{\omega_{A_1} m_A}{\rho_{A_1} \omega_{A_0}} + \frac{\omega_{B_1} m_B}{\rho_{B_1} \omega_{B_0}} + \frac{\omega_{C_1} m_C}{\rho_{C_1} \omega_{C_0}} + \cdots \tag{5}$$

[8] B. W. Gamson and K. M. Watson, *Natl. Petroleum News*, Tech. Sec., **36**, *R*554 (Aug. 2, 1944). Also "Process Engineering Data," National Petroleum Publishing Company, Cleveland (1944).

where V_0 = total volume of solution at the low temperature T_0 for the respective pure components:

V_{A0}, V_{B0}, V_{C0} = volume at temperature T_0

m_A, m_B, m_C = masses

$\rho_{A0}, \rho_{B0}, \rho_{C0}$ = densities at temperature T_0

$\rho_{A1}, \rho_{B1}, \rho_{C1}$ = densities at reference conditions

$\omega_{A0}, \omega_{B0}, \omega_{C0}$ = expansion factors at temperature T_0

$\omega_{A1}, \omega_{B1}, \omega_{C1}$ = expansion factors at the reference conditions of temperature T_1 and pressure p_1 where the density is ρ_1

The value of ω_{A_1} is determined from Fig. 109 by expressing T_{A_1} and p_{A_1} in reduced terms based on the critical properties of component A. It is evident from consideration of Fig. 109 that as T_0 is decreased ω_{A0}, ω_{B0}, $\omega_{C0}, \cdots$ approach equality and it is reasonable to assume that if the relationship were extrapolated to the absolute zero the values of ω for all compounds would become equal to ω_0. If T_0 is taken as the absolute zero,

$$V_0 = \frac{1}{\omega_0}\left(\frac{\omega_{A_1} m_A}{\rho_{A1}} + \frac{\omega_{B_1} m_B}{\rho_{B_1}} + \frac{\omega_{C_1} m_C}{\rho_{C_1}} + \cdots\right) \tag{6}$$

The volume V' of the solution at any temperature T and pressure p is obtained by applying Equation (XII-38) to the entire mixture and combining it with (6),

$$V' = \frac{V_0\omega_0}{\omega'} = \frac{1}{\omega'}\left[\left(\frac{\omega_1}{\rho_1}\right)_A m_A + \left(\frac{\omega_1}{\rho_1}\right)_B m_B + \left(\frac{\omega_1}{\rho_1}\right)_C m_C + \cdots\right] \tag{7}$$

where V' = volume of liquid solution of mass $m_A + m_B + m_C + \cdots$ at temperature T and pressure p

ω' = expansion factor for entire mixture at T and p

The expansion factor ω' of the solution is derived from Fig. 109 by expressing T and p in pseudoreduced units based on the pseudocritical properties of the mixture.

Similarly, in molal units,

$$v' = \frac{v_0'\omega_0'}{\omega'} = \frac{1}{\omega'}[(v_1\omega_1)_A N_A + (v_1\omega_1)_B N_B + (v_1\omega_1)_C N_C + \cdots] \tag{8}$$

where v' = volume per mole of mixture

v_{1A} etc. = molal volume of component A at the reference condition $1A$

ω_{1A} = expansion factor of component A at the reference condition $1A$

N_A = mole fraction of component A

By means of Equations (7) and (8) it is possible to calculate the volume of an ideal liquid system at any temperature and pressure within the range of Fig. 109, even though some of the components may be above their critical temperatures. Care must be exercised that this equation is applied only to the liquid state. Where vaporization occurs, the volumes of the liquid and vapor phases must be considered separately by methods developed in later sections.

Gamson and Watson[8] also developed methods for calculating partial molal volumes of the components of solutions if ideal systems in both the gaseous and liquid states prevail over the wide ranges of conditions.

Illustration 3. The reflux to a fractionating column which is depropanizing natural gasoline has the following composition in mole per cent:

Methane	1.5
Ethane	12.0
Propane	86.5
	100.0

To serve as a basis for the design of the pump, meter, valves, and pipe line for this stream, it is desired to calculate the density in pounds per cubic foot of this liquid solution at 100°F and an absolute pressure of 300 lb per sq in.

From Table XXVIII, page 507:

	N	T_c °R	NT_c	p_c, lb per sq in.	Np_c lb per sq in.	$v_1\omega_1$ cc per g-mole	$N(v_1\omega_1)$ cc per g-mole	M	NM
CH_4	0.015	343	5.1	673	10.6	4.36	0.07	16.04	0.24
C_2H_6	0.120	550	66.0	717	86.1	6.81	0.82	30.07	3.60
C_3H_8	0.865	666	576.1	642	555.3	9.18	7.96	44.09	38.06
		$T'_c =$	647.2	$p'_c =$	652.0	$(v_1\omega_1)' =$	8.85	$M_{avg} =$	41.90

At 100°F and 300 lb per sq in.

$$T'_r = 560/647.2 = 0.865$$

$$p'_r = 300/652 = 0.460$$

$$\omega' \text{ (Fig. 109)} = 0.096$$

From Equation (8),

$$v = \frac{8.85}{0.096} = 92.0 \text{ cc per g-mole}$$

$$\text{Density} = \frac{(41.90)(62.4)}{92.0} = 28.4 \text{ lb per cu ft}$$

Nonideal Systems. Many solutions, particularly those involving highly polar components, deviate widely from Equations (7) and (8) and at low temperature have volumes either greater or less than the sum of the volumes of the pure components. Direct experimental data are necessary in order to determine the behavior of such systems. Generally densities or specific volumes of the solutions are determined and empirically expressed as functions of composition at a constant temperature where the vapor pressures are less than 1 atm. Pressure variations in this range have a negligible effect on the volume of liquid solutions. The partial volume of each component may be derived from such data by the methods developed in Chapter VIII, page 287, and individually expressed as a function of composition.

The deviations from ideal solution behavior which are exhibited by ideal systems may be designated as deviations due to differences in molecular size. Nonideal systems show this same type of deviation and in addition deviations due to differences in molecular type or chemical dissimilarity which characterize the behavior of such systems at low temperatures. In general, it is observed that deviations from ideal behavior which are due to chemical dissimilarity decrease with increased temperature, whereas deviations due to differences in molecular size increase as the temperature is increased. The opposite temperature effects result in a temperature range of minimum deviation from the laws of ideal solutions for some nonideal systems. At low temperatures large deviations due to chemical dissimilarity are encountered which are reduced as the temperature is increased. Still further increase in temperature leads to important deviations due to dissimilarity in size. It is possible that Equation (7) is applicable without serious errors to nonideal systems in this higher-temperature range where deviations due to chemical dissimilarity are negligible in comparison to those due to differences in molecular size.

ENTHALPIES OF SOLUTIONS

Where an ideal solution is formed, the enthalpy of the solution is equal to the sum of the enthalpies of the components at the existing temperature and pressure. In the general case of a nonideal solution in either an ideal or nonideal system, a heat of mixing is involved in the formation of the solution, as discussed in Chapter VIII. The magnitude of the heat of mixing is a measure of the extent of the deviation from ideal behavior.

Enthalpies of Gaseous Mixtures. Under conditions of nonideal gas behavior the enthalpy of a gaseous mixture is best determined by first obtaining its enthalpy in the ideal state at the existing temperature and applying an isothermal correction for the effect of pressure on the enthalpy of the entire mixture based upon its pseudocritical properties.

This procedure is straightforward where only the gaseous state is involved. The enthalpy of the mixture at zero pressure is obtained as the sum of the enthalpies of the components at the existing temperature and zero pressure. The pseudoreduced conditions of the mixture are then calculated and the corresponding value of $\frac{\text{H}^* - \text{H}}{T'_c}$ for the mixture is determined directly from Figs. 105 or 106.

Enthalpies of Liquid Solutions. In dealing with solutions of liquids in an ideal system under such conditions that no component is at a reduced temperature above 0.9, it is generally satisfactory to assume that the enthalpy of the mixture is equal to the sum of the enthalpies of the components at the temperature and pressure of the mixture. For nonideal systems the heat of mixing must be added, as discussed in Chapter VIII. If some components of the liquid are at reduced temperatures above 0.9, a more nearly accurate procedure is to calculate the enthalpy of the liquid from the enthalpy of the ideal-gas mixture at the existing temperature. It is assumed that the liquid is formed by compressing the gas to the pseudo-vapor pressure of the mixture which may be considered as the vapor pressure of the equivalent pure compound whose critical properties, liquid density, and heat of vaporization are equal to the corresponding pseudo properties of the mixture. Condensation to the liquid state is considered as occurring at this constant pressure accompanied by what may be termed a *pseudo heat of condensation*, $-\lambda'$. The liquid is then compressed to the pressure of the mixture. Actually condensation occurs over a range of pressures.

At low temperatures where the saturated vapor of each component behaves as an ideal gas and the enthalpy of the vapors is independent of pressure, the heat of vaporization of a mixture of an ideal system at a temperature t is equal to the sum of the heats of vaporization of the individual components at this same temperature. Thus, the pseudo heat of vaporization of a mixture may be estimated by assuming that at the absolute zero the heats of vaporization of the components are additive. This value of Λ'_0 is corrected to any desired finite pseudoreduced temperature by Equation (VII-32), page 233. Thus for a mixture of n_A, n_B, n_C, $\cdots$ moles of components A, B, C, $\cdots$

$$\Lambda' = (1 - T'_r)^{0.38}\left[n_A \frac{\lambda_{bA}}{(1 - T_{rbA})^{0.38}} + n_B \frac{\lambda_{bB}}{(1 - T_{rbB})^{0.38}} + n_C \frac{\lambda_{bC}}{(1 - T_{rbC})^{0.38}} + \cdots\right] \tag{9}$$

where Λ' = pseudo heat of vaporization of the mixture at pseudoreduced temperature T'_r

$\lambda_{bA}, \lambda_{bB}$ = molal heats of vaporization of components A, B, at their normal boiling points

$T_{rbA_1} T_{rbB}$ = reduced temperatures of the normal boiling points of components A, B, based on their individual true critical temperatures

T'_r = pseudoreduced temperature of the mixture

In view of the other approximations involved, it is generally satisfactory to assume, for the calculation of liquid enthalpies, that the pseudoreduced vapor pressure at which condensation of the mixture is assumed to occur is a unique function of pseudoreduced temperature. This assumption is not accurate and should not be used for other purposes but leads to little error in calculating enthalpies. Such pseudoreduced vapor pressures based on the data for *n*-pentane are given in Table XXIXa. On the basis of this assumption unique values of $\left(\frac{\mathrm{H}^* - \mathrm{H}_s}{T'_c}\right)'_T$ for the gaseous state and ψ'_s for the liquid state have been calculated for saturation conditions as functions of reduced temperatures only and are plotted in Fig. 141.

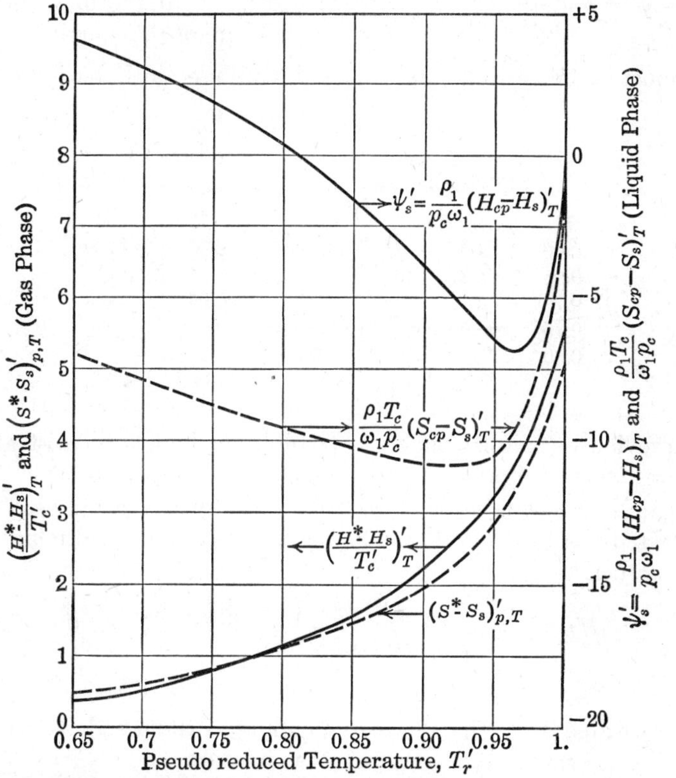

FIG. 141. Enthalpy and Entropy Corrections of Gases and Liquids for Saturation at Various Reduced Temperatures.

(Reproduced in " C.P.P. Charts ")

The enthalpy H_L of the liquid mixture comprising n_t moles at temperature t and pressure p is then

$$H_L = H^* - n_tT'_c\left(\frac{\mathrm{H}^* - \mathrm{H}_s}{T'_c}\right) - \Lambda' + n_tp'_c(v_1\omega_1)'(\psi'_s - \psi') \qquad (10)$$

where H^* = enthalpy of mixture as an ideal gas at temperature t

$\frac{H^* - H_s}{T'_c}$ = enthalpy correction due to nonideal behavior of gas from Fig. 141 corresponding to the pseudoreduced temperature of the mixture

Λ' = pseudo heat of vaporization at temperature t

ψ'_s = enthalpy correction for liquid from Fig. 141 corresponding to the pseudoreduced temperature of the mixture at saturation

$(v_1\omega_1)'$ = defined by Equation (8)

ψ' = value from Fig. 110 corresponding to the final pseudoreduced temperature and pressure of the liquid

Care must be taken that consistent units are used in Equation (10).

TABLE XXIXa

PSEUDOREDUCED VAPOR PRESSURES

T'_r	p'_r	T'_r	p'_r	T'_r	p'_r
0.60	0.0110	0.85	0.310	0.96	0.759
0.65	0.027	0.90	0.478	0.97	0.815
0.70	0.057	0.92	0.561	0.98	0.873
0.75	0.109	0.94	0.655	0.99	0.935
0.80	0.189	0.95	0.699	0.995	0.990
				1.000	1.000

As previously pointed out, it is possible for liquids to exist at temperatures above the pseudocritical temperature of the mixture. Where only a single phase is present, such liquids may be treated as a highly compressed gas to which Fig. 106 is applicable. The enthalpy of the liquid is then

$$H_L = H^* - n_t T'_c \left(\frac{H^* - H}{T'_c}\right) \quad (\text{where} \quad T'_r > 1.0) \tag{11}$$

where

$\frac{H^* - H}{T'_c}$ = value from Fig. 106 corresponding to the pseudoreduced temperature and pressure of the mixture

Equations (10) and (11) are applicable only to single-phase conditions. Where two phases exist the enthalpy of each must be considered separately.

Illustration 4. Calculate the enthalpy of the mixture of Illustration 3 when completely vaporized (*a*) at a pressure of 175 lb per sq in. and 100°F and (*b*) when existing as a liquid at 100°F and 300 lb per sq in. Express the enthalpies in Btu per pound relative to the ideal gaseous state at 60°F.

Basis: 1 lb-mole of mixture = 41.9 lb

Enthalpies: Ideal gas at 100°F

The enthalpy of the mixture in the ideal gaseous state at 100°F is calculated from the heat-capacity equations of Table XXI, page 336. Over the small temperature range involved it may be assumed that the mean heat capacities are the values at the mean temperature of 80°F or 540°R.

$CH_4 = 40[3.42 + (9.91)(0.540) - (1.28)(5.40)^2(10^{-2})](0.015) =$ 5.0 Btu
$C_2H_6 = 40[1.38 + (23.25)(0.540) - (4.27)(5.40)^2(10^{-2})](0.120) =$ 60.9 Btu
$C_3H_8 = 40[0.41 + (35.95)(0.540) - (6.97)(5.40)^2(10^{-2})](0.865) =$ 615.5 Btu
Total enthalpy = 681.4 Btu per lb-mole

(*a*) In the gaseous state at 100°F and 175 lb per sq in., use of the pseudocritical pressure of 652 lb per sq in. from Illustration 3 gives,

$$p_r' = 175/652 = 0.268; \qquad T_r' = \frac{560}{647.2} = 0.865$$

From Fig. 106,

$$(\text{H}^* - \text{H})/T_c' = 1.25$$

$$\text{H} = 681.4 - (1.25)(647.2) = -127.6 \text{ Btu per lb-mole}$$

Or, since the average molecular weight of the mixture is 41.9,

$$H = -127.6/41.9 = -3.05 \text{ Btu per lb}$$

(*b*) In the liquid state at 100°F and 300 lb per sq in., from Illustration 3,

$$T_r' = 0.865; \qquad p_r' = 0.460$$

The following heats of vaporization at the normal boiling points are recommended by Doss.[9]

	N	λ_b cal per g-mole	$T_b°R$	$T_c'°R$	T_{rb}	$1 - T_{rb}$	$\dfrac{\lambda_b}{(1 - T_{rb})^{0.38}}$	$\dfrac{N\lambda_b}{(1 - T_{rb})^{0.38}}$
CH_4	0.015	2,218	201.2	343	0.587	0.413	3,100	46.5
C_2H_6	0.120	3,515	331.8	550	0.603	0.397	5,000	600
C_3H_8	0.865	4,493	416.2	666	0.625	0.375	6,530	5650
							λ_0' cal per g-mole =	6,296

By using the pseudoreduced temperature of 0.865 calculated in Illustration 3 the foregoing data may be substituted in Equation (9) and the pseudoheat of vaporization at 100°F calculated:

$$\lambda' = (1 - 0.865)^{0.38}(6,296) = 2,940 \text{ cal per g-mole}$$

From Fig. 110, $\psi' = -1.5$

From Fig. 141, $\psi_s' = -2.2; \qquad (\text{H}^* - \text{H}_s)'/T_c' = 1.75$

In Illustration 3 $(v_1\omega_1)'$ was found to be 8.85 cc per g-mole, which is equal to (8.85)(454)/28,320 = 0.1417 cu ft per lb-mole. Substitution in Equation (10), for 1 lb-mole, gives

$$H = 681.4 - 647.2(1.75) - (2,940)(1.8) + \frac{(652)(144)(0.1417)(-2.2 + 1.5)}{778}$$
$$= 681.4 - 1132 - 5,290 - 12 = -5,753 \text{ Btu per lb-mole or } -5,753/41.9$$
$$= -137 \text{ Btu per lb}$$

[9] M. P. Doss, "Physical Constants of the Principal Hydrocarbons," 3d Ed., Texas Company, New York (1942).

ENTROPIES OF SOLUTIONS

Problems involving the expansion or compression of a mixture may be handled directly by the methods developed in Chapter XIII if the thermodynamic properties of the mixture can be calculated. Volumes and enthalpies may be obtained by the methods demonstrated in the preceding sections and entropies are calculated by a procedure parallel to that developed for enthalpies. If pseudoreduced terms are employed, Figs. 107 and 111 are applicable to mixtures. Thus, the entropy of a gaseous mixture is obtained by subtracting the correction of Fig. 107 from the entropy which it would possess were it an ideal gas at the temperature and pressure of the mixture. Similarly, the entropy of a pseudosaturated liquid mixture is obtained by subtracting a pseudo entropy of vaporization from the entropy of the pseudosaturated vapor.

On the assumption of a pseudoreduced vapor pressure corresponding to a given pseudoreduced temperature as given in Table XXIX, unique values of $(\mathrm{s}^*-\mathrm{s}_s)'_{p,T}$ for the gaseous state and $\frac{T'_c}{p'_c}\frac{\rho_1}{\omega_1}(\mathrm{s}_{cp}-\mathrm{s}_s)'_T$ for the liquid state have been calculated and are plotted in Fig. 141. The values of $(\mathrm{s}^*-\mathrm{s}_s)'_{p,T}$ for the gaseous state give the corrections in entropy per mole in going from ideal conditions to conditions of the pseudosaturated vapor at the same temperature and pressure.

Because of the entropy changes which accompany the formation of even ideal solutions, entropies of mixtures of fixed composition are most conveniently expressed relative to the mixture at some reference state rather than relative to the pure components.

CRITERIA OF COMPLEX EQUILIBRIA

The discussion on page 449 of equilibrium in a closed system is restricted to consideration of the conditions attained by the system considered as a whole. However, for complete equilibrium it is necessary that equilibrium also be maintained between all parts within the system. In the example discussed, two phases, liquid ether and a gaseous mixture of ether and nitrogen, are present, and together constitute a closed system in which the phases must reach complete equilibrium with each other. However, each individual phase considered by itself constitutes an open system which can undergo changes in composition and mass. For this reason the criteria of equilibrium must be extended to include the effects of these variables in dealing with such systems.

Chemical Potentials. In order to define a single-phase open system, it is necessary to specify mass and composition and two independent variables such as temperature, pressure, volume, and entropy; for a system of given mass and composition specification of two of the last four variables serves to fix the other two. Any infinitesimal change taking place in such a system can be mathematically expressed in terms of all the infinitesimal changes in its independent properties in accordance with the properties of a continuous function represented by Equation (XI-31), page 455. Thus for an infinitesimal change in internal energy, if volume, entropy, and composition are selected as the independent variables, the general differential equation is as follows:

$$dU = \left(\frac{\partial U}{\partial S}\right)_{V, n_1, n_2 \ldots} dS + \left(\frac{\partial U}{\partial V}\right)_{S, n_1, n_2 \ldots} dV + \left(\frac{\partial U}{\partial n_1}\right)_{V, S, n_2 \ldots} dn_1 + \left(\frac{\partial U}{\partial n_2}\right)_{S, V, n_1 \ldots} dn_2 + \cdots \quad (12)$$

where the subscripts 1, 2, $\cdots$ refer to the individual components of the phase which undergo change in mass.

Where composition is constant, all except the first three terms of Equation (12) become zero and in accordance with the equation

$$dU = T\,dS - p\,dV,$$

it is evident that $\left(\frac{\partial U}{\partial V}\right)_{S, n_1, n_2, \ldots} = -p$ and $\left(\frac{\partial U}{\partial S}\right)_{V, n_1, n_2, \ldots} = T$ when reversible changes are under consideration. The differential coefficients of mass are designated by the symbol μ. Thus,

$$(\mu_1)_{V,S} = \left(\frac{\partial U}{\partial n_1}\right)_{V, S, n_2 \ldots}; \quad (\mu_2)_{V,S} = \left(\frac{\partial U}{\partial n_2}\right)_{V, S, n_1 \ldots} \quad (13)$$

and Equation (12) may be written for reversible changes as

$$dU = T\,dS - p\,dV + (\mu_1)_{V,S}\,dn_1 + (\mu_2)_{V,S}\,dn_2 + \cdots \quad (14)$$

It may be noted from Equation (14) that a change in the total internal energy of a system is the sum of the changes of a number of energy terms, each of which comprises an intensive and an extensive factor. Thus, for mechanical work of expansion pressure is the intensive and volume the extensive factor. Similarly temperature is the intensive factor of heat energy and entropy the extensive factor. Since n_1 and n_2 are extensive factors, it follows that $(\mu_1)_{V,S}$ and $(\mu_2)_{V,S}$ represent intensive factors of energy associated with the masses of components 1, 2,$\cdots$ in solution and hence, like temperature and pressure, may be designated as potentials.

Expressions similar to Equation (14) may be developed for the other three energy functions. Thus, for reversible changes

$$dH = T\,dS + V\,dp + (\mu_1)_{S,p}\,dn_1 + (\mu_2)_{S,p}\,dn_2 + \cdots \quad (15)$$

$$dA = -S\,dT - p\,dV + (\mu_1)_{T,V}\,dn_1 + (\mu_2)_{T,V}\,dn_2 + \cdots \quad (16)$$

$$dG = -S\,dT + V\,dp + (\mu_1)_{T,p}\,dn_1 + (\mu_2)_{T,p}\,dn_2 + \cdots \quad (17)$$

From the definitions of the energy functions,

$$U = H - pV = A + TS = G + TS - pV \tag{18}$$

Differentiation yields

$$\begin{aligned} dU &= dH - p\,dV - V\,dp = dA + T\,dS + S\,dT \\ &= dG + T\,dS + S\,dT - p\,dV - V\,dp \end{aligned} \tag{19}$$

Adding $(-T\,dS + p\,dV)$ to each of the equalities, gives

$$\begin{aligned} dU - T\,dS + p\,dV = dH - T\,dS - V\,dp &= dA + p\,dV + S\,dT \\ &= dG + S\,dT - V\,dp \end{aligned} \tag{20}$$

If Equation (20) is compared with Equations (14–17) it is evident that

$$(\mu_1)_{V,S} = (\mu_1)_{S,p} = (\mu_1)_{T,V} = (\mu_1)_{T,p} = \mu_1 \tag{21}$$

and

$$\begin{aligned} \mu_1 &= \left(\frac{\partial U}{\partial n_1}\right)_{V,S,n_2\ldots} = \left(\frac{\partial H}{\partial n_1}\right)_{S,p,n_2\ldots} = \left(\frac{\partial A}{\partial n_1}\right)_{T,V,n_2\ldots} \\ &= \left(\frac{\partial G}{\partial n_1}\right)_{T,p,n_2\ldots} = \bar{G}_1 \end{aligned} \tag{22}$$

As previously pointed out μ_1 is the intensive factor of the internal energy associated with the mass of component 1, but it may be defined in terms of any of the energy functions by Equation (22). This intensive property is designated as the *chemical potential.* From the definition of partial molal quantities (page 285) and Equation (22), it is evident that *the chemical potential of a component is equal to its partial molal free energy.* The other partial derivatives in Equation (22) are equal to the chemical potential but are not partial molal quantities in the restriction of constant temperature and pressure.

If a closed system consisting of several phases, each of which is an open system, is at equilibrium, its properties as a whole must satisfy the criteria established on page 449. Thus, if the conditions of restraint are constant temperature and pressure, at equilibrium, $dG = 0$, if no means of performing useful work is present. Also, any change in the free energy of the system must equal the sum of the corresponding changes of its parts. If the properties of the different individual phases of the system are identified by prime markings and the different components of the various phases by subscripts, from Equation (17),

$$\begin{aligned} dG = &(-S'\,dT + V'\,dp + \mu_1'\,dn_1' + \mu_2'\,dn_2' + \cdots) \\ &+ (-S''\,dT + V''\,dp + \mu_1''\,dn_1'' + \mu_2''\,dn_2'' + \cdots) \\ &+ (-S'''\,dT + V'''\,dp + \mu_1'''\,dn_1''' + \mu_2'''\,dn_2''' + \cdots) + \cdots \end{aligned} \tag{23}$$

At equilibrium under conditions of constant temperature and pressure $dG = 0$, $dT = 0$, $dp = 0$, and

$$\begin{aligned} &(\mu_1'\,dn_1' + \mu_1''\,dn_1'' + \mu_1'''\,dn_1''' + \cdots) \\ &+ (\mu_2'\,dn_2' + \mu_2''\,dn_2'' + \mu_2'''\,dn_2''' + \cdots) + \cdots = 0 \end{aligned} \tag{24}$$

If the total number of moles of each component in the system remains constant,

$$dn_1' + dn_1'' + dn_1''' + \cdots = 0 \quad \text{and} \quad dn_2' + dn_2'' + dn_2''' = 0 \quad \text{etc.} \tag{25}$$

From comparison of Equations (24) and (25), it is evident that those two equalities are satisfied only if

$$\left.\begin{aligned} \mu_1' = \mu_1'' = \mu_1''' = \cdots \\ \mu_2' = \mu_2'' = \mu_2''' = \cdots \end{aligned}\right\} \tag{26}$$

In the development of Equation (26) it was assumed that the total number of moles of each component in the system remains constant. This is true in the case of a system in physical equilibrium not involving chemical reactions. Where chemical reactions take place, the total number of moles of a component in the system may vary as a result of certain types of possible changes. However, the system must be in equilibrium with respect to all possible changes, both physical changes involving no change in the number of moles of the components as well as chemical changes. Therefore, Equation (25) represents one type of change which must be considered, and for this type Equation (24) is satisfied only if (26) is true. *Thus, in a system at equilibrium the chemical potentials of any component must be equal in all phases in which it can be present.*

This criterion is of far-reaching significance and applies to equilibria under any conditions of restraint. For example, instead of considering a system restrained to constant temperature and pressure, the foregoing analysis may be applied to a system at constant volume and temperature. Such a system might be achieved by means of a container of rigid conducting walls in contact with a constant-temperature heat reservoir and divided into two parts by a semipermeable membrane. In one compartment might be placed a solvent to which the membrane is permeable and in the other compartment a solution in this solvent of a solute to which the membrane is not permeable. If the compartments are initially filled at constant pressure, solvent will diffuse into the compartment containing the solution and increase its pressure until an equilibrium is reached with different pressures in the two compartments but with equal values for the chemical potential of the solvent. This follows from Equation (XI-19) which states that the criterion of equilibrium in a system at constant temperature and volume is that $dA = 0$. On this basis an expression for dA analogous to Equation (23) may be written, and Equation (26) may be derived by the same procedure as before. Systems involving other conditions of restraint are more difficult to visualize but are subject to the same criteria regarding equilibrium between phases.

Gibbs Phase Rule. In any system comprising several phases existing at equilibrium there is a limited number of intensive properties which can be freely varied without causing a change in the number of phases or the number of components in some phase. For example, in a system comprising a pure liquid and its pure vapor at equilibria, temperature can be varied over a wide range without affecting the number of phases. However, it is impossible to vary both temperature and pressure freely without causing the disappearance of one phase or the other. The number of intensive properties which can be varied without changing the number of phases or the number of components in any phase is termed the number of *degrees of freedom* of the system.

Since in addition to temperature and pressure the chemical potential of each component is an intensive property, the total number of intensive properties subject to variation in a single phase containing C components is $C + 2$. If this phase is a part of a system in equilibrium at constant temperature and pressure, a differential expression for the free-energy change, which is the criterion of equilibrium, may be written in the form of Equation (23). Thus, for each phase,

$$dG' = f'(T, p, \mu_1, \mu_2, \cdots) = 0$$
$$dG'' = f''(T, p, \mu_1, \mu_2, \cdots) = 0$$
$$dG''' = f'''(T, p, \mu_1, \mu_2, \cdots) = 0$$

Etc.

where the functions f', f'', f''' are characteristic of the respective phases. In this manner, if ϕ phases are present, simultaneous equations may be written in terms of $C + 2$ variables. Since for determination of the variables in simultaneous equations one equation is required for each variable, it follows that the number of variables not fixed by the equations is $C + 2 - \phi$ or

$$F = C + 2 - \phi \tag{27}$$

where

F = degrees of freedom

C = number of components

ϕ = number of phases

Equation (27) is the famous phase rule of Gibbs, developed in 1875. The foregoing derivation follows that presented by Keenan[10]:

As an example of the application of the rule, the equilibrium among the liquid, vapor, and solid states may be considered. Thus, if only a pure liquid, for example water, is in equilibrium with its vapor, $C = 1$, $\phi = 2$, and $F = 1$; one intensive property, either temperature or pressure,

[10] J. H. Keenan, "Thermodynamics," John Wiley & Sons, New York (1941).

but not both may be freely varied. If, however, ice is also present in the system, $C = 1$, $\phi = 3$, and $F = 0$. Under such conditions no conditions can be varied and the specification of three phases fixes both temperature and pressure. If instead of pure water a binary solution of water and alcohol is in equilibrium with its vapor, $C = 2$, $\phi = 2$, and $F = 2$. Thus, two properties may be freely varied in such a system. For example, both temperature and pressure may be varied freely over restricted ranges with corresponding changes in composition of the phases, but all three variables cannot be independently varied. A specified composition and temperature fixes the corresponding pressure.

When used in connection with the phase rule, the number of components is the least number of independently variable chemical substances from which the system in all its variations can be produced. Elsewhere in this text the term *component* is not used in this restricted sense.

FUGACITY

The concept of the chemical potential is of value in establishing the fundamental thermodynamic requirements which must be satisfied by a complex system when equilibrium is reached. Since at equilibrium the chemical potentials of a component must be equal in all phases in which it can occur, differences in the chemical potentials must be equalized by redistribution of the component as a system approaches equilibrium. If the chemical potential is high in one phase and low in another material escapes from the first to the second phase until equality is reached. Thus, the chemical potential may be considered as a measure of what may be termed the *escaping tendency.*

Since the chemical potential is a measure of the escaping tendency, it follows that any other property which is a unique function of the chemical potential is also a measure of the escaping tendency. Such a function, termed the *fugacity,* was so defined by Lewis[11] as to simplify the mathematical relationships of the equilibrium. By definition,

$$(d\mu_A = d\bar{G}_A = RT\, d \ln f_A)_T \tag{28}$$

where

f_A = fugacity of component A at temperature T

μ_A = chemical potential of component A at temperature T

$\bar{G}_A$ = partial molal free energy of component A at temperature T

From this definition it follows that *in any system which is restrained to a constant temperature at equilibrium the fugacities of any component must be equal in all phases in which it appears.* For a pure component the par-

[11] G. N. Lewis, *Proc. Am. Acad. Arts, Sci.,* **37**, 49 (1901).

tial molal free energy $\bar{G}_1$ is equal to the molal free energy G and Equation (28) reduces to

$$(dG = RT\,d\ln f)_T \qquad (29)$$

Equations (28) and (29) do not serve to define the numerical values of f but do establish the changes in f which correspond to specified changes in G or $\bar{G}$ at constant temperature.

From Equation (k), Table XXIV, page 472,

$$\left(\frac{\partial G}{\partial p}\right)_T = v$$

Combining this with Equation (29) gives

$$\left(\frac{\partial \ln f}{\partial p}\right)_T = \frac{v}{RT} \qquad (30)$$

In applying Equation (30) to a gas v may be replaced by zRT/p. Then,

$$(\partial \ln f = z\partial \ln p)_T \qquad (31)$$

From Equation (31) it is evident that for an ideal gas, where $z = 1.0$ the fugacity is proportional to the pressure. The definition of fugacity is completed, and numerical values are assigned to it by arbitrarily setting the fugacity of an ideal gas equal to its pressure. Thus fugacity has the units of pressure and is numerically equal to pressure in the ideal gaseous state. Under other conditions pressures and fugacities are not equal, and fugacity is sometimes referred to as a corrected or thermodynamic pressure. However, from the definition it is evident that the corrections involved are a function of free-energy changes and the fugacity is useful only for equilibrium calculations where free-energy changes are the criterion.

Fugacities of Pure Gases. An expression for the fugacity of a gas at any conditions is obtained by rearranging Equation (31) in terms of the ratio of fugacity to pressure f/p which by definition is equal to 1.0 at zero pressure where ideal behavior is realized. Thus

$$d\ln f/p = z\,d\ln p - d\ln p = (z - 1)\,d\ln p \qquad (32)$$

Integration from $p = 0$ to p gives

$$\ln\frac{f}{p} = \int_0^p (z - 1)\,d\ln p = \int_0^p \frac{(z - 1)}{p}\,dp \qquad (33)$$

Equation (33) may be integrated graphically for any substance for which compressibility data are available. The integration is more accurately carried out by analytical methods if an equation of state is evaluated.

On the basis of the theorem of corresponding states a generalized integration of Equation (33) is obtained from the compressibility-factor relationship of Fig. 103. Various investigators have carried out this integration which is conveniently presented by plotting the ratio of fugacity to pressure (f/p) against reduced temperature and pressure. This ratio $f/p = \nu$ will be termed the *fugacity coefficient* and is equal to unity where the ideal-gas law is valid. This ratio is frequently termed the "activity coefficient," but it is believed preferable to reserve this latter term for expressing relationships between fugacities and composition in solutions.

Average values of the fugacity coefficients of gases are plotted in Fig. 142.[12] This chart was derived from the same data on which Fig. 103 was based. For many purposes these charts may be taken as applicable to all gases. The errors involved are generally less than 10 per cent.

The fugacity of any gas or vapor at specified conditions is readily obtained when its critical temperature and pressure are known.

Illustration 5. Calculate the fugacity of methane at 122°F and 1000 lb per sq in. abs.

From Table XXVIII, $T_c = 343°R$ and $p_c = 673$ lb per sq in.

$$\text{Reduced temperature} = \frac{122 + 460}{343} = 1.70$$

Reduced pressure = $\frac{1000}{673} = 1.49$

Fugacity coefficient (Fig. 142) = 0.94

Fugacity = (1000)(0.94) = 940 lb per sq in.

Fugacities of Pure Liquids and Solids. From the basic concept of fugacity as a measure of escaping tendency, it follows that the fugacity of a liquid or solid must be equal to that of its vapor in equilibrium with it. Thus, when a liquid is in equilibrium with its pure vapor, the fugacity of the liquid is determined by calculating the fugacity of the vapor at the equilibrium temperature and pressure.

Illustration 6. Calculate the fugacity of liquid benzene in equilibrium with its pure vapor, at a temperature of 428°F. The critical temperature of benzene is 550°F, and its critical pressure is 700 lb per sq in. The vapor pressure at 428°F is 281 lb per sq in.

$$\text{Reduced temperature} = \frac{428 + 460}{550 + 460} = 0.88$$

Reduced pressure = 281/700 = 0.40

Fugacity coefficient (Fig. 142) = 0.79

Fugacity of vapor = fugacity of liquid = (281)(0.79) = 222 lb per sq in.

The fugacity of the liquid state is a function of total pressure as expressed by Equation (30). This equation may be integrated by expressing v as

[12] B. W. Gamson and K. M. Watson, *Natl. Petroleum News*, Tech. Sec., **36**, *R*623 (September 6, 1944). Also "Process Engineering Data," National Petroleum Publishing Company, Cleveland (1944).

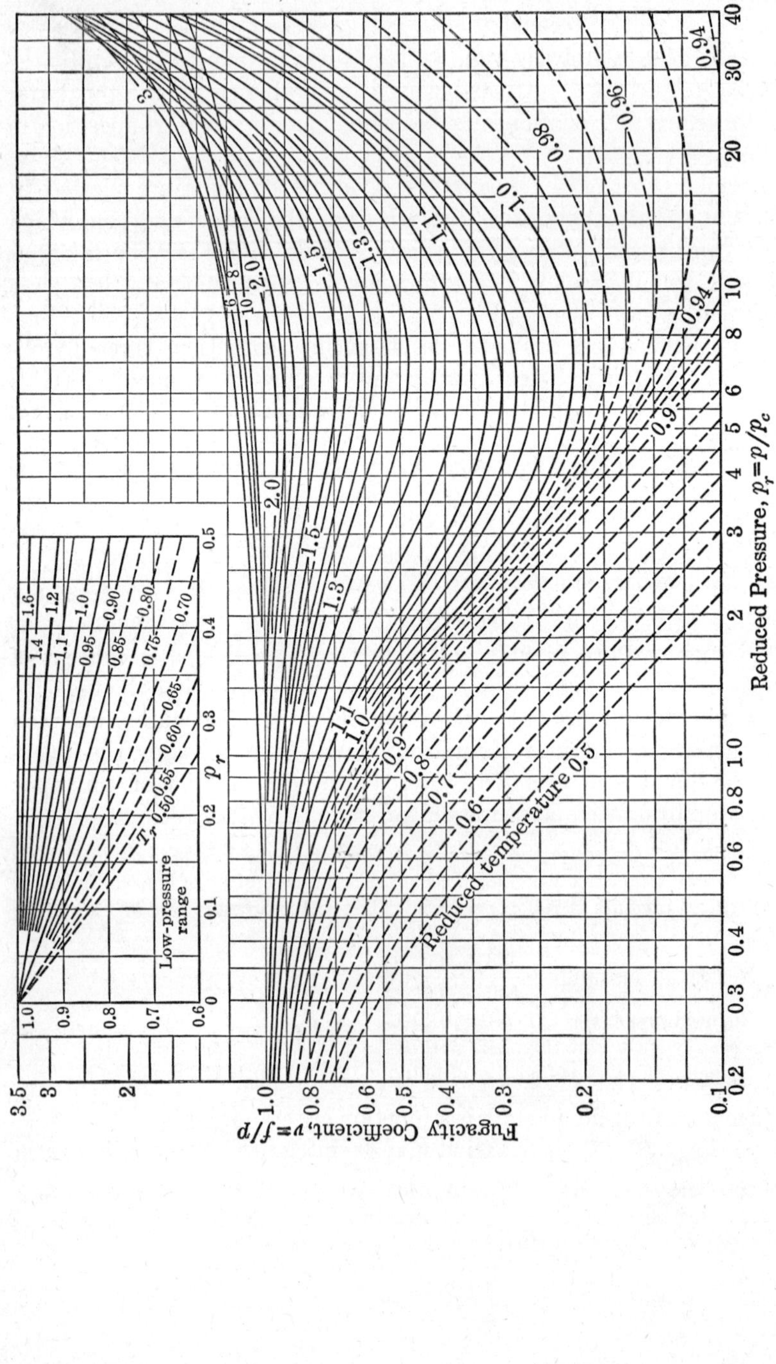

FIG. 142. Fugacity Coefficients of Gases.

(Reproduced in " C.P.P. Charts ")

a function of pressure by means of Fig. 109. Since the volume of a liquid is little affected by pressure at reduced temperatures below 0.8, an approximate integration assuming v constant at an average value is generally satisfactory in this range. Thus,

$$\ln \frac{f_\pi}{f_P} = \frac{v_m(\pi - P)}{RT} \tag{34}$$

where

f_π = fugacity at total pressure π
f_P = fugacity at normal vapor pressure P

Equation (30) permits calculation of the fugacities of liquids at any total pressure but the integrated form, Equation (34), is useful only at reduced temperatures below 0.8.

Illustration 7. Calculate the fugacity of liquid benzene at 428°F if the liquid is in an atmosphere of hydrogen such that the total gauge pressure is 2000 lb per sq in. The average density of liquid benzene at these conditions is 0.63 g per cc.

$$v = \frac{78}{(0.63)(62.4)} = 1.98 \text{ cu ft per lb-mole}$$

Substitution in Equation (34), gives

$$\ln \frac{f_\pi}{222} = \frac{1.98(2015 - 281)}{(10.71)(428 + 460)} = 0.361$$

$$\log f_\pi/222 = (0.361)(0.434) = 0.157 \quad \text{or} \quad f_\pi/222 = 1.43$$

f_π = the fugacity at 2015 lb per sq in. = (1.43)(222) = 318 lb per sq in.

Effect of Temperature on Fugacity. The following equation is derived by the method of Lewis and Randall[13] to relate fugacity to temperature. A substance in a given state at temperature T, pressure p, molal free energy G, fugacity f, and molal enthalpy H is compared to the same substance at the same temperature but at a low pressure where the free energy is G*, and the fugacity f^* is equal to the pressure. Under these conditions the enthalpy H* is independent of pressure. From Equation (29),

$$G^* - G = RT(\ln f^* - \ln f) \tag{35}$$

By differentiation with respect to temperature at constant pressure,

$$\left(\frac{\partial G^*}{\partial T}\right)_p - \left(\frac{\partial G}{\partial T}\right)_p = R \ln \frac{f^*}{f} + RT\left(\frac{\partial \ln f^*}{\partial T}\right)_p - RT\left(\frac{\partial \ln f}{\partial T}\right)_p \tag{36}$$

Since f^* is equal to pressure,

$$\left(\frac{\partial \ln f^*}{\partial T}\right)_p = 0$$

[13] Lewis and Randall, "Thermodynamics," McGraw-Hill Book Company, New York (1923).

Combining (35) and (36) results in

$$\left(\frac{\partial \text{G}^*}{\partial T}\right)_p - \left(\frac{\partial \text{G}}{\partial T}\right)_p = \frac{\text{G}^*}{T} - \frac{\text{G}}{T} - RT\left(\frac{\partial \ln f}{\partial T}\right)_p \tag{37}$$

Combining (37) with Equation (XI-4) and (l) of Table XXIV, gives

$$\left(\frac{\partial \ln f}{\partial T}\right)_p = \frac{\text{H}^* - \text{H}}{RT^2} \tag{38}$$

where

H = molal enthalpy at an elevated pressure p and temperature T

H* = molal enthalpy at the same temperature T but at a pressure sufficiently low so that the gas behaves ideally

The quantity H* − H may be obtained from the generalized correlation of Fig. 105 or may be calculated from compressibility data by the methods developed in Chapter XII. Charts such as Figs. 117 or 120 or thermodynamic tables permit direct evaluation of H* − H for either the liquid or vapor state as a function of temperature and pressure.

Fugacities in Solutions. The partial molal extensive thermodynamic properties of a component in solution are related to each other by expressions parallel to those developed for pure substances in Chapter XI. For example, by Equation (XI-4), page 446, for any system,

$$G = H - TS \tag{39}$$

By differentiation of this equation with respect to the number of moles of component *2*, all other conditions being held constant, there results

$$\left(\frac{\partial G}{\partial n_2}\right)_{T,\,p,\,n_1\ldots} = \left(\frac{\partial H}{\partial n_2}\right)_{T,\,p,\,n_1\ldots} - T\left(\frac{\partial S}{\partial n_2}\right)_{T,\,p,\,n_1\ldots} \tag{40}$$

Each of the derivatives of Equation (40) represents a partial molal quantity. Therefore,

$$\overline{\text{G}}_2 = \overline{\text{H}}_2 - T\overline{\text{S}}_2 \tag{41}$$

Equation (41) is entirely parallel in significance to (39). In a similar manner it may be shown that the other relationships of Chapter XI which involve extensive properties are valid for partial molal quantities. On this basis, from Table XXIV, Equation (k),

$$\left(\frac{\partial \overline{\text{G}}_2}{\partial p}\right)_T = \overline{v}_2 \tag{42}$$

from Equation (l),

$$\left(\frac{\partial \overline{\text{G}}_2}{\partial T}\right)_p = -\overline{\text{S}}_2 = \frac{\overline{\text{G}}_2 - \overline{\text{H}}_2}{T} \tag{43}$$

Combining (28) with (42) gives

$$\left(\frac{\partial \ln f_2}{\partial p}\right)_T = \frac{\bar{v}_2}{RT} \tag{44}$$

Combining (28) and (43) in the manner used in developing (38), gives

$$\left(\frac{\partial \ln f_2}{\partial T}\right)_p = \frac{\text{H}_2^* - \bar{\text{H}}_2}{RT^2} \tag{45}$$

In using Equation (45) care must be taken that the partial molal enthalpy $\bar{\text{H}}_2$ is expressed with respect to the same reference state as the molal enthalpy of the ideal gas, H_2^*.

Equations (28), (44), and (45) permit calculations of the fugacities of components in solutions from a variety of data such as vaporization, solubility, or distribution equilibrium measurements. For example, the fugacity of a pure solid at a specified temperature may be determined from its vapor pressure, and this value is equal to the fugacity of the solid as a solute in any saturated solution at this same temperature. The variation with temperature and pressure of the fugacity in solution can then be calculated by means of Equations (44) and (45), if partial volume and enthalpy data are available.

If Equation (VIII-34) is written for free energies, at constant temperature and pressure,

$$n_1\, d\bar{\text{G}}_1 + n_2\, d\bar{\text{G}}_2 + n_3\, d\bar{\text{G}}_3 + \cdots = 0 \tag{46}$$

or, considering 1 mole of solution, at constant temperature and pressure,

$$N_1\, d\bar{\text{G}}_1 + N_2\, d\bar{\text{G}}_2 + N_3\, d\bar{\text{G}}_3 + \cdots = 0 \tag{47}$$

Equation (47) may be written as partial derivatives with respect to the mole fraction of any selected component, temperature and pressure being kept constant

$$N_1\left(\frac{\partial \bar{\text{G}}_1}{\partial N_2}\right)_{T,p} + N_2\left(\frac{\partial \bar{\text{G}}_2}{\partial N_2}\right)_{T,p} + N_3\left(\frac{\partial \bar{\text{G}}_3}{\partial N_2}\right)_{T,p} + \cdots = 0 \tag{48}$$

Combining (28) and (48) gives

$$N_1\left(\frac{\partial \ln f_1}{\partial N_2}\right)_{T,p} + N_2\left(\frac{\partial \ln f_2}{\partial N_2}\right)_{T,p} + N_3\left(\frac{\partial \ln f_3}{\partial N_2}\right)_{T,p} + \cdots = 0 \tag{49}$$

If Equation (49) is applied to a binary solution, $dN_1 = -dN_2$ and

$$N_1\left(\frac{\partial \ln f_1}{\partial N_1}\right)_{T,p} = N_2\left(\frac{\partial \ln f_2}{\partial N_2}\right)_{T,p} \tag{50}$$

This important relation between the fugacities of the components of a solution is termed the Gibbs–Duhem equation.

Fugacities in Ideal Solutions. In an ideal solution, either gaseous or liquid, the fugacity of a component is proportional to its mole fraction.

$$f_2 = N_2 f_2' \tag{51}$$

where

f_2 = fugacity of component 2 in the solution

f_2' = fugacity of pure component 2 at the temperature and pressure of the solution

N_2 = mole fraction of component 2

Lewis and Randall[13] define the ideal solution solely by Equation (51) and point out that all other properties attributed to ideal solutions such as additivity of volumes and enthalpies follow from this definition.

For gaseous mixtures at low pressures Equation (51) is equivalent to Dalton's law while for liquid solutions whose vapors are ideal gases it is equivalent to Raoult's law. However, at conditions resulting in large deviations from ideal-gas behavior, fugacities are not equal to partial pressures, and Equation (51) may be looked upon as an improved statement of Dalton's and Raoult's laws which is generally applicable over a much wider range of conditions.

The assumption that gases form ideal solutions is of a much higher order of accuracy than the ideal-gas law and may be applied with accuracy satisfactory for many purposes to gaseous mixtures at pseudoreduced pressures less than 0.8. If the pseudoreduced temperature is low, of the order of 1.0 or less, difficulty is encountered in handling the high-boiling components of the mixture, which, if they existed alone at the temperature and pressure of the mixture, would be liquefied. In such cases the term f'_2 of Equation (51) may be estimated from the broken-line extrapolations of the curves of Fig. 142 into the two-phase region. Thus, hypothetical fugacity coefficients for the pure gases at pressures above their vapor pressures are obtained. This procedure is not satisfactory where extended extrapolation is required. For mixtures at higher temperatures, above the critical temperatures of all components, the assumption of ideal solutions is more satisfactory and may be extended to higher pressure ranges.

Illustration 8. A mixture of gases has the following composition expressed in mole per cent:

Methane	17
Ethane	35
Propane	48
	100

Assuming an ideal gaseous solution, calculate from Fig. 142 the fugacity of each component when the mixture is at an absolute pressure of 300 lb per sq in. and a temperature of 100°F.

Solution:

	Methane	*Ethane*	*Propane*
T_c°R (Table XXVII)	343	550	666
p_c lb per sq in. (Table XXVII)	673	717	642
T_r	1.63	1.02	0.84
p_r	0.446	0.418	0.466
ν (Fig. 142)	0.98	0.87	0.73
$f' = 300\,\nu$	294	261	219
$f = Nf'$	50.0	91.5	105

ACTIVITY

For treatment of problems involving solutions and chemical equilibria it is convenient to define another thermodynamic property which is directly related to fugacity and hence also to free energy and the chemical potential. This property, called *activity*, a, is defined as the ratio of the fugacity of a component in a given state to its fugacity in an arbitrarily defined standard state at the same temperature. Thus,

$$a = \left(\frac{f}{f^\circ}\right)_T \tag{52}$$

where

a = activity
f = fugacity in the given state
f° = fugacity in the standard state at the same temperature

Combining Equations (28) and (52) gives

$$RT \ln a = \bar{G} - \bar{G}^\circ \tag{53}$$

where

$\bar{G}$ = partial molal free energy in given state at temperature T
$\bar{G}^\circ$ = partial molal free energy in the standard state at the same temperature T

The concept of activity is particularly useful in dealing with liquid solutions. For example, if the standard state of a component which is in solution is taken as the pure component at the temperature and pressure of the solution, the activity becomes a function of the concentration or fraction of the component in the solution. Thus, the activity provides a basis for the thermodynamic expression of concentrations or compositions in terms directly related to free energy by Equation (53).

Standard States. The choice of the standard state necessary to complete the definition of activity is arbitrary since it affects only the numerical magnitude of the function and not its relationship to other properties. It is, however, desirable to select standard states such that the resulting numerical scales will be convenient and correspond to common expressions of composition in ideal systems. Different standard states may be selected for the same substance for use in different types of relationships, the choice being dictated by convenience. It is important however that the standard state of each component be kept the same throughout any one series of relationships or calculations. It is evident that *the numerical value of activity is without significance unless the standard state is specified.*

The following standard states have been found convenient for certain types of calculations and are in more or less general use.

Components of Gaseous Mixtures. (*a*) It is frequently convenient to define the standard state as the state of unit fugacity, or $f^\circ = 1.0$. With this choice of standard state the activity of a gaseous component is equal to its fugacity and has the same numerical value. In mixtures of ideal gases the activity of each component is equal to its partial pressure.

(*b*) In other cases it is more convenient to define the standard state as the pure component at the temperature and the pressure of the mixture. With this choice of standard state the activities become equal to the mole fractions in mixtures which form ideal solutions. It may be noted that with this definition the standard state changes with change in pressure whereas standard state (*a*) is independent of the pressure of the system.

Pure Liquids and Solids. When a liquid or solid is involved in a process in its pure state, it is customary to designate its activity as unity under a specified pressure. Three choices have been used in the specification of the pressure of the standard state.

(*a*) The pure component under a pressure of 1.0 atm is widely used as the standard state for nonvolatile substances.

(*b*) For volatile substances whose vapor pressures exceed 1.0 atm it is convenient to define the standard state as the pure substance under its own vapor pressure.

(*c*) In certain calculations it is convenient to define the standard state as the pure substance under the pressure of the system. With this definition the standard state varies with change in pressure whereas states (*a*) and (*b*) are independent of pressure.

Components in Nonelectrolytic Solutions. The standard state for either a solid or liquid component in a nonelectrolytic solution may be taken as the pure component in either the solid or liquid state under one of the three pressure designations enumerated in the preceding paragraph. Pressure designations (*a*) and (*b*) have the advantage of defining the standard state independent of pressure. Use of designation (*c*) leads to the convenient relationship that activities are equal to mole fractions in ideal solutions under constant pressure.

An alternate standard state which is sometimes desirable for sparingly soluble materials is so defined that the mole fraction is equal to the activity as the mole fraction approaches zero. Thus,

$$\lim_{N_2=0} \frac{a_2'}{N_2} = 1.0 \tag{54}$$

It may be experimentally demonstrated that as infinite dilution is approached the fugacity of a solute becomes proportional to its mole fraction,

$$f_2 = k_2' N_2 \quad \text{as} \quad N_2 \longrightarrow 0 \quad \text{or} \quad \left(\frac{\partial f_2}{\partial N_2}\right)_{N_2=0} = k_2' \tag{55}$$

Combining (52), (54), and (55) for an infinitely dilute solution gives

$$a_2' = N_2 = \frac{f_2}{k_2'} = \frac{f_2}{f_2^{\circ\prime}}$$

or

$$f_2^{\circ\prime} = k_2' \tag{56}$$

Thus, the standard state, according to the definition of Equation (56), is a hypothetical state in which the fugacity is equal to k_2', the factor relating fugacity to mole fraction at infinite dilution. It should be noted that the standard state is not the state of infinite dilution but is defined by reference to this state. A standard state defined in this manner will be termed a *standard state referred to infinite dilution.* The pressure at which such a standard state is defined may be either (*a*) 1 atm, (*b*) the vapor pressure of the solvent, or (*c*) the pressure of the system. If designation (*c*) is employed the activity is equal to the mole fraction at all concentrations in an ideal solution at constant pressure exactly as when the pure component is chosen as the standard state.

It is important to recognize that the standard state referred to infinite dilution is a hypothetical pure state of unit activity derived by linear extrapolation of the ideal fugacity-composition relationship at infinite dilution. This may be obtained by drawing a tangent to the fugacity–mole-fraction plot at zero mole fraction of solute. The intersection of this tangent with the abscissa of unit mole fraction $N_2 = 1.0$ gives the fugacity $f_2^{\circ\prime}$ of the standard state. Since partial molal volumes, enthalpies, and heat capacities are independent of composition in an ideal solution, it follows that *these properties of the hypothetical reference state are equal to the corresponding actual partial properties of the solute in infinite dilution.*

An alternate standard state referred to infinite dilution is so defined that molality is equal to activity as the molality approaches zero. Thus,

$$\text{Lim}\ \frac{a_2''}{m_2} = 1.0 \quad \text{as} \quad m_2 \longrightarrow 0 \tag{57}$$

As infinite dilution is approached the fugacity becomes proportional to molality, or

$$f_2 = k_2''m_2 \quad \text{as} \quad m_2 \longrightarrow 0 \quad \text{or} \quad \left(\frac{\partial f_2}{\partial m_2}\right)_{m_2=0} = k_2'' \tag{58}$$

Combining (52), (57), and (58) for an infinitely dilute solution gives

$$a_2'' = m_2 = \frac{f_2}{k_2''} = \frac{f_2}{f_2^{\circ''}} \quad \text{or} \quad f_2^{\circ''} = k_2'' \tag{59}$$

The hypothetical standard state can be obtained by drawing a tangent to the fugacity–molality plot at zero molality of solute. The intersection of this tangent with the abscissa of unit molality $m_2 = 1$ gives the fugacity $f_2^{\circ''}$ of the standard state.

Gases in Liquid Solutions. The activity of a gas dissolved in a liquid may be referred to a standard state of unit fugacity, making activities equal to fugacities. More frequently it is desirable to refer the activities to infinite dilution so that the activity equals the mole fraction as they both approach zero. The pressure of the standard state may be designated by any of the conventions discussed in the preceding paragraphs.

For gases below or not greatly above their critical points a special standard state is useful in which the standard state of the gas in the liquid solution is taken as the hypothetical pure component in the liquid state at the temperature and pressure of the solution. The fugacity of such a hypothetical reference state may be estimated by extrapolation of the vapor–pressure curve if the temperature is above the critical. This standard state is extensively used for vaporization equilibrium calculations.

As previously mentioned, the choice of the standard state is based on convenience and does not affect the relationships involved. However, it is important that the selected standard state be clearly defined and consistently adhered to in any specific problem. The concept of the activity tends to be confusing, and its value and use will be better understood as the applications of the following sections are developed.

Activity Coefficients. The concept of the activity is useful because it provides a thermodynamically defined quantity which with proper choice of standard state is a simple measure of composition in ideal solutions. Because of its thermodynamic definition it is possible to develop exact relationships between the activities of components in complex systems at equilibrium under varying conditions. Where ideal solutions are involved or approximated, these relationships between activities are directly translatable into ordinary compositions. If ideal behavior cannot be assumed, it is necessary to introduce an empirical factor which relates activity to composition. This factor is termed the *activity coefficient* γ, which is the ratio of activity to a numerical expression of composition. The numerical value of the activity coefficient is dependent both upon the standard state of the activity and upon the units of expression of composition. Both of these factors may be arbitrarily selected, and the activity coefficient has no significance unless both are specified.

The most commonly used activity coefficients relate activities to mole fractions or molalities. The following symbols are used for the three coefficients of most importance in dealing with systems at low pressures.

$$\gamma_2 = a_2/N_2 \tag{60}$$

$$\gamma_2' = a_2'/N_2 \tag{61}$$

$$\gamma_2'' = a_2''/m_2 \tag{62}$$

where

a_2 = activity of component 2 referred to the pure component

a_2' = activity of component 2 referred to infinite dilution where $a_2' = N_2$

a_2'' = activity of component 2 referred to infinite dilution where $a_2'' = m_2$

The activity coefficient of a component in a nonideal solution is a function of composition, temperature, and pressure, which ordinarily must be derived from direct experimental data on the specific system. A variety of data on systems at equilibrium may be used for this purpose. For example, the composition of the vapor in equilibrium with a liquid solution may be determined. The fugacities of the components in the liquid solution are then equal to those of the same components in the gaseous phase and may be calculated by the methods of the preceding sections. By determining such vapor–liquid equilibrium data over a range of compositions the fugacities of the liquid components may be evaluated. The activities and activity coefficients are then calculated from the fugacities of the standard states. In such investigations it is necessary to evaluate experimentally the fugacities of all but one component. The fugacity of the remaining component can then be calculated by Equation (49) or (50) if a single known value is available in the range of concentrations covered by the data.

If the unknown component is a nonvolatile material of unknown fugacity or if it is desired to express activities in terms referred to infinite dilution, it is more convenient to rewrite Equations (49) and (50) in terms of activities or activity coefficients. The limits of integration of these expressions are fixed by the concentration of the state of unit activity or of unit activity coefficient, respectively, even though the fugacity is unknown. Since $a = f/f°$, Equation (47) may be written for changes in activity due to changes in composition at constant temperature and pressure.

$$N_1 d \ln a_1 + N_2 d \ln a_2 + N_3 d \ln a_3 + \cdots = 0 \tag{63}$$

or integration, for a binary solution, gives

$$\ln a_2 = -\int_{N_2^\circ}^{N_2} \frac{N_1}{N_2} d \ln a_1 \tag{64}$$

where

N_2° = mole fraction of the standard state where $a_2 = 1.0$

Equations (63) and (64) are convenient where the pure liquid or solid component 2 is chosen as the reference state. Since $a_2 = N_2\gamma_2$ Equation (63) may also be written in terms of activity coefficients. Thus,

$$N_1 d \ln N_1 + N_1 d \ln \gamma_1 + N_2 d \ln N_2 + N_2 d \ln \gamma_2 + \cdots = 0 \tag{65}$$

also

$$dN_1 + dN_2 + dN_3 + \cdots = 0$$

Then since $N_1 d \ln N_1 = dN_1$, it follows that

$$N_1 d \ln N_1 + N_2 d \ln N_2 + N_3 d \ln N_3 + \cdots = 0 \tag{66}$$

and

$$N_1 d \ln \gamma_1 + N_2 d \ln \gamma_2 + N_3 d \ln \gamma_3 + \cdots = 0 \tag{67}$$

Integration of (67) for a binary solution gives

$$\ln \gamma_2 = -\int_{N_2'}^{N_2} \frac{N_1}{N_2} d \ln \gamma_1 + C \tag{68}$$

where C is the value of $\ln \gamma_2$ at N_2'.

Equation (67) may also be written in the form of (50) for binary solutions:

$$N_1 \left(\frac{\partial \ln \gamma_1}{\partial N_1} \right)_{T,p} = N_2 \left(\frac{\partial \ln \gamma_2}{\partial N_2} \right)_{T,p} \tag{69}$$

Equations (63) through (67) may also be written in terms of common logarithms and then integrated graphically by plotting $\log a_1$, or $\log \gamma_1$, against N_1/N_2. The equations in terms of the activity coefficients are preferable for graphical manipulation, because in general the coefficient varies less with change in composition than does the activity.

For solutions of more than two components the procedure is more complicated since an additional integral is introduced for each additional component. Such integrations are best carried out with respect to a constant number of moles of all components other than the one under consideration. Thus,

$$-\ln a_2 = \int_{n_2^\circ}^{n_2} \frac{n_1}{n_2}\, d\ln a_1 + \int_{n_2^\circ}^{n_2} \frac{n_3}{n_2}\, d\ln a_3 + \cdots \tag{70}$$

where n_1, n_3, $\cdots$ are constant.

Equations (63–70) are rigorous thermodynamic relations which are valid for conditions of constant temperature. They are of great value in minimizing the experimental data necessary to evaluate the properties of a system and for detecting inconsistent or erroneous measurements.

Where activities are referred to molalities at infinite dilution the activity coefficients γ'' are not constant or equal to unity, even in ideal solutions except in the dilute range. Their principal value is in dealing with dilute aqueous solutions in which ionization occurs. Equations (69) and (70) are not applicable to activity coefficients defined in this manner.

From the definition of activity it is evident that the activities of a solute when expressed on the basis of the three different standard states used in Equations (60), (61), and (62) must be related to each other by constant factors, independent of concentration. Thus,

$$f_2 = a_2 f_2^\circ = a_2' f_2^{\circ\prime} = a_2'' f_2^{\circ\prime\prime} \tag{71}$$

The conversion of activities from one standard state to another for a given system is obtained by rearranging Equation (71) to give

$$a_2 = a_2'\left(\frac{f_2^{\circ\prime}}{f_2^\circ}\right) = a_2''\left(\frac{f_2^{\circ\prime\prime}}{f_2^\circ}\right) \tag{72}$$

It is thus evident that the ratio of activities relative to different standard states is given by the ratio of the fugacities of the standard states according to Equation (72) and that these conversion factors are constant over the entire range of concentrations.

A further relation between activities in aqueous solutions results from consideration of the infinitely dilute solution where, since 1000 g of water is equivalent to 55.51 g-moles, it follows that,

$$m_2 = 55.51\, N_2 \quad \text{or} \quad dm_2 = 55.51\, dN_2 \text{ as } m_2 \longrightarrow 0 \tag{73}$$

and where, from Equations (55, 56),

$$\left(\frac{\partial f_2}{\partial N_2}\right)_{N_2=0} = f_2^{\circ\prime} \tag{74}$$

and where, from Equations (58, 59),

$$\left(\frac{\partial f_2}{\partial m_2}\right)_{m_2=0} = f_2^{\circ\prime\prime} \tag{75}$$

Combining (73–75) gives $\quad f_2^{\circ\prime} = 55.51\, f_2^{\circ\prime\prime}$ (76)

Combining (72) and (76) gives

$$a_2'' = 55.51\, a_2' \tag{77}$$

A comparison of activities and of activity coefficients referred to the three different standard states is illustrated diagrammatically in Fig. 143 in which fugacities, activities, and activity coefficients of a solute in a nonideal solution are plotted against mole fraction. In this example, $f_2^\circ = 0.5$ and $a_2 = 1.0$ at $N_2 = 1.0$. The value of $f_2^{\circ\prime}$ is obtained by drawing a tangent to the fugacity curve at zero mole fraction and extending it to $N_2 = 1.0$ where $f_2^{\circ\prime} = 1.3$. The value of $f_2^{\circ\prime\prime}$ is obtained by constructing a tangent to a fugacity–molality curve at zero molality and extending it to $m_2 = 1.0$, or from Equation (76), $f_2^{\circ\prime\prime} = \dfrac{1.3}{55.51} = 0.0234.$ The ratios of activities referred to the different standard states are given by Equation (72) as

$$a_2 = \left(\frac{1.3}{0.5}\right) a_2' = \left(\frac{0.0234}{0.5}\right) a_2'' = 2.6\, a_2' = 0.0468\, a_2''$$

For mole fractions of zero and unity the following values are obtained:

	At $N_2 = 0$	At $N_2 = 1.0 (m_2 = \infty)$
a_2	0	1.0
a_2'	0	0.384
a_2''	0	21.35
γ_2	2.60	1.0
γ_2'	1.0	0.384
γ_2''	1.0	0

Where fugacities of component 2 are unknown the ratio of standard fugacities in Equation (72) can be obtained from the activity coefficient of component 2 at infinite dilution $(\gamma_2)_0$. By combination of Equations (52) and (60) and Equations (61), (57), and (73), and, since at infinite dilution γ_2' and γ_2'' are each equal to unity, Equation (72) becomes

$$a_2 = (\gamma_2)_0\, a_2' = (\gamma_2)_0\, a_2'' M_1/1000 \tag{77a}$$

Illustration 9. From the data for the vapor pressure of water above sugar ($C_{12}H_{22}O_{11}$) solutions taken from the International Critical Tables and summarized in Table XXX, calculate the corresponding:

(*a*) Fugacities of the water in solution.

(*b*) Activities a_1 and activity coefficients γ_1 of the water referred to pure liquid water.

(*c*) Activities a_2 and activity coefficients γ_2 of the sugar referred to the pure solid.

(*d*) Activities a_2' and activity coefficients γ_2' of the sugar referred to mole fractions at infinite dilution.

(*e*) Activities a_2'' and activity coefficients γ_2'' of the sugar referred to molality at infinite dilution.

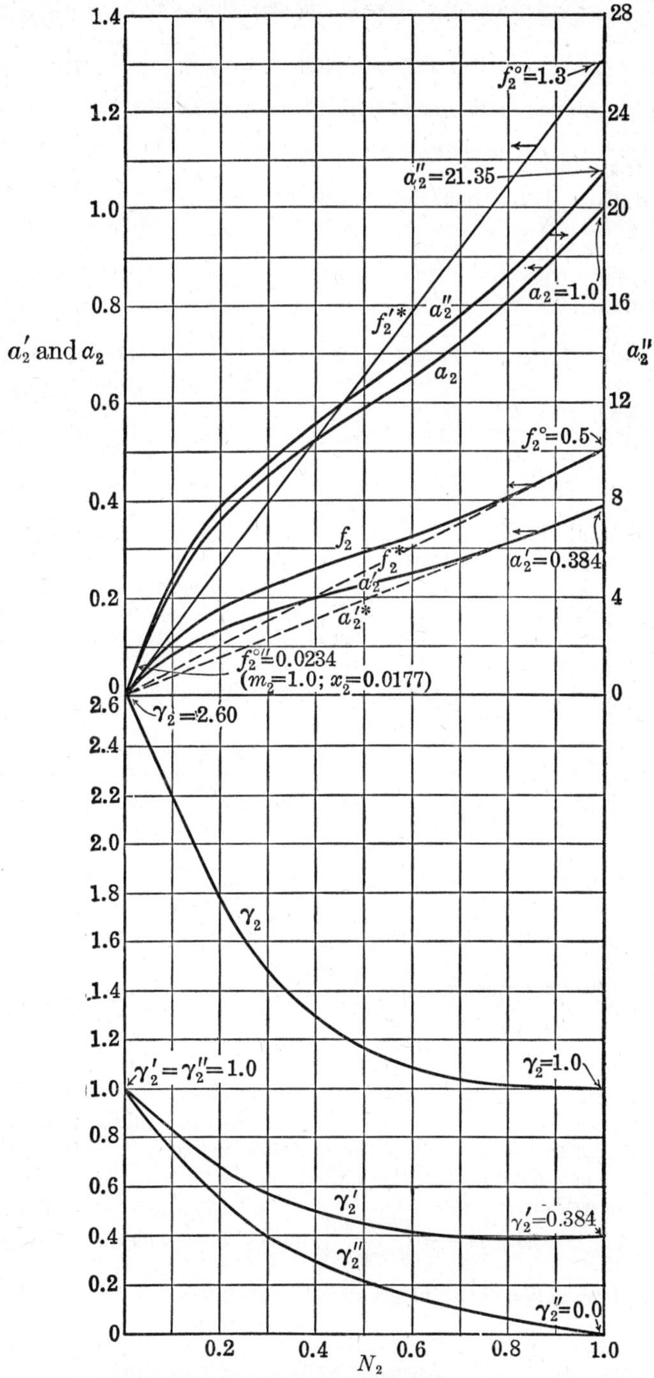

FIGS. 143*a* and *b*. Comparison of Activities and Activity Coefficients for Different Standard States.

TABLE XXX

ACTIVITY OF WATER IN SUGAR SOLUTIONS AT 25°C

Molality	*Mole Fractions* $C_{12}H_{22}O_{11}$	H_2O	*Vapor pressure* mm Hg				
m	N_2	N_1	p_1	a_1	γ_1	$\log \gamma_1$	N_1/N_2
0	0	1.0	23.756	1.0	1.0	0	∞
0.1	0.00180	0.99820	23.714	0.99823	1.0000	0.0000	554.56
0.4	0.00715	0.99285	23.585	0.99280	0.99995	−0.00002	138.86
1.0	0.01770	0.98230	23.302	0.98089	0.99856	−0.00063	55.497
2.0	0.03478	0.96522	22.762	0.95816	0.99269	−0.00319	27.752
3.0	0.05127	0.94873	22.166	0.93307	0.98349	−0.00723	18.505
4.0	0.06722	0.93278	21.521	0.90592	0.97120	−0.01269	13.877
5.0	0.08263	0.91737	20.846	0.87750	0.95654	−0.01930	11.102
6.0	0.09755	0.90245	20.20	0.85031	0.94222	−0.02585	9.251
6.18 (Sat)	0.10018	0.89982	20.08	0.84526	0.93937	−0.02716	8.9820

Solution: (*a*) At the low pressures involved it may be assumed that the fugacities of the water vapor f_1 are equal to partial pressures p_1.

(*b*) The activities a_1 of the water in the solution are obtained by dividing p_1 by 23.756 the vapor pressure and fugacity of the pure water at 25°C which is chosen as the standard state. At the low pressures involved the pressure designation of the standard state is of no consequence since pressures of this order have a negligible effect on the fugacity of the liquid. The activity coefficients γ_1 are equal to a_1/N_1.

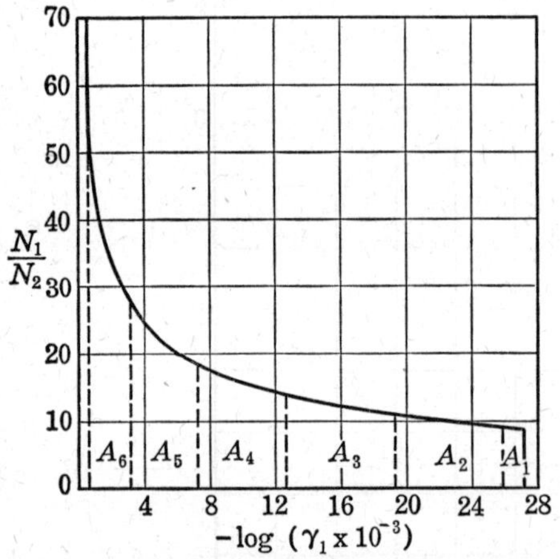

FIG. 144. **Evaluation of the Activity Coefficients of Sucrose**

(*c*) The activity coefficients of the sugar γ_2 are determined by graphical integration of Equation (68). Values of $-\log \gamma_1$ and N_1/N_2 from Table XXX are plotted in Fig. 144. Since the fugacity of the sugar in the saturated solution is equal to that of the pure solid, the activity a_2 of the sugar in the saturated solution is 1.0 referred to the pure solid. The corresponding activity coefficient $\gamma_2 = 1.0/N_2 = 9.9820$ where $N_1/N_2 = 8.9820$. These values fix the lower limit of the integral which may be written

$$\log \gamma_2 = -\int_{8.9820}^{N_1/N_2} \frac{N_1}{N_2}\, d \log \gamma_1 + \log 9.9820$$

The incremental evaluation of the integral is indicated in Fig. 144, and the resultant values of γ_2 are given in Table XXXI. Corresponding values of a_2 are obtained by multiplying N_2 by γ_2.

(*d*) In Fig. 145 values of γ_2 are plotted as ordinates on a logarithmic scale against

N_2 and the resulting curve extrapolated to $N_2 = 0$. Values of a_2' and a_2'' are then calculated from Equation (72). The corresponding activity coefficients are by definition $\gamma_2' = a_2'/N_2$ and $\gamma_2'' = a_2''/m$. These results are summarized in Table XXXI and Fig. 145.

TABLE XXXI

ACTIVITIES OF SUGAR IN AQUEOUS SOLUTIONS AT 25°C

m	N_2	γ_2	a_2	a_2'	γ_2'	a_2''	γ_2''
0	0	3.2800	0	0	1.000	0	1.000
0.1	0.00180	3.2899	0.00592	0.00180	1.000	0.1000	1.000
0.4	0.00715	3.3696	0.02409	0.00734	1.027	0.4074	1.019
1.0	0.01770	3.8539	0.06821	0.02080	1.175	1.155	1.155
2.0	0.03478	4.7222	0.16424	0.0501	1.400	2.779	1.390
3.0	0.05127	5.7860	0.29665	0.0904	1.764	5.020	1.673
4.0	0.06722	7.0175	0.47172	0.1438	2.139	7.983	1.996
5.0	0.08263	8.3880	0.69310	0.2113	2.557	11.730	2.346
6.0	0.09755	9.7225	0.94843	0.2899	2.964	16.05	2.675
6.18	0.10018	9.9820	1.00000	0.3049	3.043	16.92	2.738

From inspection of Fig. 145 it is evident that the activity coefficients γ_2 and γ_2' are widely different in numerical values but are related by a constant factor. The coefficient γ_2'' approaches γ_2' at low concentrations but is not related to the other two coefficients by a constant factor.

It may be noted that the graphical integration shown in Fig. 144 becomes uncertain at low concentrations where N_1/N_2 approaches infinity. Special graphical methods have been developed by Lewis and Randall[13] to circumvent this difficulty and permit accurate integrations with zero concentration as one limit.

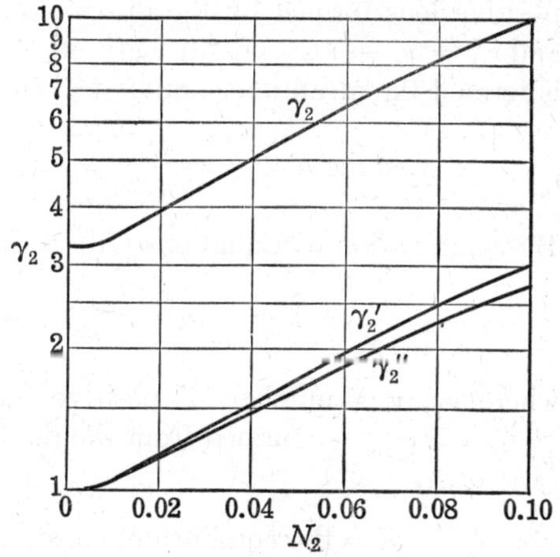

FIG. 145. Activity Coefficients of Sucrose in Aqueous Solutions at 25°C.

Although determination of activities and activity coefficients from vapor-pressure data is most direct, in principle the results are frequently less accurate than those obtained by other methods involving different types of equilibrium.

Lewis and Randall[13] discuss in detail the evaluation of activity coefficients from data on solubilities, distribution coefficients, freezing points, boiling points, and electromotive-force measurements.

Ionic Activities. The expression of activities in solution of electrolytes is complicated by the dissociation of the components into ions. This effect may be disregarded, and the *apparent molecular activity*, frequently termed the activity of the undissociated electrolyte, may be determined from vapor-pressure data by the procedure of Illustration 9. For example, the molecular activities of sodium chloride in water may be determined in this manner from data in the literature. It is found, however, that the molecular-activity coefficient $\gamma_2 = a_2/N_2$ varies widely with concentration and approaches zero instead of a constant value at zero concentration. This behavior results from the fact that the effective mole fraction of the binary solute is increased by dissociation of one molecule into two ions, each of which behaves as a solute, and the actual molal concentration of the solute approaches twice the stoichiometric value based on the undissociated salt. Thus, for determining molecular activities in a binary solution Equation (64) may be written

$$d \log a_2 = -\frac{n_1}{n_2} d \log a_1 = -\frac{1000}{M_1 m_2} d \log a_1 \tag{78}$$

Actually, however, the number of moles of effective solute is equal to $n_2(\nu_- + \nu_+)$ or $n_2\nu_\pm$ where ν_- and ν_+ are the numbers of negative and positive ions formed by the dissociation of one molecule of the solute and $\nu_\pm = \nu_- + \nu_+$. On this basis an expression may be written for what is termed the mean ionic activity $a_{\pm 2}$.

$$d \log a_{\pm 2} = -\frac{n_1}{n_2 \nu_{2\pm}} d \log a_1 = -\frac{1000}{M_1 m_2 \nu_\pm} d \log a_1 \tag{79}$$

By combination of Equations (78) and (79) it follows that

$$a_2 = k(a_\pm)^{\nu_\pm} = \frac{a_+^{\nu_+} a_-^{\nu_-}}{K} \tag{80}$$

where ν_+, ν_- = number of positive and negative ions, respectively, formed from the dissociation of one molecule,

$$\nu_\pm = \nu_+ + \nu_-$$

K = the equilibrium constant of ionic dissociation

k = a proportionality factor, the value of which is determined by the standard states chosen for a_2 and $a_{\pm 2}$

Rearrangement of Equation (80) gives,

$$a_\pm = \left(\frac{a_+^{\nu_+} a_-^{\nu_-}}{kK}\right)^{\frac{1}{\nu_\pm}} \tag{80a}$$

The standard state of ionic activity is conventionally based upon the *mean ionic molality* which takes the same form as Equation (80a); thus:

$$m_{\pm 2} = m_2(\nu_+^{\nu_+} \nu_-^{\nu_-})^{\frac{1}{\nu_\pm}} \tag{81}$$

where m_2 = stoichiometric molality

For example, calcium chloride dissociates into one positive and two negative ions or $\nu_+ = 1$, $\nu_- = 2$, $\nu_\pm = 3$, and $m_{\pm 2} = m_2(4)^{\frac{1}{3}}$.

The standard state for the expression of the activities of electrolytes is so chosen that $a_{\pm 2} = m_{\pm 2}$ as infinite dilution is approached. Thus,

$$\lim_{m_{\pm 2}=0} a_{\pm 2} = m_{\pm 2} = m_2(\nu_+^{\nu_+}\nu_-^{\nu_-})^{\frac{1}{\nu_\pm}} \tag{82}$$

The mean ionic-activity coefficient $\gamma_{\pm 2}$ is defined by

$$\gamma_{\pm 2} = \frac{a_{\pm 2}}{m_{\pm 2}} \tag{83}$$

It is evident that $\gamma_{\pm 2} = 1.0$ at zero concentration.

For the expression of the apparent molecular activities of electrolytes, a special standard state is commonly selected in order that the following relationship shall be valid.

$$a_2''' = (a_{\pm 2})^{\nu_\pm} = (f_2^\circ/f_2^{\circ\prime\prime\prime})a_2 \tag{84}$$

where a_2''' = the apparent molecular activity referred to what may be termed the electrolytic standard state having a fugacity $f_2^{\circ\prime\prime\prime}$.

a_2 = the apparent molecular activity referred to the pure component having a fugacity f_2°. The apparent molecular activity coefficient γ_2''' may be defined as follows:

$$\gamma_2''' = \frac{a_2'''}{m_2} \tag{85}$$

It should be noted that this definition of standard state does not require that γ_2''' be unity at infinite dilution.

By combination of Equations (81), (82), and (84) mean ionic activities may be determined from vapor-pressure measurements which permit calculation of the apparent molecular activity a_2 referred to the pure component or some other convenient state. Thus,

$$k_2\gamma_{\pm 2} = \frac{(a_2)^{\frac{1}{\nu_\pm}}}{m_2} \tag{86}$$

where

$$k_2 = \frac{(\nu_+^{\nu_+}\nu_-^{\nu_-})^{\frac{1}{\nu_\pm}}}{(f_2^\circ/f_2^{\circ\prime\prime\prime})^{\frac{1}{\nu_\pm}}} \tag{87}$$

Values of $k_2\gamma_{\pm 2}$ from Equation (86) are plotted against m_2 and extrapolated to zero molality where by definition $\gamma_{\pm 2} = 1.0$. This intercept is equal to k_2 from which values of $\gamma_{\pm 2}$ at all concentrations may be calculated by Equation (86). Corresponding values of $a_{\pm 2}$ and a_2''' are then obtained from Equations (83) and (84).

It is of interest to develop the relationship between $\gamma_{\pm 2}$ and the apparent molecular-activity coefficients $\gamma_2''' = a_2'''/m_2$ and $\gamma_2 = a_2/N_2$.

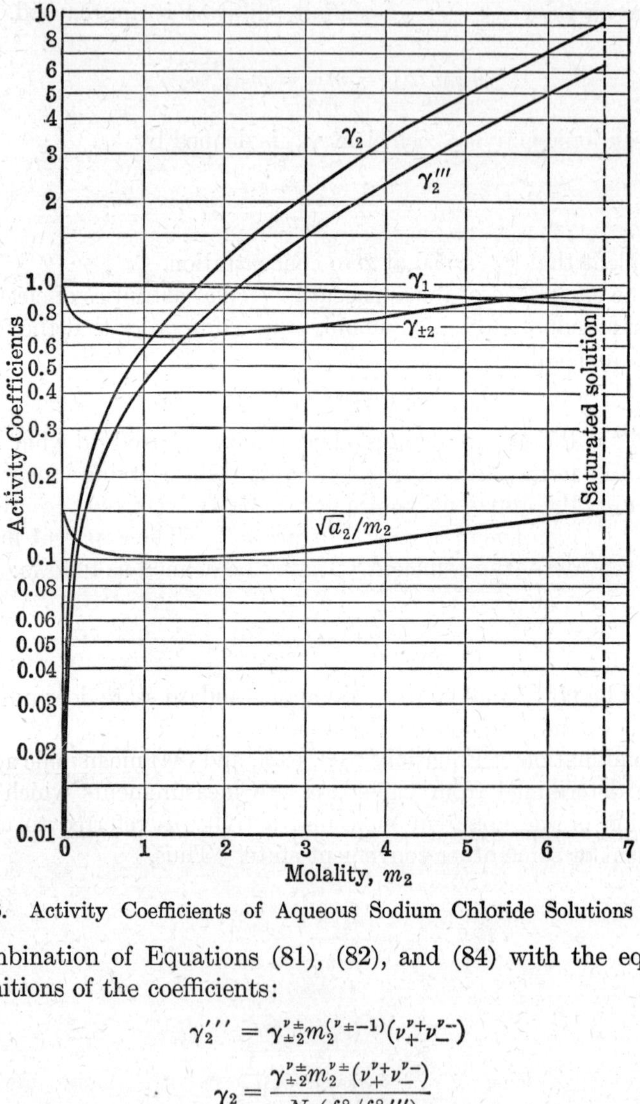

FIG. 146. Activity Coefficients of Aqueous Sodium Chloride Solutions at 25°C.

By combination of Equations (81), (82), and (84) with the equations of definitions of the coefficients:

$$\gamma_2''' = \gamma_{\pm 2}^{\nu_\pm} m_2^{(\nu_\pm - 1)} (\nu_+^{\nu_+} \nu_-^{\nu_-}) \tag{88}$$

$$\gamma_2 = \frac{\gamma_{\pm 2}^{\nu_\pm} m_2^{\nu_\pm} (\nu_+^{\nu_+} \nu_-^{\nu_-})}{N_2 (f_2^\circ / f_2^{\circ\prime\prime\prime})} \tag{89}$$

It is evident from Equations (88) and (89) that both γ_2''' and γ_2 must approach zero at zero concentration where $\gamma_{\pm 2} = 1.0$ and $m_2 = 0$. The relationship between the various activity coefficients is shown graphically in Fig. 146 for NaCl in water at 100°C. Values of a_2 and γ_2 referred to pure NaCl(s) may be calculated from vapor-pressure data by the method of Illustration 9. In accordance with Equation (86) the quantity $\sqrt{a_2}/m_2$ is plotted against m_2 and extrapolated to $m_2 = 0$. The value of k_2 is equal to this intercept, since $\gamma_{\pm 2} = 1.0$. The corresponding values of γ_2'' are obtained from Equation (88). This method is presented to clarify the relationships between the various standard states and activity coefficients. Actually it is not a practical means of evaluating activity coefficients with a high degree of accuracy because of the difficulty of obtaining precise vapor-pressure data for very dilute solutions. More accurate experimental methods are discussed in detail by Lewis and Randall.[13]

PROBLEMS

1. A mixture of hydrocarbon gases has the following composition in mole per cent:

CH_4	38.38
C_2H_6	7.56
C_3H_8	7.05
n-C_4H_{10}	11.29
n-C_5H_{12}	35.72
	100.00

Calculate the density of this mixture in pounds per cubic foot in the gaseous state at a temperature of 200°F and an absolute pressure of 400 lb per sq in.

2. Calculate the density in grams per cubic centimeter of the mixture of problem 1 in the liquid state at a temperature of 100°F and an absolute pressure of 1300 lb per sq in.

3. Calculate the enthalpy in Btu per pound of the mixture of problem 1 in the gaseous state at the specified conditions. As the reference state of zero enthalpy use the ideal gaseous state at 60°F for the propane and lighter constituents. For butane and the heavier constituents use the saturated liquid state at 60°F as the reference. It may be assumed that the enthalpies of the components are additive under these reference conditions. Heats of vaporization of the three light gases are given in Illustration 4. Doss[9] recommends the following values in calories per gram at the normal boiling point: n-butane 92.0; n-pentane 85.5.

4. Calculate the enthalpy in Btu per pound of the mixture in the liquid state at the conditions of problem 2, using the reference states specified in problem 3.

5. From the data of Table XXVIII and Fig. 142 calculate the fugacity in pounds per square inch of pure ethane gas at a temperature of 200°F and an absolute pressure of 1500 lb per sq in.

6. Calculate the fugacities of the following gases;

(*a*) Air at 60°F and 100 atm.

(*b*) Ammonia at 80°C and 40 atm.

(*c*) Carbon dioxide at 150°F and a gauge pressure of 2,000 lb per sq in.

7. Calculate the fugacity of the liquid ethylene in contact with its saturated vapor at 0°C and 40.6 atm.

8. Calculate the fugacity of liquid chlorine in contact with a mixture of hydrogen and its own vapors at a temperature of 122°F and an absolute pressure of 1,000 lb per sq in. The vapor pressure of liquid chlorine at this temperature is 14.1 atm, and its density is 1.557 g per cc at −33.6°C.

9. Assuming that an ideal solution is formed, calculate the fugacities of the components of the mixture of problem 1 under the specified conditions.

10. The gas from the converter of a synthetic ammonia plant has the following composition in mole per cent:

$$N_2 = 20.2$$
$$H_2 = 60.8$$
$$NH_3 = 19.0$$

Calculate the fugacity of each component of the mixture at 800°F and an absolute pressure of 4,500 lb per sq in., assuming that an ideal solution is formed.

11. The International Critical Tables give the following data for the lowering of the vapor pressure of water at 0°C by urea $[CO(NH_2)_2]$:

m	$100\,R$
1	1.52
2	1.49
4	1.46
6	1.45
10	1.43

$$R = \frac{(P_0 - p)}{mP_0}$$

where m = molality; P_0 = vapor pressure of the pure water; p = vapor pressure of H_2O above the solution. The solubility of urea at 0°C is 67.1 g per 100 g of H_2O.

Calculate and plot as functions of molality the activity of the water a_1 and of the urea a_2 referred to the solid at 0°C, and estimate the activities of the urea a_2' and a_2'' referred to infinite dilution. Also calculate and plot the corresponding activity coefficients, γ_1, γ_2, γ_2', and γ_2''.

12. The following data on the vapor-pressure lowering of aqueous NaI solutions at 100°C are expressed in the terms defined in problem 11:

m	$100\,R$	m	$100\,R$
0.7	3.25	7	4.91
1.5	3.62	9	5.00
4.0	4.46	11	4.89
5.0	4.67	17	4.25
6.0	4.82	20	3.85

The solubility of NaI at 100°C is 20.24 moles per 1000 g of H_2O.

Calculate and plot as functions of the molality: a_1; γ_1; a_2 (referred to solid NaI); γ_2; $a_{\pm 2}$; $\gamma_{\pm 2}$; a_2'''; γ_2'''.

13. The following data from the International Critical Tables show the partial pressures in millimeters of Hg of toluene and acetic acid in solutions at a temperature of 69.94°C:

x_1 = mole per cent toluene	p_1 (toluene)	p_2 (acetic acid)
0	0	136
12.50	54.8	120.5
23.10	84.8	110.8
31.21	101.9	103.0
40.19	117.8	95.7
48.60	130.7	88.2
53.49	137.6	83.7
59.12	145.2	78.2
66.20	155.7	69.3
75.97	167.3	57.8
82.89	176.2	46.5
90.58	186.1	30.5
95.65	193.5	17.2
100	202	0

Calculate the activity coefficients γ_1, γ_2 from these data. Plot the activity coefficients as ordinates on a logarithmic scale against the mole per cent toluene on a uniform scale and check these curves for consistency with the Gibbs–Duhem equation at compositions of 10, 30, 50, 70, and 90 per cent toluene.

CHAPTER XV

PHYSICAL EQUILIBRIUM

The concepts of fugacity and activity are of particular value in problems of equilibrium in both physical and chemical processes. Where only pure components or ideal solutions are involved, such problems are readily solved by applications of the principles developed in the preceding chapters. If a system under consideration involves a nonideal solution, the evaluation of the activity coefficients of the components is frequently the most difficult problem. Some direct experimental data combined with empirical or semiempirical relationships are generally necessary.

VAPOR–LIQUID EQUILIBRIUM

The evaluation of activity coefficients in solutions of miscible liquids is of particular importance for correcting data on vapor–liquid equilibrium. In working with such mixtures it is convenient to choose the pure components at the pressure of the solution as the standard states. With this convention activity is equal to mole fraction in an ideal solution at constant temperature and pressure. Then,

$$\gamma_1 = \frac{a_1}{N_1} = \frac{f_1}{N_1 f_1^\circ} \tag{1}$$

where

γ_1 = activity coefficient of component 1
a_1 = activity of component 1 in solution referred to the pure component
f_1 = fugacity of component 1 in solution
f_1° = fugacity of pure component 1 at the temperature and pressure of the solution

If it is assumed that the vapors form an ideal solution, the fugacity f_1 of Equation (1) is readily calculated as in Illustration 8 (Chapter XIV) from the composition, temperature, and pressure of the vapor in equilibrium with the solution. The fugacity of the standard state is calculated from the vapor pressure of the pure compound and the total pressure of the solution by the method of Illustration 7, page 623. In this manner, if sufficient data are available, activity coefficients may be evaluated in any given system as a function of composition, temperature, and total pressure. The results may be verified for thermodynamic consistency by means of the Gibbs–Duhem Equation (XIV-67). This equation must be satisfied by activity coefficients at a constant temperature and pressure.

Application of these methods to binary solutions is complicated by the fact that when a vapor phase is present such a system possesses only two degrees of freedom and it is not possible to vary composition at constant temperature and pressure. For rigorous application of the Gibbs–Duhem equation, fugacities f_1 determined at a constant temperature and varying total pressures should be corrected to a constant-pressure basis by means of Equation (XIV-45). However, this correction is small for moderate pressure changes and generally may be neglected except when one is working with solutions near their critical points where partial molal volumes and pressures are both high. Similarly, at moderate pressures and small liquid volumes the effect of pressure on the fugacity of the liquid is negligible, and f_1° may be taken as the fugacity of the liquid under its own vapor pressure. Under conditions such that the vapors may be treated as ideal gases still further simplification is possible and Equation (1) may be written:

$$\gamma_1 = \frac{p_1^*}{x_1 P_1^*} = \frac{y_1 \pi^*}{x_1 P_1^*} \tag{2}$$

where

x_1 = mole fraction of component 1 in the liquid phase
y_1 = mole fraction of component 1 in the vapor phase
p_1^* = partial pressure of component 1 in vapor
P_1^* = vapor pressure of pure component 1 at the temperature of the solution
π^* = total pressure of the system

Component 1 is usually taken as the component which in the pure state has the lower boiling point.

In Fig. 147 are plotted the activity coefficients of benzene and toluene (system I) of water and *n*-butanol (system II) of isopropyl ether and isopropyl alcohol (system III) and of acetone and chloroform (system IV) all at 760 mm of Hg. For the nearly ideal benzene–toluene system, the activity coefficients of both components are the same and equal to 1.0. The corresponding vapor–liquid composition curves of these four systems are plotted in Fig. 148 and the boiling-point curves in Fig. 149.

It should be noted that the activity coefficients of Fig. 147 are restricted to a constant total pressure and therefore correspond to varying temperatures. Similar curves for the same midrange conditions

* Indicates a range of pressures sufficiently low so that the ideal-gas law is applicable.

might be plotted for activity coefficients at constant temperature and varying pressure. In general, the differences between such plots are negligible for systems whose boiling points vary only a few degrees with change in composition. The constant pressure plots are in general more useful because most vaporization operations are conducted at substan-

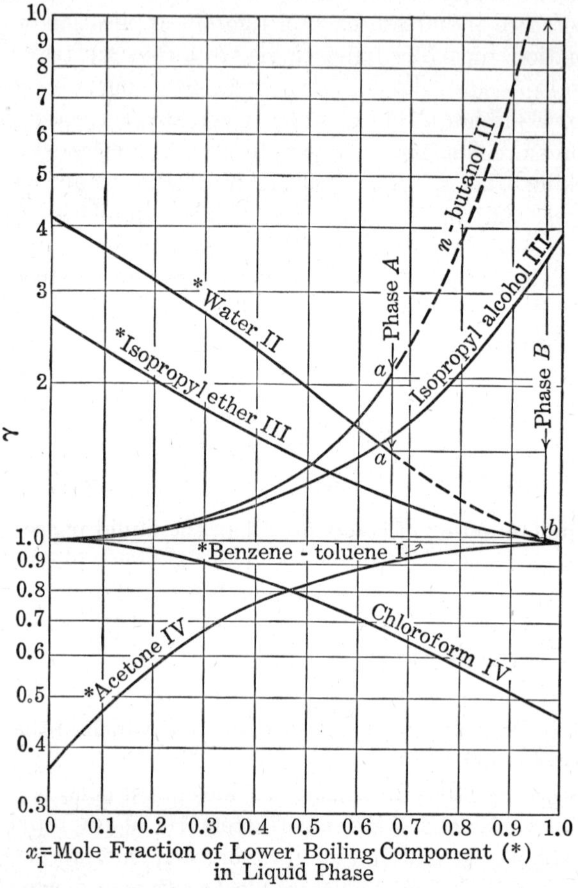

FIG. 147. Activity Coefficients of Typical Binary Systems.

tially constant pressure. On the other hand, activity coefficients at constant temperature are more accurately rationalized by the Gibbs–Duhem equation.

From Fig. 149 it may be noted that the isopropyl alcohol–isopropyl ether system forms a solution having a minimum boiling point at 78 per cent ether, whereas the acetone–chloroform system forms a solution having a maximum boiling point at 40 per cent acetone, and the water–

butanol system has a minimum boiling point at 75 per cent water. Such solutions which have maximum or minimum boiling points and which evolve vapors of the same composition as the liquid are termed *azeotropic solutions or azeotropes.* By comparison of Figs. 149 and 147 it is seen that for a system forming an azeotrope of minimum boiling point the activity coefficients of the separate components are each greater than unity over the entire range of compositions, whereas for

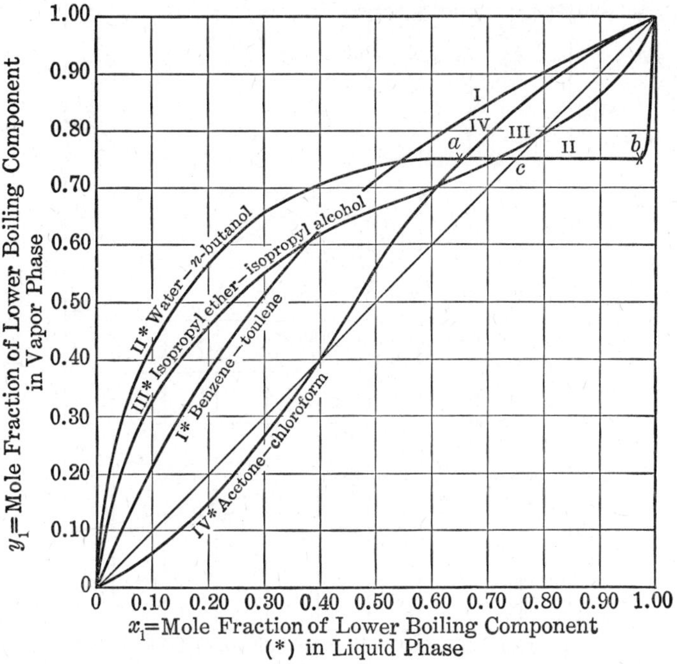

FIG. 148. Vapor–Liquid Equilibria of Typical Binary Systems.

the system forming an azeotrope of maximum boiling point the activity coefficients are less than unity. This behavior is general for many nonideal systems. Activity coefficients greater than 1.0 are much the more common.

If the effects of the small variations in temperature are neglected, the slopes of the curves of Fig. 147 for any system at constant temperature and low pressure are related to each other by Equation (XIV-69), and if one curve is known the other is readily calculated. For the prediction of both activity curves from a minimum of experimental data a general relation between activity coefficients and mole fraction is necessary.

Free Energy of Mixing. Thermodynamically consistent expressions for the activity coefficients of components in solutions are most conveniently derived by consideration of the free-energy changes accompanying the formation of a solution from its components in their standard states. This *free-energy change of mixing* ΔG_M is given by the following general equation where the standard state is taken as the pure

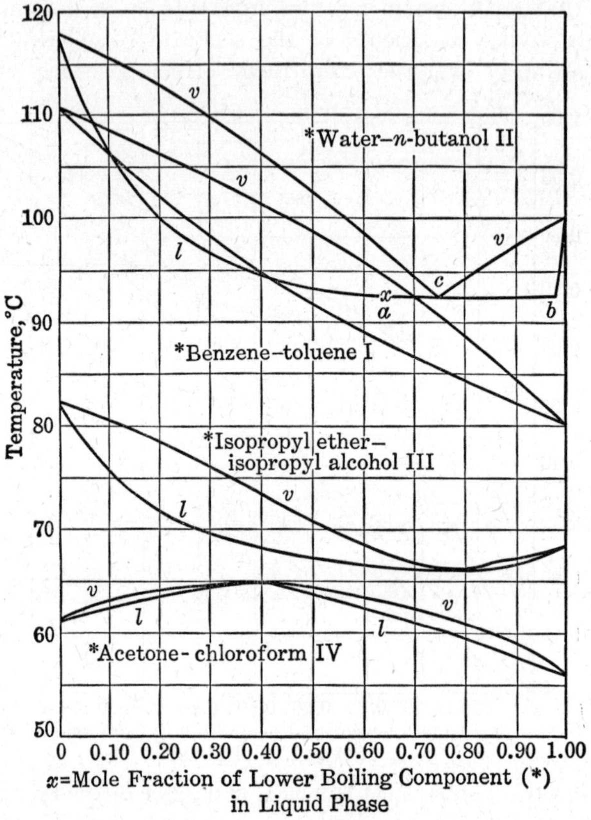

FIG. 149. Saturation Temperatures of Typical Binary Systems at 1.0 Atm Pressure.

components at the pressure and temperature of the system:

$$\Delta G_M = \Sigma n_i \overline{G}_i - \Sigma n_i G_i^\circ, \tag{3}$$

where

$$\Sigma n_i \overline{G}_i = n_1 \overline{G}_1 + n_2 \overline{G}_2 + n_3 \overline{G}_3 + \cdots = G$$

$$\Sigma n_i G_i^\circ = n_1 G_1^\circ + n_2 G_2^\circ + n_3 G_3^\circ + \cdots$$

Combining (3), (XIV-53) and (XIV-60) and using x_i for mole fraction instead of N_i for the liquid phase gives

$$\Delta G_M = RT \Sigma n_i \ln x_i + RT \Sigma n_i \ln \gamma_i \tag{4}$$

where

$$\Sigma n_i \ln x_i = n_1 \ln x_1 + n_2 \ln x_2 + n_3 \ln x_3 + \cdots$$

For an ideal solution all activity coefficients are equal to unity and the term $RT\Sigma n_i \ln \gamma_i$ reduces to zero. This term is designated by Scatchard and Hamer[1] as the *excess free energy* G^E of the solution. Thus, the free energy G of any solution may be written as

$$G = \Sigma n_i \mathrm{G}_i^\circ + RT\Sigma n_i \ln x_i + G^E \tag{5}$$

where

$$G^E = RT\Sigma n_i \ln \gamma_i = \text{the excess free energy}$$

The chemical potential of any component is obtained by differentiating Equation (5), thus for component 1,

$$\mu_1 = \overline{\mathrm{G}}_1 = \mathrm{G}_1^\circ + RT\frac{\partial \Sigma n_i \ln x_i}{\partial n_1} + \frac{\partial G^E}{\partial n_1} \tag{6}$$

The expansion of the second term combined with the principle of Equation (XIV-66) gives

$$RT\frac{\partial \Sigma n_i \ln x_i}{\partial n_1} = RT\left[\ln x_1 + n_1\frac{\partial \ln x_1}{\partial n_1} + n_2\frac{\partial \ln x_2}{\partial n_1} + \cdots\right] = RT \ln x_1 \tag{7}$$

Then,

$$\mu_1 = \overline{\mathrm{G}}_1 = \mathrm{G}_1^\circ + RT \ln x_1 + \frac{\partial G^E}{\partial n_1} \tag{8}$$

By comparison of Equations (8) and (XIV-53) it may be seen that

$$RT \ln \gamma_1 = \frac{\partial G^E}{\partial n_1} \tag{9}$$

Similarly,

$$RT \ln \gamma_2 = \frac{\partial G^E}{\partial n_2} \tag{10}$$

Equations (9) and (10) permit derivation of thermodynamically consistent expressions for the activity coefficients of all components of a solution from an empirical expression relating the excess free energy to the composition of the solution.

It was pointed out by Wohl[2] that the empirical equations which have been commonly used for correlation of activity coefficients represent special cases of the following general equation for the excess free energy:

$$\frac{\mathrm{G}^E}{RT(\Sigma q_i x_i)} = \Sigma_{ih} z_i z_h a_{ih} + \Sigma_{ihj} z_i z_h z_j a_{ihj} + \Sigma_{ihjl} z_i z_h z_j z_l a_{ihjl} + \cdots \tag{11}$$

where

x_i = mole fraction of component i
q_i = effective molal volume of component i
z_i = effective volume fraction of component i

[1] G. Scatchard and W. J. Hamer, *J. Am. Chem. Soc.*, **57**, 1805 (1935).
[2] K. Wohl, *Trans. Am. Inst. Chem. Engrs.* **42**, 215–250 (1946).

Subscripts i, h, j, l each may correspond to any component of the mixture in the terms of the indicated summations.

a_{ih} etc. = empirical constants corresponding to the indicated groups of components in the summations.

Equation (11) is designated a *four-suffix* empirical equation as characterized by the last term. The first summation term may be considered as indicating the effect on the excess free energy of the interactions between unlike molecules of the components in pairs. The second term indicates the contributions from interactions of groups of three molecules while the third takes into account interactions of groups of four. If the last summation is omitted the remainder is a *three-suffix equation*, whereas if the last two terms are omitted it becomes a two-suffix equation. Each additional summation introduces added constants to permit improved representation of increasingly complex relationships.

Each summation in Equation (11) represents the sum of all possible combinations of molecules in unlike groups of the indicated size. Thus, in a mixture of components 1, 2, 3, · · ·, component 1 is first taken as i, and all possible groups are expanded by successively designating each component of the mixture as h, j, and l. For example, for a ternary system the second summation is expanded to 24 terms as follows for all possible triple combinations among two and three unlike species.

$$\begin{aligned}\Sigma_{ihj}z_iz_hz_ja_{ihj} &= z_1z_1z_2a_{112} + z_1z_1z_3a_{113} + z_1z_2z_1a_{121} + z_1z_2z_2a_{122} \\ &+ z_1z_2z_3a_{123} + z_1z_3z_1a_{131} + z_1z_3z_2a_{132} + z_1z_3z_3a_{133} + z_2z_1z_1a_{211} \\ &+ z_2z_1z_2a_{212} + z_2z_1z_3a_{213} + z_2z_2z_1a_{221} + z_2z_2z_3a_{223} + z_2z_3z_1a_{231} \\ &+ z_2z_3z_2a_{232} + z_2z_3z_3a_{233} + z_3z_1z_1a_{311} + z_3z_1z_2a_{312} + z_3z_1z_3a_{313} \\ &+ z_3z_2z_1a_{321} + z_3z_2z_2a_{322} + z_3z_2z_3a_{323} + z_3z_3z_1a_{331} + z_3z_3z_2a_{332} \qquad (12)\end{aligned}$$

Since the order of multiplication is immaterial, terms such as $z_1^2z_2a_{112}$ and $z_1z_2z_1a_{121}$ are identical and may be grouped together. Thus, Equation (12) reduces to seven terms:

$$\begin{aligned}\Sigma_{ihj}z_iz_hz_ja_{ihj} &= z_1^2z_23a_{112} + z_1^2z_33a_{113} + z_1z_2^23a_{122} + z_1z_2z_36a_{123} \\ &+ z_1z_3^23a_{133} + z_2^2z_33a_{223} + z_2z_3^23a_{233} \qquad (13)\end{aligned}$$

The other summations may be expanded in a similar manner for any system of any number of components.

The effective molal volume q in Equation (11) serves to relate the effective volume fractions z to the mole fractions x. Thus,

$$z_1 = \frac{x_1}{x_1 + \dfrac{q_2}{q_1}x_2 + \dfrac{q_3}{q_1}x_3 + \cdots} \qquad (14)$$

and

$$z_2 = \frac{x_2 \dfrac{q_2}{q_1}}{x_1 + \dfrac{q_2}{q_1} x_2 + \dfrac{q_3}{q_1} x_3 + \cdots} \tag{15}$$

Activity Coefficients in Binary Systems. For a binary system of components 1 and 2, a three-suffix equation of the form of (11) may be expanded as follows:

$$\frac{\text{G}^{\text{E}}}{RT(q_1x_1 + q_2x_2)} = z_1z_2 2a_{12} + z_1^2z_2 3a_{112} + z_1z_2^2 3a_{122} \tag{16}$$

or, since $z_1 + z_2 = 1.0$,

$$\frac{\text{G}^{\text{E}}}{RT} = \left[x_1 + \frac{q_2}{q_1} x_2\right] z_1z_2[z_1q_1(2a_{12} + 3a_{112}) + z_2q_1(2a_{12} + 3a_{122})] \tag{17}$$

This equation may be written with the following abbreviations:

$$A' = q_1(2a_{12} + 3a_{122}) \tag{18}$$

$$B' = q_2(2a_{12} + 3a_{112}) \tag{19}$$

Then

$$\frac{\text{G}^{\text{E}}}{RT} = \left[x_1 + \frac{q_2}{q_1} x_2\right] z_1z_2\left[z_1\left(\frac{q_1}{q_2}\right) B' + z_2A'\right] \tag{20}$$

The term $\ln \gamma_1$ may be evaluated by partial differentiation of Equation (20) with respect to n_1 in accordance with the relationship of Equation (9). This is done by first replacing x and z by n by means of the following relations:

$$x_1 = \frac{n_1}{n_1 + n_2}; \quad x_2 = \frac{n_2}{n_1 + n_2}; \quad z_1 = \frac{n_1}{n_1 + \dfrac{q_2}{q_1} n_2}; \quad z_2 = \frac{n_2 \dfrac{q_2}{q_1}}{n_1 + \dfrac{q_2}{q_1} n_2} \tag{21}$$

$$x_1 + x_2 = 1.0; \qquad z_1 + z_2 = 1.0$$

Also on the molal basis, $\dfrac{G^E}{n_1 + n_2} = \text{G}^{\text{E}}$

By substitution in Equation (20),

$$\frac{G^E}{RT} = \frac{n_1^2n_2B' + n_1n_2^2\left(\dfrac{q_2}{q_1}\right)^2 A'}{\left(n_1 + \dfrac{q_2}{q_1} n_2\right)^2} \tag{22}$$

Upon differentiation with respect to n_1, where $A = 2.303A'$ and $B = 2.303B'$,

$$\frac{\partial G^E}{2.303\partial n_1(RT)} = \log \gamma_1 = \frac{n_2^2\left(\frac{q_2}{q_1}\right)^2\left[-n_1A + 2n_1B\left(\frac{q_1}{q_2}\right) + n_2A\left(\frac{q_2}{q_1}\right)\right]}{\left(n_1 + n_2\frac{q_2}{q_1}\right)^3} \tag{23}$$

Replacing values of z_1 and z_2 gives

$$\log \gamma_1 = z_2^2\left[A + 2\left(B\frac{q_1}{q_2} - A\right)z_1\right] \tag{24}$$

Similarly,

$$\log \gamma_2 = z_1^2\left[B + 2\left(A\frac{q_2}{q_1} - B\right)z_2\right] \tag{25}$$

These general equations developed by Wohl[2] involve three constants, A, B, and q_1/q_2, which must be empirically determined for each system under consideration. By the use of various assumptions regarding the ratio of the effective molal volumes q_2/q_1 the number of empirical constants may be reduced to two.

Margules[3] in effect assumed that $q_2/q_1 = 1.0$. With this assumption Equations (24) and (25) reduce to the familiar Margules equations as modified by Carlson and Colburn.[4]

$$\log \gamma_1 = x_2^2[A + 2(B - A)x_1] = (2B - A)x_2^2 + 2(A - B)x_2^3 \tag{26}$$

$$\log \gamma_2 = x_1^2[B + 2(A - B)x_2] = (2A - B)x_1^2 + 2(B - A)x_1^3 \tag{27}$$

Scatchard[5] and co-workers[(6,7)] took the effective molal volumes as equal to the actual molal volumes of the pure components. On this basis $q_2/q_1 = V_2/V_1$, and Equations (24) and (25) reduce to what is termed the Scatchard–Hamer equations. Thus,

$$\log \gamma_1 = z_2^2\left[A + 2\left(B\frac{V_1}{V_2} - A\right)z_1\right] \tag{28}$$

$$\log \gamma_2 = z_1^2\left[B + 2\left(A\frac{V_2}{V_1} - B\right)z_2\right] \tag{29}$$

In these equations z represents the volume fraction based on the volumes of the pure components and is the true volume fraction if volume changes in mixing are negligible.

If it is assumed that $q_1/q_2 = A/B$ Equations (24) and (25) reduce to

[3] M. Margules, *Sitzber. Akad. Wiss. Wien. Math. naturw. Klasse*, II, 104, 1243–78 (1895).

[4] H. C. Carlson and A. P. Colburn, *Ind. Eng. Chem.*, **34**, 581 (1942).

[5] G. Scatchard, *Chem. Rev.*, **8**, 321 (1931).

[6] G. Scatchard and S. S. Prentiss, *J. Am. Chem. Soc.*, **56**, 1486 (1934).

[7] G. Scatchard and W. J. Hamer, *J. Am. Chem. Soc.*, **57**, 1805 (1935).

the form developed by van Laar[8,9] as rearranged by Carlson and Colburn.[4]

$$\log \gamma_1 = Az_2^2 = \frac{Ax_2^2}{\left(\frac{A}{B}x_1 + x_2\right)^2} \tag{30}$$

$$\log \gamma_2 = Bz_1^2 = \frac{Bx_1^2}{\left(x_1 + \frac{B}{A}x_2\right)^2} \tag{31}$$

Differentiation proves that Equations (24–31) satisfy the general relation of Equation (XIV-69). The constants A and B are characteristic of each particular system. In the van Laar equations (30) and (31) it is evident that A is the terminal value of $\log \gamma_1$, where $x_1 = 0$, and B is the terminal value of $\log \gamma_2$, where $x_2 = 0$.

Equations (24–31) are basically restricted to conditions of constant temperature. However, as previously pointed out, little error is generally encountered in applying them to activity coefficients at constant pressure in close boiling systems.

If the validity of the van Laar Equations (30) and (31) is accepted, determination of γ_1 and γ_2 at a single known composition permits evaluation of A and B and calculation of the complete γ curves. Measurement of a single set of equilibrium liquid and vapor compositions, together with knowledge of the vapor pressures of the pure components, suffices for calculation of γ_1 and γ_2 from Equation (2). Where an azeotrope is formed the composition of only one phase need be measured since the liquid and vapor compositions are the same. By rearrangement of Equations (30) and (31) the constants A and B are obtained directly.

$$A = \log \gamma_1 \left[1 + \frac{x_2 \log \gamma_2}{x_1 \log \gamma_1}\right]^2 \tag{32}$$

$$B = \log \gamma_2 \left[1 + \frac{x_1 \log \gamma_1}{x_2 \log \gamma_2}\right]^2 \tag{33}$$

Illustration 1. The azeotrope of the ethanol–benzene system has a composition of 44.8 mole per cent ethanol with a boiling point of 68.24°C at 760 mm Hg. At 68.24°C the vapor pressure of pure benzene is 517 mm Hg, and that of ethanol is 506 mm Hg. Calculate the van Laar constants for the system, and evaluate the activity coefficients for a solution containing 10 mole per cent ethanol.

Solution: Ethanol is designated as component 1. At the azeotropic composition, from Equation (2), since $y = x$,

$$\gamma_1 = \tfrac{760}{506} = 1.502; \qquad \log \gamma_1 = 0.177$$

$$\gamma_2 = \tfrac{760}{517} = 1.47; \qquad \log \gamma_2 = 0.167$$

[8] J. J. van Laar, *Z. phys. Chem.*, **72**, 723 (1910).
[9] J. J. van Laar, *Z. phys. Chem.*, **185**, 35 (1929).

From Equation (32),

$$A = 0.177\left[1 + \frac{(0.552)(0.167)}{(0.448)(0.177)}\right]^2 = 0.830$$

Similarly, from Equation (33),

$$B = 0.577$$

For a solution containing 10 per cent ethanol, from Equations (30) and (31),

$$\log \gamma_1 = \frac{(0.830)(0.90)^2}{\left[\frac{(0.830)(0.10)}{0.577} + 0.90\right]^2} = 0.6168; \qquad \gamma_1 = 4.14$$

$$\log \gamma_2 = 0.0109; \qquad \gamma_2 = 1.026$$

By this procedure the activity coefficients of the entire range of compositions may be calculated.

It was pointed out by Carlson and Colburn[4] that the van Laar constants may be evaluated without any liquid–vapor equilibrium composition data if a series of isothermal vapor-pressure or isobaric boiling-point measurements are available for solutions of known composition over the entire composition range. It is frequently desirable to work from such data because of the difficulty of securing a reliable analysis of vapor mixtures and the ease of making up liquid solutions of a desired composition. Since, in Equation (2),

$$p_1^* = \pi^* - p_2^* = \pi^* - \gamma_2 x_2 P_2^* \tag{34}$$

as x_2 approaches 1.0 and γ_2 is approximately 1.0,

$$\lim_{x_1=0} \gamma_1 = \frac{\pi^* - x_2 P_2^*}{x_1 P_1^*} \tag{35}$$

Similarly,

$$\lim_{x_2=0} \gamma_2 = \frac{\pi^* - x_1 P_1^*}{x_2 P_2^*} \tag{36}$$

In working from isothermal total vapor-pressure data the observed values of π^* are substituted in Equation (35) with the corresponding mole fractions, and vapor pressures and apparent values of γ_1 are calculated. By plotting $\log \gamma_1$ against x_1 and extrapolating to $x_1 = 0$ the value of A is determined. Similarly, when working from isobaric boiling-point data, values of the vapor pressures of the pure components at the observed temperatures are substituted in Equation (35) with the corresponding total pressure and mole fractions. The calculated apparent values of γ_1 are extrapolated to $x_1 = 0$ as before. This latter procedure involves the assumption that the activity coefficients are independent of temperature over the range covered.

Illustration 2. In Table XXXII are data for the boiling points of ethanol–benzene solutions at 750 mm Hg and the vapor pressures of pure ethanol and benzene at these temperatures. Calculate the van Laar constants from these data, assuming the activity coefficients to be independent of temperature.

Solution: The values of apparent γ_1 and γ_2 shown in Table XXXII are calculated from Equations (35) and (36) and plotted in Fig. 150 against x_1. Extrapolating the γ_1 curve to $x_1 = 0$ gives the value of $\gamma_1 = 5.7$ or $A = \log \gamma_1 = 0.756$. Similarly, γ_2 is determined to be 4.06 by extrapolating the γ_2 curve to $x_1 = 1.0$ and $B = \log \gamma_2 = 0.608$. These constants are in fair agreement with those derived from the azeotrope composition.

TABLE XXXII

BOILING POINTS OF ETHANOL–BENZENE SOLUTIONS AT 750 MM HG

Mole Fraction Ethanol x_1	*Mole Fraction* Benzene x_2	*Boiling Point* t_B°C	*Vapor Pressure of Pure Components* Mm of Hg at t_B°C *Ethanol* P_1	*Benzene* P_2	*Apparent* γ_1	*Apparent* γ_2	*True* γ_1	*True* γ_2
0.0	1.0	79.7	804 mm	750 mm	—	1.00	5.70	1.00
0.04	0.96	75.2	671	648	4.74	1.16	4.20	1.00
0.11	0.89	70.8	560	562	4.07	1.38	3.66	1.05
0.28	0.72	68.3	507	518	2.66	1.63	2.11	1.16
0.43	0.57	67.8	497	509	2.15	1.84	1.58	1.34
0.61	0.39	68.3	507	518	1.77	2.18	1.22	1.84
0.80	0.20	70.1	545	549	1.47	2.86	1.05	2.64
0.89	0.11	72.4	598	592	1.285	3.33	1.02	3.14
0.94	0.06	74.4	650	632	1.165	3.64	1.00	3.58
1.00	0.0	78.1	750	711	1.00	—	1.00	4.06

Data from the International Critical Tables, III, pp. 217–21, 313.

Improved accuracy in the determination of van Laar constants from boiling-point or total-pressure data is obtained by first calculating the constants from Equations (34) and (35) as in Illustration 2. Values of γ_2 are then calculated from the van Laar equation with these approximate values of A and B substituted into Equation (34) instead of $\gamma_2 = 1.0$ being assumed as in the first approximation. Similarly, calculated values of γ_1 are substituted in Equation (35), and the extrapolation of the two equations to zero concentration is repeated to obtain second approximations of A and B. In the second approximation of Illustration 2 the same values of A and B were obtained. The corrected values of γ_1 and γ_2 are plotted in Fig. 150.

The preferred method for evaluating the constants of the van Laar equation depends on the nature of the system and the data available. Reliable liquid–vapor composition data in the dilute ranges are most desirable. A generally applicable method for obtaining such data by equilibrium condensation has been developed by Colburn, Schoenborn, and Shilling.[10] Where such data are not available the composition of

[10] A. P. Colburn, E. M. Schoenborn, and G. D. Shilling, *Ind. Eng. Chem.*, **35**, 1250 (1943).

the azeotrope generally furnishes the best basis if the azeotrope is in the middle of the composition range, between $x_1 = 0.25$ and $x_1 = 0.75$. If the azeotrope composition is outside this range, boiling-point or vapor-pressure data provide a better basis.

Once the activity coefficients of a system are established Equation (2) permits calculation of equilibrium compositions and vapor-pressure and

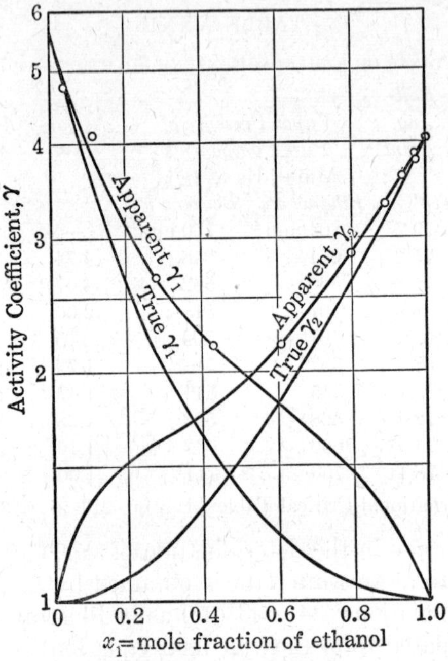

FIG. 150. Apparent and True Activity Coefficients of the Ethanol–Benzene System at 750 mm Hg.

boiling-point curves at low pressures where the ideal-gas law may be assumed. For a binary system of components 1 and 2,

$$\pi^* = p_1^* + p_2^* = y_1\pi^* + y_2\pi^* \tag{37}$$

Combining (2) and (37), since $x_2 = 1.0 - x_1$, gives

$$\pi^* = x_1 P_1^* \gamma_1 + (1 - x_1) P_2^* \gamma_2 \tag{38}$$

Combining (2) and (38) gives

$$y_1 = \frac{x_1 P_1^* \gamma_1}{x_1 P_1^* \gamma_1 + (1 - x_1) P_2^* \gamma_2} = \frac{\dfrac{P_1^* \gamma_1}{P_2^* \gamma_2}\left(\dfrac{x_1}{1 - x_1}\right)}{1 + \dfrac{P_1^* \gamma_1}{P_2^* \gamma_2}\left(\dfrac{x_1}{1 - x_1}\right)} \tag{39}$$

Equations (38) and (39) permit direct calculation of isothermal vapor-pressure and vapor–liquid composition curves. For deriving the more valuable isobaric boiling-point curves and the corresponding isobaric liquid–vapor composition curves it is necessary that P_1^* and P_2^* be known as functions of temperature. Equation (39) may then be solved by a trial and error method for the boiling points and vapor compositions corresponding to selected liquid compositions. As a first approximation the boiling point is assumed; this fixes a trial value of the *relative volatility* P_1^*/P_2^* which is substituted in Equation (39), and a first approximation y_1 is calculated. Then, from Equation (2), P_1^* is equal to $y_1\pi^*/x_1\gamma_1$, and a corrected boiling point is obtained as the temperature corresponding to this value of P_1^*. Since P_1^*/P_2^* varies but little with temperature, the value of y_1 calculated from Equation (39) with the corrected temperature being used is generally satisfactory.

Illustration 3. From the van Laar constants of Illustration 1, calculate the boiling point at 750 mm Hg of a solution containing 28 mole per cent ethanol in benzene and the equilibrium composition of the vapor evolved. The vapor pressures of the pure components may be estimated from the data of Illustration 2.

Solution: As a first approximation assume $t_B = 70°C$. Then,

$$\frac{P_1^*}{P_2^*} = \frac{543}{548} = 0.99$$

From Equations (30) and (31), the values of A and B from Illustration (1) give

$$\gamma_1 = 2.19; \qquad \gamma_2 = 1.18$$

By substitution in Equation (39), a first approximation of y_1 is obtained:

$$y_1 = \frac{(0.99)\left(\dfrac{2.19}{1.18}\right)\left(\dfrac{0.28}{0.72}\right)}{1 + (0.99)\left(\dfrac{2.19}{1.18}\right)\left(\dfrac{0.28}{0.72}\right)} = \frac{0.715}{1.715} = 0.416$$

Substitution of this value in Equation (2) gives

$$P_1^* = (y_1\pi^*)/x_1\gamma_1 = \frac{(0.416)(750)}{(0.28)(2.19)} = 509 \text{ mm Hg}$$

This vapor pressure corresponds to a temperature of 68.4°C, the corrected boiling point. At this temperature $P_2^* = 520$ mm: $P_1^*/P_2^* = \frac{509}{520} = 0.978$. Substitution in Equation (39) gives

$$y_1 = \frac{0.706}{1.706} = 0.413$$

In this manner complete boiling-point and liquid–vapor composition curves may be derived. The differences between the values calculated in this manner from the data of Table XXXII are attributable to the inadequacy of the van Laar equation and the uncertainty of experimental data of this type.

Partially Miscible Liquids. As the chemical dissimilarity of the components of a solution increases, the deviations from unity of the activity coefficients increase until finally miscibility is incomplete and two liquid phases result. This behavior is illustrated by the aqueous solutions of the alcohols. The ethanol–water system shows considerable deviation from ideal behavior with activity coefficients greater than 1.0. The water–*n*-propanol system has higher activity coefficients but complete miscibility at its normal boiling points. The water–butanol system has still higher activity coefficients and is only partially miscible at its normal boiling temperatures.

The properties of the water–butanol system at atmospheric pressure are shown in Figs. 147, 148, and 149. It may be noted that two liquid phases exist in the composition range from *a* to *b* corresponding to 66 to 98 mole per cent water. Between these limits the fugacity of each component is constant and equal in the two phases. As a result the boiling point is constant, and the vapor evolved is of constant composition *c*. The apparent activity coefficients in this range are inversely proportional to the mole fractions based on the combined phases. Although the vapor evolved from any liquid mixture of composition between *a* and *b* is of constant composition *c*, the liquid and vapor compositions are equal only at point *c*, which may be termed the *hetero-azeotropic* composition. However, in each of the individual phases the activity and composition are constant as indicated in Fig. 147. Since the fugacities in the two phases are equal at equilibrium, in the region of the two liquid phases,

$$a_1 f_1^\circ = a_1' f_1^{\circ\prime} \tag{40}$$

where the primed quantities represent one liquid phase and the unprimed the other. Since the standard state is the pure liquid component, $f_1^\circ = f_1^{\circ\prime}$ and

$$x_1\gamma_1 = x_1'\gamma_1' \quad \text{or} \quad \frac{x_1}{x_1'} = \frac{\gamma_1'}{\gamma_1} \tag{41}$$

The ratio x_1/x_1' is the distribution coefficient of component 1.

Carlson and Colburn[4] pointed out that Equation (41) permits evaluation of the constants of the van Laar equation for a binary system from liquid–liquid solubility data. By combining Equations (41) with (30) and (31), the following equations were developed which may be solved simultaneously for A and B:

$$\frac{A}{B} = \frac{\left(\dfrac{x_1}{x_2} + \dfrac{x_1'}{x_2'}\right)\left(\dfrac{\log x_1'/x_1}{\log x_2/x_2'}\right) - 2}{\dfrac{x_1}{x_2} + \dfrac{x_1'}{x_2'} - \dfrac{2x_1x_1' \log x_1'/x_1}{x_2x_2' \log x_2/x_2'}} \tag{42}$$

$$A = \frac{\log x_1'/x_1}{\dfrac{1}{\left(1 + \dfrac{Ax_1}{Bx_2}\right)^2} - \dfrac{1}{\left(1 + \dfrac{Ax_1'}{Bx_2'}\right)^2}} \tag{43}$$

It was found that the van Laar equation with constants derived from liquid-solubility data in this manner did not accurately represent activity coefficients in solutions far removed from the two-phase region. For example, for the water–n-butanol system the constant A based on liquid solubility was found to be 0.334, whereas that based on vapor-composition data in the dilute range is 0.61. Similarly, B is 1.60 from solubility and 1.34 from vapor compositions. Equations (42) and (43) are useful principally as a means of obtaining rough approximations for activity coefficients where no vapor-composition data are available.

As the chemical dissimilarity of the components of a system increases, the deviation of the activity coefficients from unity increases and the range of mutual solubility decreases. This effect corresponds to lengthening the horizontal portions between a and b on the curves of Figs. 147, 148, and 149. As complete immiscibility is approached, point a approaches zero, point b approaches 100 per cent composition, and the activity coefficients and van Laar constants become very large. To serve as a rough guide to the relative magnitudes of the activity coefficients corresponding to various degrees of miscibility Colburn[11] prepared Table XXXIII from Equations (42) and (43) for symmetrical systems in which the mutual solubilities of the two components are equal.

TABLE XXXIII

ACTIVITY COEFFICIENTS AT VARIOUS MUTUAL SOLUBILITIES IN SYMMETRICAL SYSTEMS

Solubility Limits x_s *Mole Fraction*	*Van Laar Constants A and B*	γ at $x = 0$	γ at $x = x_s$
0.5	0.875	7.5	1.68
0.2	1.0	10.0	5.0
0.1	1.2	15.8	9.8
0.05	1.43	27	20
0.02	1.77	59	50
0.01	2.03	106	94

Ternary Systems. Few reliable data are available from which activity coefficients in ternary systems can be calculated. The general problem of the correlation of activity coefficients in such systems has been reviewed by Wohl[2] who has developed a series of equations analogous to those for binary systems which have already been discussed. The accuracy of any of these equations depends to a large extent on the number of experimentally determined constants involved. The most useful form requiring minimum data is probably the ternary van Laar equation which involves only six constants, all of which can be evaluated from data on the three individual binary systems. The ternary van Laar

[11] A. P. Colburn, private communication (1942).

equation for a system of components 1, 2, and 3 was developed by Wohl by applying the van Laar type of relationship that $q_1/q_2 = A_{1-2}/A_{2-1}$ to a two-suffix form of Equations (13–15). Thus,

$$\log \gamma_1 = \frac{x_2^2 A_{1-2}\left(\dfrac{A_{2-1}}{A_{1-2}}\right)^2 + x_3^2 A_{1-3}\left(\dfrac{A_{3-1}}{A_{1-3}}\right)^2}{\left(x_1 + x_2\dfrac{A_{2-1}}{A_{1-2}} + x_3\dfrac{A_{3-1}}{A_{1-3}}\right)^2} + \frac{x_2 x_3 \dfrac{A_{2-1}A_{3-1}}{A_{1-2}A_{1-3}}\left(A_{1-2} + A_{1-3} - A_{3-2}\dfrac{A_{1-3}}{A_{3-1}}\right)}{\left(x_1 + x_2\dfrac{A_{2-1}}{A_{1-2}} + x_3\dfrac{A_{3-1}}{A_{1-3}}\right)^2} \tag{44}$$

In Equation (44) the constants A_{1-2} and A_{2-1} correspond to the van Laar constants A and B of the binary system of components 1 and 2. Similar equations for $\log \gamma_2$ and $\log \gamma_3$ are obtained by changing the subscripts on all terms in Equation (44) in accordance with the following schedule of rotation:

Subscript in Equation (44) for	*Subscripts in Equation for*	
$\log \gamma_1$	$\log \gamma_2$	$\log \gamma_3$
1	2	3
2	3	1
3	1	2

Thus, in the equation for $\log \gamma_2$, x_1 of Equation (44) becomes x_2; x_2 becomes x_3; x_3 becomes x_1; and A_{1-2} becomes A_{2-3} etc. Similarly, in the equation for $\log \gamma_3$, x_3 becomes x_2, and A_{1-2} becomes A_{3-1}.

By means of Equation (44) complete activity coefficient relationships may be estimated from the coefficients of the three binary systems which in turn may be approximated from the binary liquid-solubility data by Equations (42) and (43). Colburn[11] tested this general method on the methyl cyclohexane–n-heptane–aniline system and found good agreement with ternary solubility data. However, it is impossible to predict the errors which may be encountered in other systems.

The most generalized equation for ternary systems recommended by Wohl is the four-suffix form and involves 14 constants, 11 of which may be evaluated from data on the three binary systems.

Knowledge of the activity coefficients in a ternary system permits calculation of the distribution of any one of the components between the two liquid phases in a region of partial miscibility by means of Equation (41) which is valid for any component regardless of the number of others present. These relationships are of fundamental importance in problems of solvent extraction and extractive or azeotropic distillation.

Effect of Temperature on Activity Coefficients. An equation for the variation of the activity coefficient with temperature at constant composition results from combination of Equation (XIV-45) and (1):

$$\left(\frac{\partial \ln \gamma_1}{\partial T}\right)_p = \left(\frac{\partial \ln f_1}{\partial T}\right)_p - \left(\frac{\partial \ln f_1^\circ}{\partial T}\right)_p$$
$$= \frac{(\text{H}_1^* - \overline{\text{H}}_1) - (\text{H}_1^* - \text{H}_1^\circ)}{RT^2} = -\frac{(\overline{\text{H}}_1 - \text{H}_1^\circ)}{RT^2} \tag{45}$$

The quantity $\overline{\text{H}}_1 - \text{H}_1^\circ$ is the partial molal enthalpy relative to the pure components at the temperature of the solution or the differential heat of solution. If $\overline{\text{H}}_1 - \text{H}_1^\circ$ is positive, γ is reduced by an increase in temperature. Carlson and Colburn[4] observed that in general if the activity coefficients of a system are greater than 1.0 the heat of solution is negative, whereas if the coefficients are less than 1.0 the heat of solution is positive. Thus, in either case increasing the temperature causes the activity coefficients to approach 1.0. Conversely, lowering the temperature tends to increase deviation from ideal behavior, and many systems miscible at high temperatures separate into two phases at low temperatures.

Use of Equation (45) for calculating the effects of temperature is complicated by the fact that values of $(\overline{\text{H}} - \text{H}_1^\circ)$ vary considerably with temperature and are rarely known over any extensive range.

An approximate method for estimating the effect of temperature on activity coefficients is obtained by extension of a method suggested by Carlson and Colburn for estimating the effect of temperature on azeotrope composition. It is assumed that the ratio of the activity coefficients γ_1/γ_2 is independent of temperature at a given composition. It follows from Equation (2) that at the azeotropic composition where $x_1 = y_1$,

$$\left(\frac{\gamma_1}{\gamma_2}\right)_z = \frac{P_2^*}{P_1^*} \tag{46}$$

The variation of the azeotropic composition with pressure or temperature is estimated from Equation (46) by plotting γ_1/γ_2 against x_1 from the data at the known temperature. On the same scale the ratio of the vapor pressures P_2^*/P_1^* is plotted against temperature. The azeotropic composition at a selected temperature is then the composition at which γ_1/γ_2 is equal to the value of P_2^*/P_1^* at this temperature. This procedure is demonstrated in Fig. 151 developed by Carlson and Colburn[4] for the ethyl acetate–ethyl alcohol system. The γ_1/γ_2 curve was derived by evaluating the van Laar equation from the composition of the azeotrope at atmospheric pressure and 71.8°C. To estimate the composition of the azeotrope at 91.4°C the broken lines on the diagram are followed from this temperature vertically to the P_2^*/P_1^* curve, then horizontally to the γ_1/γ_2 curve, and vertically to the composition axis giving a value of

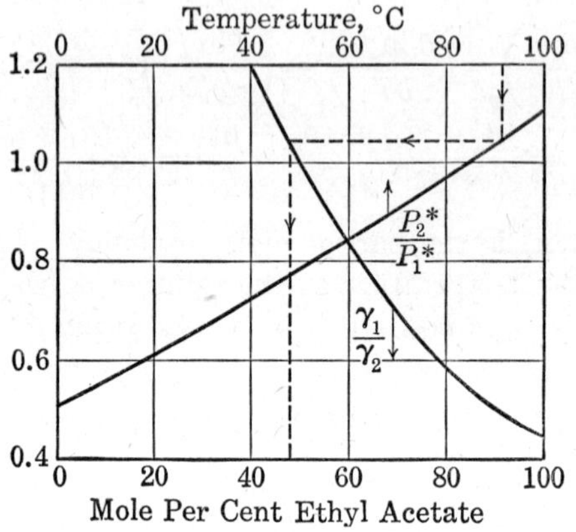

FIG. 151. Effect of Temperature on the Composition of the Ethyl Acetate–Ethyl Alcohol Azeotrope.

48 mole per cent ethyl acetate. In Table XXXIV are values estimated[4] in this manner compared with the experimental measurements of Merriman.[12]

The total pressures P_z corresponding to the various azeotrope compositions and temperatures may be approximated by the following empirical equation,

$$P_z = \frac{P_z'(x_1P_1 + x_2P_2)}{(x_1'P_1' + x_2'P_2')} \tag{47}$$

where

P_z, P_1, P_2 = vapor pressures of the azeotrope, component 1, and component 2 at temperature t

P_z', P_1', P_2' = vapor pressures of the azeotrope, component 1, and component 2 at the reference temperature t'

x_1, x_2 = mole fractions of components 1 and 2 in the azeotrope at temperature t

x_1', x_2' = mole fractions of components 1 and 2 in the azeotrope at temperature t'

From the azeotrope compositions and pressures calculated in this manner corresponding constants for the van Laar equation may be derived by means of Equations (32) and (33). In this manner complete

[12] R. W. Merriman, *J. Chem. Soc.*, **103**, 1801–16 (1913).

activity coefficient and vapor–liquid composition curves may be estimated over limited ranges of temperatures or pressures. This method has not been sufficiently explored to permit estimation of the errors which may be encountered.

TABLE XXXIV

EFFECT OF PRESSURE ON THE ETHYL ACETATE–ETHYL ALCOHOL AZEOTROPE[2]

Temperature, °C	P_2^*/P_1^*	*Mole Fraction Ethyl Acetate Calculated*	*Mole Fraction Ethyl Acetate Measured*	*Total Pressure* mm Hg
18.7	0.598	0.787	0.734	77.4
40.5	0.725	0.677	0.660	220
56.3	0.827	0.602	0.601	423
71.8	0.921	0.539	0.539	760
83.1	0.994	0.498	0.490	1121
91.4	1.049	0.480	0.451	1476

HIGH-PRESSURE VAPOR-LIQUID EQUILIBRIUM

It has been pointed out that in any heterogeneous system it is a general criterion of equilibrium that the chemical potential μ_1 of any component must be equal in every phase in which it is present. For systems restricted to constant temperature this requirement of equality of chemical potential may be replaced by the requirement of equality of the more convenient function, fugacity. Since fugacity is equal to the product of activity times the fugacity of the standard state, the criterion of equilibrium at constant temperature is expressed by the following equation.

$$(a_1)_L(f_1^\circ)_L = (a_1)_{L'}(f_1^\circ)_{L'} = (a_1)_v(f_1^\circ)_v = (a_1)_s(f_1^\circ)_s = \cdots \quad (48)$$

where

$(a_1)_L$, $(a_1)_{L'}$, etc. = activities of component 1 in the phases L, L', v, $s \cdots$

$(f_1^\circ)_L$, $(f_1^\circ)_{L'}$, etc. = standard states for activity of component 1 in the phases L, L', v, $s \cdots$

For a system comprising a single liquid phase L and a vapor phase v Equation (48) may be written

$$\frac{(a_1)_v}{(a_1)_L} = \frac{(f_1^\circ)_L}{(f_1^\circ)_v} = K_1 \quad (49)$$

where K_1 is termed the *vaporization equilibrium* constant of component 1.

It is evident that the value of the equilibrium constant is determined by the definitions of the standard states for both phases. It is convenient to choose as the standard state for each component in the vapor phase the pure component in the gaseous state at the temperature and pressure of the system. Similarly, the standard state for each component in the liquid phase is chosen as the pure component in the liquid state at the temperature and pressure of the system. These definitions of standard states have the advantage that where the liquid and vapor approximate ideal solutions activities are equal to mole fractions. It follows from these definitions of the standard states that K_1 is a function of both the temperature and pressure of the system as well as of the nature of the component.

The vaporization equilibrium constant as defined by Equation (49) provides a thermodynamically rigorous relationship between the activities of a component in vapor and liquid phases which are in equilibrium. However, before this relationship is useful it is necessary to relate activities to mole fractions through the introduction of activity coefficients. Thus, in the vapor phase,

$$(a_1)_v = \frac{(f_1)_v}{(f_1^\circ)_v} = \phi_{v1} y_1 \tag{50}$$

where

$(a_1)_v$ = activity of component 1 in the vapor phase
y_1 = mole fraction of component 1 in the vapor phase
ϕ_{v1} = activity coefficient of component 1 in the vapor phase

The activity coefficient ϕ_v is dependent on the properties of both component 1 and the other components with which it is mixed and also upon the temperature and pressure. Under conditions of moderate pressures and elevated temperatures gaseous solutions tend to be ideal if the components are not too widely different in properties. In such cases ϕ_v may be taken as 1.0. Maximum deviations from 1.0 occur in the region of the critical point of the solution.

In the liquid phase the relationship between activity and mole fraction is complicated by the two types of deviation from the laws of ideal solutions which are discussed on page 609. Because of the widely different contributing causes and relationships it is convenient to consider these two types of deviation separately and assign an activity coefficient to express each. Thus,

$$(a_1)_L = \phi_{L1} \gamma_1 x_1 \tag{51}$$

where

$(a_1)_L$ = activity of component 1 in the liquid phase
ϕ_{L1} = activity coefficient of component 1 in the liquid phase which takes into account deviations from the laws of ideal solutions resulting from differences in molecular size or volatility
γ_1 = activity coefficient of component 1 in the liquid state which takes into account deviations from the laws of ideal solutions resulting from differences in molecular or chemical type
x_1 = mole fraction of component 1 in the liquid phase

For ideal systems $\gamma_1 = 1.0$ at all conditions. For nonideal systems γ_1 may be determined by the methods of the preceding sections which are applicable at low pressures where the effects of differences in molecular size or volatility are small and ϕ_{L_1} may be taken as 1.0. As the temperature and pressure in such a system are increased γ_1 tends to approach 1.0 while deviations of ϕ_{L_1} from 1.0 tend to increase.

Combining Equations (49), (50), and (51) gives

$$\frac{y_1}{x_1} = \frac{\phi_{L_1}\gamma_1}{\phi_{v_1}} K_1 = K_1' \tag{52}$$

It was suggested by Gamson and Watson[13] that the ratio of the mole fractions in the vapor and liquid phases be termed the *vaporization ratio* K'_1 of component 1. It is evident from Equation (52) that the vaporization ratio is equal to the thermodynamic equilibrium constant only in ideal solutions where the activity coefficients are each equal to 1.0.

Equation (52) provides a sound basis for calculations involving vapor–liquid equilibria. In many cases it is satisfactory to assume that the coefficients ϕ_{L_1} and ϕ_{v_1} are unity, and, as previously pointed out, $\gamma_1 = 1.0$ for all ideal systems. Under other conditions it is necessary to evaluate the coefficients as functions of the properties of the component and the system, either from experimental data or generalized methods of calculation.

At low temperatures and pressures where Dalton's law and the ideal-gas law are applicable ϕ_L and ϕ_v become unity and Equation (52) reduces to

$$K_1' = \frac{y_1^*}{x_1^*} = \gamma_1 K_1 = \frac{\gamma_1 P_1^*}{\pi^*} \tag{53}$$

where P_1^* is the vapor pressure of component 1 at the temperature of the system.

Vaporization Equilibrium Constants. The definition of the standard states of unit activity which determine the value of the vaporization equilibrium constant is complicated by the fact that for a heavy component of a mixture the standard state for the vapor phase is a hypothetical state in which the pure component cannot actually exist as a gas. Similarly, for a light component the standard state for the liquid phase is a hypothetical state in which the pure component cannot actually exist as a liquid. The fugacities of the standard states under these conditions must be arrived at by arbitrary extrapolation of the fugacities

[13] B. W. Gamson and K. M. Watson, *Nat. Petroleum News*, Tech. Sec., **36**, *R*623 (Sept. 6, 1944). Also "Process Engineering Data," National Petroleum Publishing Company, Cleveland (1944).

of the pure components under real conditions. However, these arbitrary extrapolations in no way interfere with the fundamental validity of Equations (49) and (52). If the method of extrapolation is changed, the values of K_1 are affected, but corresponding changes result in the activity coefficients so that it is only necessary that the values of ϕ_{v1}, ϕ_{L1}, γ_1, and K_1 all be referred to consistently defined standard states.

The concept of the vaporization equilibrium constant was introduced by Souders, Selheimer and Brown[14] who expressed it by the following equation resulting from combination of Equations (XIV-34) and (49) with the definition of the fugacity coefficient:

$$K_1 = \frac{(f_1^\circ)_L}{(f_1^\circ)_v} = \frac{P_1 \nu_{P_1} e^{\frac{v_m(\pi - P_1)}{RT}}}{\pi \nu_{\pi_1}} \tag{54}$$

P_1 = normal vapor pressure of component 1 at the temperature T of the system

ν_{P_1} = fugacity coefficient of the vapor of component 1 at pressure P_1 and the temperature of the system

π = total pressure of the system

v_m = mean molal volume of liquid component 1 between P_1 and π at the temperature of the system

ν_{π_1} = fugacity coefficient of component 1 in the vapor state at the temperature and pressure of the system

In order to evaluate ν_{π_1} extrapolation in Fig. 142 is required where π is greater than P_1.

As previously mentioned this extrapolation may be arbitrary, but it is desirable that it be so carried out that activities in the vapor phase are equal to mole fractions over as wide a range as possible. In the higher-pressure ranges where deviations from the laws of the ideal solution become large it is desirable that the extrapolation be such as to result in simple expressions for the activity coefficients. As a basis for such an extrapolation Gamson and Watson[13] replotted the data of Fig. 142 in the form shown in Fig. 152 with $f/p_c = \nu p_r$ as ordinates. The lines corresponding to low reduced temperatures were extrapolated on the basis of the available experimental data on the fugacities of hydrocarbons in vapor phase mixtures in the region where the vapors form approximately ideal solutions. At higher pressures the extrapolations were extended to approach the horizontal. Actual fugacities of components under these conditions tend to reach a maximum with increase in pressure and then diminish as the pressure is further increased. However, for any compo-

[14] M. Souders, Jr., C. W. Selheimer, and G. G. Brown, *Ind. Eng. Chem.*, **24**, 517 (1932).

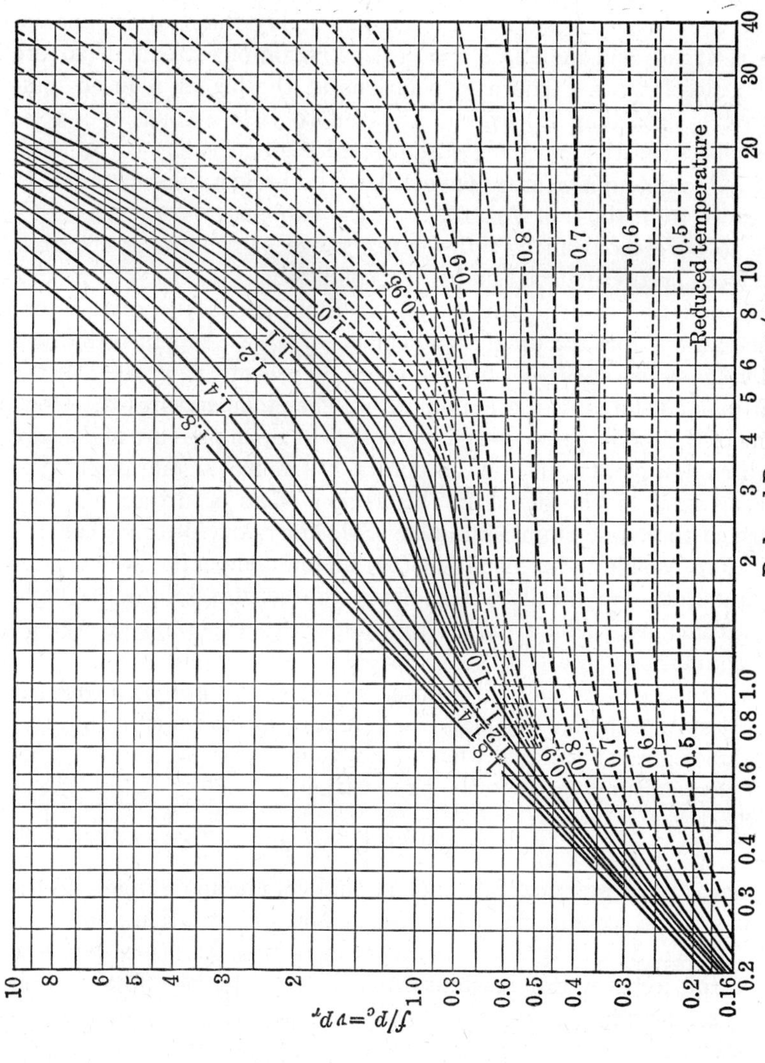

FIG. 152. Generalized Fugacities of Gases (low-temperature–high-pressure range).

(Reproduced in "C.P.P. Charts")

nent at a given temperature the pressure at which this maximum is reached and the variation with pressures beyond that of the maximum fugacity are entirely dependent on the composition of the mixture in which the component is present. The horizontal extrapolations of Fig. 152 are believed desirable in order to simplify the relationships between activities and mole fractions in this region of extreme departure from ideal solutions. The extrapolations at the higher reduced temperatures approaching 1.0 were carried out to yield a continuous relationship between the extrapolated lines and the real values corresponding to reduced temperatures above 1.0. The broken lines on Fig. 142 are derived from those of Fig. 152, and the two charts may be used together to cover conveniently the entire range of conditions and thus define $(f_1^\circ)_v$ for any material under any conditions.

The evaluation or definition of $(f_1^\circ)_L$ in Equation (54) is complicated by uncertainty as to the proper value of v_m. At low temperatures where molal volumes are small and little affected by pressure v_m is taken as the actual molal volume of the liquid. At higher temperatures the molal volume of the liquid becomes highly dependent upon pressure and above the critical temperature of the component has no real significance. The vapor pressure of the liquid in this region offers a further problem. Because of these uncertainties Souders, Selheimer, and Brown calculated equilibrium constants by Equation (54) at low temperatures and pressures and then graphically extrapolated the constants into the high-temperature and pressure range to conform to experimental vapor–liquid equilibrium data.

In order to be more specific, Gamson and Watson[13] proposed that the standard state of a component in the liquid phase be taken as a hypothetical incompressible liquid whose molal volume is expressed as a function of temperature by the following equation:

$$v_m = (v_1\omega_1)(5.7 + 3.0T_r) \tag{55}$$

where $(v_1\omega_1)$ is the product of the molal liquid volume and ω from Fig. 109 at any selected conditions. By comparison of Equation (55) with (XII-39), page 502, it is seen that the term $(5.7 + 3.0T_r)$ corresponds to $1/\omega$ for the hypothetical incompressible liquid reference state. This value is a linear extrapolation of $1/\omega$ from the absolute zero and thus defines the liquid standard state as having a constant volumetric coefficient of expansion equal to that of the real liquid at the absolute zero. At low reduced temperatures values of v_m calculated from Equation (55) differ little from actual molal volumes. As the critical point is approached v_m is smaller than the actual molal volume except at high pressures.

The definition of $(f_1^\circ)_L$ at all conditions is completed by means of the vapor-pressure Equation (III-16), page 73, which extrapolates logically above the critical point. Thus, from (55) and the numerator of (54),

$$\log (f_1^\circ)_L = \log P_1 + \log \nu_{P_1} + \frac{(v_1\omega_1)(5.7 + 3.0T_r)(\pi - P_1)}{2.303RT} \tag{56}$$

where

$$\log P_1 = -\frac{A}{T_r} + B - e^{-20(T_r - b)^2} \tag{57}$$

Equations (54), (56), and (57) together with Figs. 142 and 152 completely define the fugacities of the liquid and vapor reference states and the equilibrium constant in an arbitrary but logical manner which is definite, reproducible, and applicable to any material at any conditions.

Illustration 4. Calculate the vaporization equilibrium constant of methyl chloride at 200°F and an absolute pressure of 600 lb per sq in. The density of liquid methyl chloride is 0.920 g per cc at 18°C, the critical temperature is 750°R, the critical pressure 967 lb per sq in., and the molecular weight 50.5.

Solution: At 200° and 600 lb per sq in., $T_r = 660/750 = 0.880$; $p_r = 600/967 = 0.620$

At 18°C (64.4°F), $T_r = 524.4/750 = 0.699$

From Fig. 142, $\nu_\pi = 0.68$; $(f^\circ)_v = (0.68)(600) = 408$ lb per sq in.

From Fig. 109 at 18°C, $\omega_1 = 0.1146$

$(v_1\omega_1) = (50.5)(0.1146)/(0.920) = 6.290$ cc per g-mole = 0.1007 cu ft per lb-mole

From Table IIa, page 73:

$$\log P_1 = -\frac{2.7195}{0.880} + 7.4185 - e^{-20(0.880-0.052)^2}$$

$$= -3.0903 + 7.4185 - 10^{-5.95} = 4.3282$$

$$P_1 = 21{,}290 \text{ mm or } 21{,}290/51.70 = 412 \text{ lb per sq in.}$$

At 412 lb per sq in., $p_r = 412/967 = 0.426$

From Fig. 142, $\nu_P = 0.77$

Substitution in Equation (56) gives

$$\log (f^\circ)_L = \log 412 + \log 0.77 + \frac{(0.1007)[5.7 + (3.0)(0.880)][600 - 412]}{(2.303)(10.71)(660)}$$

$$= 2.6149 - 0.1135 + 0.0097 = 2.5111$$

$$(f^\circ)_L = 324.4 \text{ lb per sq in.}$$

From Equation (54),

$$K = 324.4/408 = 0.795$$

In this manner charts may be developed which express the vaporization equilibrium constant of a specific compound as a function of temperature and pressure. Series of such charts have been developed by several investigators for many of the lower-boiling hydrocarbons. An alternate method of graphical presentation which is convenient when working with mixtures of homologs of a single series is to restrict each

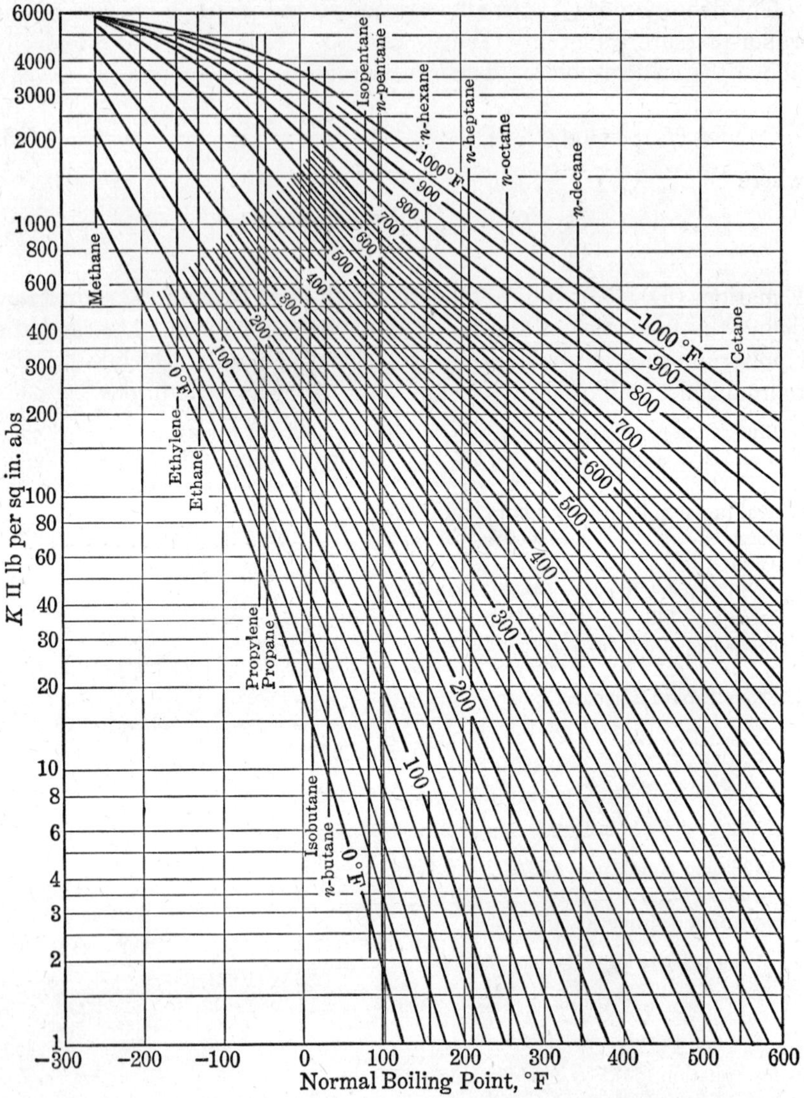

FIG. 153. Vaporization Equilibrium Constants of Hydrocarbons. (Absolute Pressure Π = 100 lb per square inch.)
(Reproduced in "C.P.P. Charts")

chart to a constant pressure and plot curves each of which expresses the equilibrium constant of a particular compound as a function of temperature at the pressure of the chart. With this scheme a series of charts each corresponding to a different pressure is required.

Another method of presenting equilibrium-constant data for an ho-

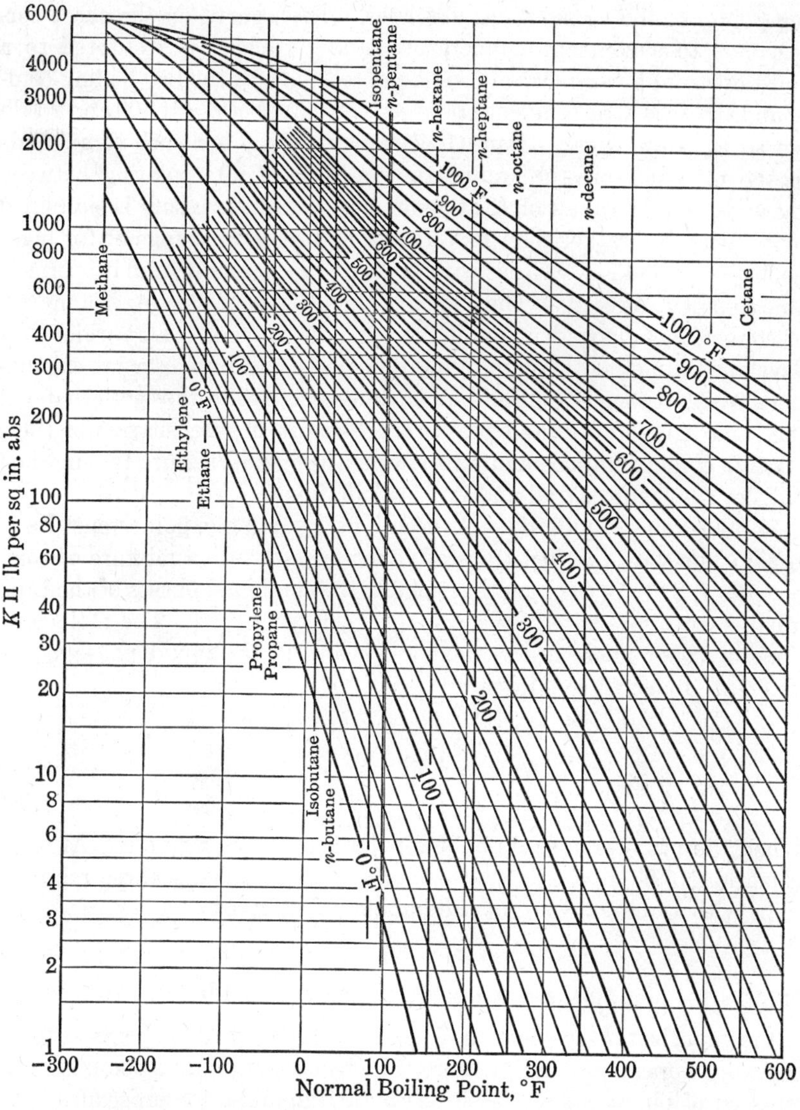

FIG. 154. Vaporization Equilibrium Constants of Hydrocarbons. (Absolute Pressure Π = 200 lb per square inch.)
(Reproduced in "C.P.P. Charts")

mologous series is shown in Figs. 153 and 154 for the paraffin hydrocarbons. These charts were calculated by Equations (54) and (56) from the data of Table XXVIII. Although based on the data for paraffins they may be used for olefins with little error. The products of $K\pi$ in pounds per square inch are plotted as ordinates with the normal boil-

507

ing points of the homologs as abscissas. Each curve on the chart corresponds to a constant temperature, and the chart is restricted to a constant total pressure π. It will be noted that for the low-boiling compounds there is little difference between the values of $K\pi$ at 100 and 200 lb per sq in., from Figs. 153 and 154, respectively. For such compounds in the moderate-pressure range the problem of interpolating between charts to obtain values of K at some intermediate pressure is simplified by plotting $K\pi$ rather than K. In the high-pressure range or for high-boiling compounds $K\pi$ varies with pressure more than K, and it is preferable to plot K directly. The choice of the method of plotting is a matter of convenience dependent upon the type of problem. Kirkbride[15] has developed a method of simplifying dew-point and bubble-point calculations by plotting all equilibrium constants relative to those of ethane at the same temperature and pressure. Because of insensitivity of these relative equilibrium constants to temperature and pressure this method reduces the labor of trial-and-error calculations.

Bubble-Point Equilibria. A liquid at its bubble point must be in equilibrium with the first trace of vapor formed. In a mixture of components 1, 2, 3 $\cdots$ the mole fractions in the combined phases of the total mixture are represented by $z_1, z_2, z_3, \cdots$, respectively. The mole fractions in vapor phase are $y_1, y_2, y_3, \cdots$, and in the liquid phase $x_1, x_2, x_3, \cdots$. Under all conditions

$$z_1 + z_2 + z_3 + \cdots = 1.0 \tag{58}$$

$$y_1 + y_2 + y_3 + \cdots = 1.0 \tag{59}$$

$$x_1 + x_2 + x_3 + \cdots = 1.0 \tag{60}$$

When the mixture is at its bubble point $x_1 = z_1$, $x_2 = z_2$, etc. Also, by Equation (52), $y_1 = K'_1 x_1$, $y_2 = K'_2 x_2$, etc. Combining these relationships with Equation (59), at the bubble point, gives

$$K'_1 z_1 + K'_2 z_2 + K'_3 z_3 + \cdots = 1.0 \tag{61}$$

or

$$\Sigma K'z = 1.0 \quad \text{and} \quad \Sigma(K'\pi z)/\pi = 1.0$$

If the vaporization coefficients K' are known for all components as functions of temperature and pressure, Equation (61) permits complete calculation of bubble-point conditions by a trial-and-error procedure.

If it is desired to calculate the bubble-point pressure of a mixture at a specified temperature, a pressure is assumed as a first approximation. Corresponding values of $K'_1, K'_2, K'_3, \cdots$ or $K'_1\pi, K'_2\pi, K'_3\pi, \cdots$ are determined, and the left-hand side of Equation (61) is evaluated. If the summation equals 1.0 the assumed pressure is correct. If not, other values are assumed, and the corresponding summations are plotted as a

[15] C. G. Kirkbride, *Petroleum Refiner*, **24**, 99 (1945).

function of pressure. The pressure at which this curve crosses 1.0 is the correct solution. The vapor composition is then calculated from Equation (52), the equilibrium constants being used at the correct pressure. A similar procedure is followed for calculating the bubble-point temperature at a specified pressure.

At low pressures Equations (53) and (61) may be combined to permit direct calculation of bubble-point pressures.

$$\gamma_1 P_1^* z_1 + \gamma_2 P_2^* z_2 + \gamma_3 P_3^* z_3 + \cdots = \pi^* \tag{62}$$

Dew-Point Equilibria. A vapor at its dew point is in equilibrium with the incipient liquid having a composition $x_1, x_2, x_3 \cdots$. Then, since $y_1 = z_1$ and $x_1 = y_1/K_1'$, Equation (60) becomes

$$\frac{z_1}{K_1'} + \frac{z_2}{K_2'} + \frac{z_3}{K_3'} + \cdots = 1.0 \tag{63}$$

or

$$\Sigma\left(\frac{z}{K'}\right) = 1.0 \quad \text{and} \quad \pi\Sigma\left(\frac{z}{K'\pi}\right) = 1.0$$

Dew-point conditions may be evaluated by a trial-and-error solution of Equation (63) by the same procedure described for bubble points. At low pressures Equations (53) and (63) may be combined to permit direct solution for dew-point pressures.

$$\pi^* = \frac{1}{\dfrac{z_1}{\gamma_1 P_1^*} + \dfrac{z_2}{\gamma_2 P_2^*} + \dfrac{z_3}{\gamma_3 P_3^*}} \tag{64}$$

Partial Vaporization. If a solution comprising F moles is partially vaporized to form L moles of liquid and V moles of vapor, all components will be in equilibrium concentrations in the two phases, as required by Equation (52). From an over-all material balance, following the method of Katz and Brown,[16] there results

$$V + L = F \tag{65}$$

Similarly, for any component,

$$V y_1 + L x_1 = F z_1 \tag{66}$$

where z_1 = mole fraction of component 1 in the F moles of total mixture. Combining (52) and (66) and solving for x_1 yields

$$x_1 = \frac{F z_1}{V K_1' + L} = \frac{F}{V}\left(\frac{z_1}{K_1' + \dfrac{L}{V}}\right) \tag{67}$$

[16] D. L. Katz and G. G. Brown, *Ind. Eng. Chem.*, **25**, 1373 (1933).

Since

$$x_1 + x_2 + x_3 + \cdots = \Sigma x = 1.0$$

$$\Sigma \frac{z}{K' + \dfrac{L}{V}} = \frac{V}{F} \tag{68}$$

Similarly, solving (66) for y_1 gives

$$y_1 = \frac{Fz_1}{V + \dfrac{L}{K_1'}} = \frac{F}{V}\left(\frac{z_1}{1 + \dfrac{L}{VK_1'}}\right) \tag{69}$$

and

$$\Sigma \frac{z}{1 + \dfrac{L}{VK'}} = \frac{V}{F} \tag{70}$$

Either Equation (68) or Equation (70) may be used in a trial-and-error solution to establish the relationship among temperature, pressure, and percentage vaporization. Values of the quantity sought are assumed, and the corresponding summations are evaluated. The correct value is that producing a summation equal to V/F. The relationship between V/F and L/V is obtained by rearranging Equation (65):

$$\frac{L}{V} = \frac{F}{V} - 1.0 \tag{71}$$

If Equation (70) is used in the summation, values of vapor compositions are obtained directly.

$$y_1 = \frac{z_1 \Big/ \left(1 + \dfrac{L}{VK_1'}\right)}{\Sigma\left[z \Big/ \left(1 + \dfrac{L}{VK'}\right)\right]} \tag{72}$$

Equation (72) may be used with little error even if the summation does not exactly equal V/F. Corresponding values of x_1 are then obtained from Equation (66).

$$x_1 = \frac{Fz_1 - Vy_1}{L} = \left(1 + \frac{V}{L}\right)z_1 - \frac{V}{L}y_1 \tag{73}$$

If Equation (68) is used for the summation, values of x are obtained directly by an expression similar to Equation (72). It is evident that in a partially vaporized mixture at equilibrium the vapor phase exists at its dew point while the liquid phase exists at its bubble point.

Illustration 5. A hydrocarbon mixture contains 25 mole per cent propane, 40 per cent n-butane, and 35 per cent n-pentane. Calculate the bubble-point temperature, the dew-point temperature, and the temperature of 45 mole per cent vaporization at an absolute pressure of 200 lb per sq in. Also, calculate the compositions of the vapor formed at the bubble point, of the liquid formed at the dew point, and of the liquid and vapor resulting from 45 mole per cent vaporization. It may be assumed that ideal solutions are formed.

Solution: (*a*) *Bubble-Point Temperature and Vapor Composition.* Values of K, which are assumed equal to K', are obtained from Fig. 154 for substitution in Equation (61).

Assumed Temperature,		180°F			200°F			220°F			
Component	z	$K'\pi$	K'	$K'z$	$K'\pi$	K'	$K'z$	$K'\pi$	K'	$K'z$	y
C_3H_8	0.25	374	1.82	0.455	420	2.10	0.525	462	2.31	0.578	0.496
C_4H_{10}	0.40	164	0.82	0.328	196	0.98	0.392	226	1.13	0.452	0.359
C_5H_{12}	0.35	74	0.37	0.130	92	0.46	0.161	110	0.55	0.193	0.145
Σ	1.00			0.926			1.078			1.223	1.000

By graphical interpolation the temperature where $\Sigma K'z = 1.0$ is found to be 190°F. For a rigorous calculation of vapor compositions y the equilibrium constants should be evaluated at this temperature. A good approximation is obtained by calculating $K_1'z_1/\Sigma K'z$ at 180° and 200° and interpolating linearly.

(*b*) *Dew-point Temperature and Liquid Composition.* Substitution in Equation (63), gives

Assumed Temperature,		220°F			240°F			260°F			
Component	z	$K'\pi$	K'	z/K'	$K'\pi$	K'	z/K'	$K'\pi$	K'	z/K'	x
C_3H_8	0.25	462	2.31	0.108	510	2.50	0.100	546	2.72	0.092	0.104
C_4H_{10}	0.40	226	1.13	0.354	254	1.27	0.315	290	1.45	0.276	0.329
C_5H_{12}	0.35	110	0.55	0.637	132	0.66	0.531	156	0.78	0.449	0.567
Σ	1.00			1.099			0.946			0.817	1.000

Graphical interpolation fixes the temperature at which $\Sigma K/z = 1.0$ at 233°F. A good approximation to the liquid composition is obtained by calculating $(z_1/K_1')/\Sigma z/K'$ at 230° and 240° and interpolating.

(*c*) *Temperature and Composition of 45 Per Cent Vaporization.*

$$V/F = 0.45$$

From Equation (71), $L/V = 1/0.45 - 1.0 = 1.22$

or $V/L = 0.82$

Equation (70) is used for the trial summations.

Assumed Temperature,		200°F			220°F			240°F		
Component	z	K'	$1+L/VK'$	$\frac{z}{1+L/VK'}$	K'	$1+L/VK'$	$\frac{z}{1+L/VK'}$	K'	$1+L/VK'$	$\frac{z}{1+L/VK'}$
C_3H_8	0.25	2.10	1.581	0.158	2.31	1.528	0.164	2.50	1.488	0.168
C_4H_{10}	0.40	0.98	2.245	0.178	1.13	2.080	0.192	1.27	1.961	0.204
C_5H_{12}	0.35	0.46	3.655	0.096	0.55	3.220	0.109	0.66	2.850	0.123
Σ	1.00			0.432			0.465			0.495

By graphical interpolation the temperature at which $\Sigma z/(1 + L/VK') = 0.45$ is determined as 211°F. Satisfactory values of y are obtained by interpolating between values calculated from Equation (72) at 200 and 220°F. Corresponding values of x are calculated from Equation (73).

Component	z	y	$1.82z$	$0.82y$	x
C_3H_8	0.25	0.358	0.455	0.294	0.161
C_4H_{10}	0.40	0.413	0.728	0.339	0.389
C_5H_{12}	0.35	0.229	0.637	0.187	0.450
Σ	1.00	1.000			1.000

Activity Coefficients. In Illustration 5 it is assumed that all activity coefficients are equal to 1.0 and that the vaporization coefficients are equal to the equilibrium constants. This assumption is satisfactory for many calculations for ideal systems at moderate pressures if the components are not widely different in properties. In working with nonideal systems under similar conditions it is satisfactory to assume that ϕ_L and ϕ_v are unity. Values of γ may be derived by the methods described in the preceding sections. At elevated pressures where the ideal-gas law is not valid, the following relation is used in preference to Equation (2).

$$\gamma_1 = \frac{y_1}{x_1 K_1} \tag{74}$$

At high pressures, approaching the critical of the mixture, the assumption that ϕ_L and ϕ_v are unity is unsatisfactory for both ideal and nonideal systems. Serious errors are also encountered at moderate pressures for components widely different in volatility from the average of the mixture. An example is methane dissolved at moderate pressure in absorption oil. Much effort has been directed toward the evaluation of these corrections, but no entirely satisfactory and general method has yet been developed.

On the basis of the pseudocritical concept Gamson and Watson[13] developed general methods for calculating fugacities in both the liquid and vapor states at all conditions. The accuracy of these methods is limited by the lack of sound definitions of the pseudocritical properties and by the inaccuracies inherent in the generalized compressibility-factor relationship. These difficulties require the introduction of empirical correction factors which bring the calculated results into fair agreement with the available data over wide ranges of conditions. The method of calculation is too tedious to be of practical value in the direct calculation of vapor–liquid equilibria but may be used to derive values of ϕ_L and ϕ_v from Equations (50) and (51) which for a particular system can be related graphically to the properties and conditions of the respective phases.

Graphical correlation of ϕ_L and ϕ_v, either from experimental data or from the generalized method of calculation of Gamson and Watson, is complicated by the large number of variables entering into the relationships. The most important are:

(a) T_{ci}/T'_c = the ratio of the critical temperature of the component to the pseudocritical temperature of the *phase* under consideration

(b) p'_r = the pseudoreduced pressure of the phase under consideration

(c) T'_r = the pseudoreduced temperature of the phase under consideration

For a system composed of homologs, ϕ_v and ϕ_L may be expressed in terms of only these three variables over limited ranges of variation of the average properties of the mixtures. In order to obtain more general relationships additional variables must be considered, even when dealing with homologs.

Consideration of the activity coefficients considerably complicates vaporization calculations. Since the activity coefficients are functions of the properties of the individual phases, it is necessary to assume trial phase compositions, and then evaluate approximate activity coefficients and vaporization coefficients. The phase compositions are then calculated and corrected by a repetition of the procedure.

SOLUBILITY

The distribution of a component among the phases of any system at equilibrium under a constant temperature is expressed by Equation (48). As has been pointed out, this general relationship may be used to estimate activity coefficients and vapor–liquid equilibria from solubility or distribution data. Conversely, these same methods may be used to estimate distribution and solubility data from vapor pressure or vapor–liquid equilibrium measurements.

Solids and Liquids. For solutions of nonelectrolytes Hildebrand[17] recommends that the activity in the liquid phase of a solid solute be referred to the pure solute in a hypothetical liquid state at the temperature and pressure of the system. The activity in the solid state is taken as unity for the pure solid at the temperature and pressure of the system. With these conventions Equation (48) for the solubility of a solid reduces to

$$\frac{x_2\gamma_{2L}}{s_2\gamma_{2s}} = \frac{(f_2^\circ)_s}{(f_2^\circ)_L} = K_2 \tag{75}$$

[17] J. H. Hildebrand, "Solubility of Non-Electrolytes," Reinhold Publishing Corporation, 1936.

where

s_2, γ_{2s} = the mole fraction and activity coefficient of component 2 in the solid phase

x_2, γ_{2L} = the mole fraction and activity coefficient of component 2 in the liquid phase

$(f_2^\circ)_s$ = fugacity of component 2 in the solid state at the temperature and pressure of the system

$(f_2^\circ)_L$ = fugacity of component 2 in the hypothetical liquid state at the temperature and pressure of the system

K_2 = solubility constant of component 2, referred to the pure component

An expression for the solubility constant as a function of temperature is obtained by combining Equations (XIV-38) and (75):

$$\left(\frac{\partial \ln K_2}{\partial T}\right)_p = \frac{\text{H}_{2L}^\circ - \text{H}_{2s}^\circ}{RT^2} \tag{76}$$

where H_{2L}° and H_{2s}° are the molal enthalpies of pure component 2 in the hypothetical liquid and solid states, respectively, at the temperature and pressure of the system. If it is assumed that the heat of fusion is independent of temperature $\text{H}_{2L}^\circ - \text{H}_{2s}^\circ$ may be taken as the normal heat of fusion and Equation (76) integrated with $K_2 = 1.0$ at the melting point T_m where $(f_2^\circ)_s = (f_2^\circ)_L$. Then[17]

$$\ln K_2 = -\frac{\lambda_f}{R}\left(\frac{1}{T} - \frac{1}{T_m}\right) = -\frac{\lambda_f(T_m - T)}{RTT_m} \tag{77}$$

where

λ_f = molal heat of fusion of component 2

T = temperature of the system

T_m = melting point of component 2

Similarly, the effect of pressure upon the solubility equilibrium constant is obtained by combining Equations (XIV-30) and (75) to give

$$\left(\frac{\partial \ln K_2}{\partial p}\right)_T = -\frac{(v_{L2}^\circ - v_{s2}^\circ)}{RT} \tag{78}$$

or

$$\ln K_2 = -\frac{(v_{L2}^\circ - v_{s2}^\circ)_{\text{avg}}}{RT}(\pi - P_m) \tag{79}$$

where

v_{s2}° = molal volume of the pure solid solute at the temperature and pressure of the system

v_{L2}° = molal volume of the hypothetical liquid solute at the temperature and pressure of the system

P_m = equilibrium pressure corresponding to melting point T

It may be noted from Equations (77–79) that for an ideal solution where $K_2 = x_2$ the solubility of a solid decreases with increase in molal heat of fusion, increases when the solid contracts upon melting, and when expressed as mole fraction is independent of the nature of the solvent; on a weight basis the solubility decreases with increase in molecular weight of the solvent for ideal solutions.

Equation (75) furnishes a sound basis for the evaluation of the corresponding activity coefficients. These coefficients approach unity in systems which do not involve polar components, large differences in molecular size, or large differences between T_m and T. If the activity coefficients are assumed to be unity, the *ideal solubility*[17] of a pure solid is equal to K_2, or $x_2 = K_2$.

Equation (77) may be used for predicting the solubility of a solid from its melting-point and heat-of-fusion data or conversely for calculating heats of fusion from solubility data where conditions are such that ideal solutions are approximated. The usefulness of Equation (75) is restricted by the extensive experimental data required for evaluation of the activity coefficients.

Illustration 6. From Fig. 10, page 113, it may be seen that the solubility of naphthalene in benzene is 60 per cent by weight at 45°C and 21 per cent at 0°C. Assuming the solution to be ideal, calculate heats of fusion of naphthalene from these two solubility values. The melting point of naphthalene is 80.1°C.

Solution:

Basis: 100 g of solution (component 2 is naphthalene).

		0°C			45°C		
	Molecular Weight	g	g-moles	*Mole Fraction*	g	g-moles	*Mole Fraction*
Naphthalene	128	21	0.164	0.14	60	0.469	0.478
Benzene	78	79	1.013	0.86	40	0.512	0.522
		100	1.177	1.00	100	0.981	1.000
Log K_2 = log x_2			−0.854			−0.321	

$$\text{At } 0°\text{C}, \lambda_f = \frac{(2.303)(0.854)(1.99)(273.1)(353.2)}{(353.2 - 273.1)} = 4710 \text{ cal per g-mole}$$

$$\text{at } 45°\text{C}, \lambda_f = \frac{(2.303)(0.321)(1.99)(318.1)(353.2)}{(353.2 - 273.1)} = 4710 \text{ cal per g-mole}$$

These results are in fair agreement with the observed value of 4,550.

The fair agreement between the heats of fusion calculated at the two temperatures of Illustration 6 indicates that the solutions of benzene in naphthalene closely approximate ideal behavior. In other systems involving less similar components much larger discrepancies can be expected and the activity coefficients of Equation (75) must be evaluated for satisfactory results. These deviations in a number of systems are discussed by Hildebrand.[17]

For solutes having melting points far removed from the temperature of the system or for electrolytes which dissociate into ions in solution, it is not advantageous to use the hypothetical standard state of Equation (75) for the liquid phase. A more convenient convention is to express the activity of the solute in the liquid referred to the infinitely dilute solution at the pressure of the system. As is mentioned on page 629 this convention corresponds to the selection of a hypothetical standard state the fugacity of which is the value which would exist if the solution behaved ideally at unit mole fraction (or unit molality) as it does at zero concentration. The enthalpy and volume of this hypothetical standard state are therefore equal to the partial molal enthalpy and volume, respectively, at zero concentration. The activities in the solid phase are best referred to the pure solid at the pressure of the system as the standard state. On this basis the solubility of a solid is expressed by

$$\frac{a'_{2L}}{a_{2s}} = \frac{f^\circ_{2s}}{f^{\circ\prime}_{2L}} = \frac{x_2\gamma'_{2L}}{s_2\gamma_{2s}} = K'_2 \tag{80}$$

where

s_2, γ_{2s} = mole fraction and activity coefficient of component 2 in the solid phase

x_2, γ'_{2L} = mole fraction and activity coefficient of component 2 in the given solution

f°_{2s} = fugacity of the pure solid at the temperature and pressure of the system

$f^{\circ\prime}_{2L}$ = fugacity of the hypothetical standard state for the liquid phase referred to infinite dilution

K'_2 = solubility equilibrium constant of component 2 referred to infinite dilution

The effect of temperature on the solubility constant is given by an equation expressing the effect of temperature on the fugacities of the standard states. From Equations (XIV-45) and (80),

$$\left(\frac{\partial \ln K'_2}{\partial T}\right)_p = \frac{(\overline{H}^\circ_{2L} - H^\circ_{2s})}{RT^2} \tag{81}$$

where

$\overline{H}^\circ_{2L}$ = partial molal enthalpy of component 2 at infinite dilution at the temperature and pressure of the system

H°_{2s} = molal enthalpy of component 2 in the solid state at the temperature and pressure of the system

The quantity $\overline{H}^\circ_{2L} - H^\circ_{2s}$ is the differential heat of solution at infinite dilution.

Similarly, by combining Equations (XIV-44) and (80) the effect of pressure on solubility is derived,

$$\left(\frac{\partial \ln K_2'}{\partial p}\right)_T = -\frac{(\bar{v}_{2L}^\circ - v_{2s}^\circ)}{RT} \tag{82}$$

where

$\bar{v}_{2L}^\circ$ = partial molal volume at infinite dilution at the temperature and pressure of the solution

v_{2s}° = molal volume of the pure solid at the temperature and pressure of the system

It may be noted from Equation (81) that the solubility of a solid is increased by increased temperature if its differential heat of solution is positive, corresponding to absorption of heat in dissolution. Similarly, the solubility is increased by increase in pressure if the partial molal volume in dilute solution is less than the volume of the solid.

The practical value of Equations (80–82) is restricted by the great variations with concentration of the activity coefficients in liquid solutions. Except in dilute solutions these coefficients must be evaluated from empirical data which are seldom available.

Equations (80–82) are directly applicable to electrolytes with activities expressed in either molecular or ionic terms. However, the use of mean ionic activities and molalities for these materials has the advantage that the activity coefficients approach unity in dilute solutions, as demonstrated in Fig. 145, whereas the molecular activity coefficients become very small.

Solubility of Gases. The solubility in a liquid of a gas, either in the pure state or from a mixture, is expressed thermodynamically by Equation (49). If the temperature of the system is below or not greatly above the critical temperature of the gas, the standard state for the gaseous component in the liquid phase is conveniently taken as the pure component in a hypothetical incompressible liquid state at the temperature and pressure of the system, exactly as described for handling vapor–liquid equilibria. If it is assumed that ideal solutions are formed in the gaseous phase and the standard state for the gaseous phase is taken as the state of unit fugacity, Equation (49) combined with (51) reduces to

$$\frac{y_2 \pi \nu_{2\pi}}{x_2 \gamma_2 \phi_{2L}} = (f_2^\circ)_L = K_2 \tag{83}$$

where

π = total pressure of the system

$\nu_{2\pi}$ = fugacity coefficient of pure gaseous component 2 at the temperature and pressure of the system

γ_2, ϕ_{2L} = activity coefficients in the liquid state as defined by Equation (51), page 664

Comparison of Equation (83) with (V-5), page 147, in which the partial pressure p_2 is equal to πy_2 results in a thermodynamic expression for Henry's constant H_2. Thus

$$H_2 = \frac{K_2 \gamma_2 \phi_{2L}}{\nu_{2\pi}} \tag{84}$$

Equations (83) and (84) may be used to predict what may be termed *ideal solubilities* if γ_2 and ϕ_{2L} are assumed to be unity; $\nu_{2\pi}$ is derived from Fig. 142, and $(f_2^\circ)_L$ is calculated from Equation (56). As pointed out by Hildebrand,[17] this method gives fair results in systems not involving highly polar materials if the temperature is below the critical temperature of the gas. At higher temperatures the calculation of $(f_2^\circ)_L$ becomes uncertain. Arbitrary specification of $(f_2^\circ)_L$ in such ranges requires consideration of corresponding values of ϕ_{2L} which must be based on experimental data. Such data are necessary in any case for the prediction of γ_2 in systems involving high polarity or chemical reaction and association.

Illustration 7. Calculate the ideal solubility of chlorine in carbon tetrachloride at a pressure of 1.0 atm and a temperature of 0°C. The density of liquid chlorine is 1.561 g per cc at −34.6°C. The vapor pressure at 0°C is 3.65 atm. T_c is 417°K and p_c is 76.1 atm.

Solution: At −34.6°C, $T_r = 238.6/417 = 0.571$

From Fig. 109, $\omega_1 = 0.1266$

$$(v_1\omega_1) = (71)(0.1266)/(1.561) = 5.758 \text{ cc per g-mole}$$

At 0°C and 3.65 atm, $T_r = 0.655$ and $p_r = 3.65/76.1 = 0.048$. Then, from Fig. 142, $\nu_P = 0.95$.

From Equation (56), at 0°C,

$$\log (f_2^\circ)_L = \log 3.65 + \log 0.95 + \frac{5.758\,[5.7 + (3.0)(0.655)](1.0 - 3.65)}{(2.303)(82.06)(273.2)}$$

$$= 0.5623 - 0.0223 - 0.0023 = 0.5377$$

$$(f_2^\circ)_L = K_2 = 3.45$$

From Equation (83), neglecting ν_π at the low pressure involved and assuming that γ_2 and ϕ_{2L} are unity, we have

$$x_2 = (1.0)(1.0)/(3.45) = 0.290$$

The experimentally determined value is 0.298.

It may be noted that with this definition of the standard state in the liquid-phase gas solubility calculations are identical with the vapor–liquid equilibrium calculation demonstrated in preceding sections.

For systems involving gases far above their critical temperatures or where dissociation or chemical reaction is involved, the standard state used in Equation (83) for the liquid phase loses its advantage. It is frequently more convenient to express the activities of the solute gas referred to infinite dilution. With this convention Equation (83) becomes

$$\frac{y_2 \nu_{2\pi} \pi}{x_2 \gamma'_{2L}} = (f_2^{\circ\prime})_L = K'_2 \tag{85}$$

where γ'_{2L}, $(f_2^{\circ\prime})_L$ and K'_2 all correspona to activities referred to infinite dilution. For such systems the single coefficient $(\gamma'_2)_L$ may be used to take into account all types of deviation from ideality. In dilute solutions or when dealing with nonpolar nonreacting solutes it is frequently satisfactory to assume that $(\gamma'_2)_L$ is unity. With this assumption a single experimental solubility value serves to determine K'_2. The variation of the solubility constant with temperature is derived from Equation (XIV-45)

$$\left(\frac{\partial \ln K'_2}{\partial T}\right)_p = \frac{\text{H}^*_{2G} - \overline{\text{H}}^{\circ}_{2L}}{RT^2} \tag{86}$$

where

H^*_{2G} = enthalpy of component 2 in the ideal gaseous state at temperature T

$\overline{\text{H}}^{\circ}_{2L}$ = partial molal enthalpy of component 2 in solution at infinite dilution at the temperature and pressure of the system

The effect of pressure on K'_2 is derived from Equation (XIV-44),

$$\left(\frac{\partial \ln K'_2}{\partial p}\right)_T = \frac{\overline{v}^{\circ}_{2L}}{RT} \tag{87}$$

where

$\overline{v}^{\circ}_{2L}$ = partial molal volume of component 2 in the liquid phase at infinite dilution

Equation (86) may be used to calculate the effect of temperature on solubility from heat-of-dissolution data or conversely to calculate the differential heat of dissolution at infinite dilution $-(\text{H}^*_{2G} - \overline{\text{H}}^{\circ}_{2L})$ from solubility data at two temperatures if the ideal solution behavior is approximated.

Illustration 8. At a partial pressure of 1.0 atm pure CO_2 is in equilibrium with aqueous solutions containing 0.096 mole per cent CO_2 at 10°C and 0.0294 mole per cent CO_2 at 60°C.

(*a*) Assuming the solution to behave ideally, calculate the mole fraction of CO_2 in an aqueous solution saturated with CO_2 at 60°C and 100 atm. The effect of pressure on K'_2 may be neglected.

(*b*) Calculate the average differential heat of dissolution of CO_2 in water at infinite dilution in the range of 10–60°C.

Solution:

(*a*) For CO_2, $T_c = 304.3°\text{K}$; $p_c = 73.0$ atm

At 60°C, 1 atm, $T_r = 1.10; \quad p_r = 0.014$

From Fig. 142, $\nu_\pi = 0.995$

Substitution in Equation (85), if it is assumed that $\gamma'_{2L} = 1.0$, gives

$$K'_2 = (1.0)(0.995)/(0.000294) = 3380$$

At 60°C, 100 atm, $T_r = 1.10; \quad p_r = 1.37; \quad \nu_\pi = 0.63$

From Equation (85),

$$x_2 = (100)(0.63)/3380 = 0.0186 \text{ or } 1.86 \text{ per cent}$$

(b) At 10°C, and 1.0 atm, $T_r = 0.93; \quad p_r = 0.014; \quad \nu_\pi = 0.995$

$$K'_2 = (1.0)(0.995)/(0.00096) = 1036$$

Substitution in Equation (86), after integration, gives

$$(\text{H}^*_{2G} - \overline{\text{H}}^\circ_{2L}) = -\left[(1.99)(2.303)\log\frac{3380}{1036}\right]\Big/\left(\frac{1}{333.2} - \frac{1}{283.2}\right) = 4{,}431 \text{ cal per g-mole}$$

The differential heat of dissolution of the ideal gas at infinite dilution is therefore −4,431 cal per g-mole.

The experimentally observed mole percentage of CO_2 in water at 60° and 100 atm is approximately 1.8 per cent.[18] Assumption of Henry's law in this case would predict a value of 2.94 per cent, an error of over 60 per cent as compared to an error of less than 5 per cent for Equation (85). The good agreement with Equation (85) indicates that the activity coefficients in the liquid phase are substantially unity as assumed, in spite of the chemical reaction occurring between carbon dioxide and water and the accompanying ionization. This may be interpreted as indicating that only a small fraction of the dissolved CO_2 reacts with the water. Where higher solute concentrations or more extensive reactions are involved, the assumption of ideal behavior may lead to large errors which require evaluation of the activity coefficients. The solubility of gases with accompanying chemical reaction is discussed in Chapter XVI, page 742.

Gas–Solid Adsorption Equilibria. Equilibrium conditions in adsorption of gases on solids are discussed in Chapter V, pages 149–160, and the enthalpy of such systems in Chapter VIII, pages 297–300. From Equation (VIII-48),

$$(w_2 + 1)H = H_1 + H_2w_2 + \Delta H = \overline{H}_1 + w_2\overline{H}_2 \qquad (88)$$

or

$$\Delta H = \overline{H}_1 - H_1 + w_2(\overline{H}_2 - H_2) = \Delta\overline{H}_1 + w_2\Delta\overline{H}_2 \qquad (89)$$

where

$\Delta\overline{H}_1$ = the differential heat of adsorption per unit mass of adsorbent
$\Delta\overline{H}_2$ = the differential heat of adsorption per unit mass of adsorbate
ΔH = the integral heat of adsorption per unit mass of adsorbent
w_2 = mass of adsorbate per unit mass of adsorbent

From Equation (89), at constant mass of adsorbent,

$$\Delta\overline{H}_2 = \left(\frac{\partial\Delta H}{\partial w_2}\right)_{w_1=1,T} \qquad (90)$$

[18] "International Critical Tables," McGraw-Hill Book Company, New York.

Since the differential heat of adsorption varies with the mass adsorbed, the total heat of adsorption in adsorbing a mass w_2 of gas is obtained by integration, thus

$$\Delta H = \int_0^{w_2} \Delta \overline{H}_2 \, dw_2 \tag{91}$$

From Equation (g) page 472, applied to gas adsorption under conditions of constant composition,

$$\left(\frac{\partial \Delta \overline{S}}{\partial \Delta \overline{V}}\right)_{T,\, w_2} = \left(\frac{\partial p}{\partial T}\right)_{w_2} \tag{92}$$

Since adsorption at constant temperature and composition also takes place at constant pressure $T\,\Delta S = \Delta H$ and Equation (92) may be directly integrated,

$$\overline{S}_2 - S_{2g} = \left(\frac{\partial p}{\partial T}\right)_{w_2} (\overline{V}_2 - V_{2g}) = \frac{\overline{H}_2 - H_{2g}}{T} \tag{93}$$

If the partial molal volume of the adsorbed gas is neglected and ideal-gas behavior is assumed,

$$\overline{H}_2 - H_{2g} = -RT^2 \left(\frac{\partial \ln p}{\partial T}\right)_{w_2} = R \left(\frac{\partial \ln p}{\partial \dfrac{1}{T}}\right)_{w_2} = \Delta \overline{H}_2 \tag{94}$$

and

$$\Delta H = \frac{R}{M_2} \int_0^{w_2} \left(\frac{\partial \ln p}{\partial \dfrac{1}{T}}\right)_{w_2} dw_2 \tag{95}$$

Illustration 9. In the adsorption of NO_2 gas by silica gel the following equilibrium isotherm equations were derived from the experimental work of Foster:[19]

At 15°C, $w_2 = 0.330\, p_2^{1.333}$

At 25°C, $w_2 = 0.115\, p_2^{1.490}$

At 35° C, $w_2 = 0.022\, p_2^{1.870}$

where p_2 = partial pressure of NO_2 in millimeters of Hg

w_2 = grams of NO_2 adsorbed per 100 g of silica gel

Calculate the differential heat of adsorption per gram-mole of NO_2 when $w_2 = 5.0$. From a cross-plot of ln p against $1/T$ at $w_2 = 5.0$,

$$\left(\frac{\partial \ln p}{\partial \left(\dfrac{1}{T}\right)}\right)_{w_2 = 5} = -3760°K$$

From Equation (94),

$$\Delta \overline{H}_2 = \overline{H}_2 - H_{2g} = -3760(1.987) = -7490 \text{ cal per g-mole } NO_2$$

[19] Gordon Foster, Ph.D. thesis, University of Wisconsin (1944).

To obtain the integral heat of solution of 5 g of NO_2 on 100 g of silica gel it is necessary to obtain the differential heat of solution for different values of w_2 from 0 to 5 and then integrate according to Equation (95).

MINIMUM WORK AND ENTROPY OF SEPARATION

When two dissimilar substances are mixed together to form a solution the escaping tendency and fugacity of each is diminished. As a result, the operation of mixing is accompanied by a decrease in free energy of the system. If the mixing could be carried out in a reversible manner, the process would be capable of performing a corresponding maximum quantity of useful work. Conversely, in order to separate the components of a solution, useful work must be done on the system. The useful work expended in separation is a minimum when the operation is conducted reversibly and in this case is equal to the increase in free energy accompanying the separation. The total work expended in the separation is the sum of the useful work plus the work of expansion which accompanies any changes in volume.

The employment of a reversible method for separating the components of a solution can conceivably be devised by the use of a semipermeable membrane which will permit one component to flow through under pressure but not the other. Such membranes are rarely used in industrial processes because of the extreme slowness of separation but are widely distributed in biological processes, such as metabolism and osmosis. Some electrochemical processes approach reversibility in the separation of components from solution. Actual separations normally consume energy far in excess of the reversible work requirements, which serve as a criterion for judging the thermodynamic efficiencies of separation processes.

The free energy of a solution is related to the partial free energies of its components by an expression similar to Equation (VIII-32), page 286. Thus, when a solution is formed from the standard states of its pure components, the free-energy change is represented by

$$\Delta G = (n_A\overline{G}_A + n_B\overline{G}_B + n_C\overline{G}_C + \cdots) - (n_A G_A^\circ + n_B G_B^\circ + n_C G_C^\circ + \cdots) = -w_f \tag{96}$$

Combining (XIV-53) and (96),

$$\Delta G = RT(n_A \ln a_A + n_B \ln a_B + n_C \ln a_C + \cdots) = -w_f \tag{97}$$

or

$$\Delta G = RT(n_A \ln \gamma_A N_A + n_B \ln \gamma_B N_B + n_C \ln \gamma_C N_C + \cdots) = -w_f \tag{98}$$

Where all activities are referred to the pure components, Equation (98) represents the free energy of formation of the solution from its components, each in its specified standard state.

Illustration 10. From the data of Fig. 146, calculate the minimum useful work required to separate 100 lb of a 1.0 molal solution of sodium chloride at 100°C into pure water and solid salt at the same temperature and pressure.

Solution:
Basis: 100 lb of solution.

Per cent NaCl by weight	= (58.5/1058.5)100	= 5.52
Lb-moles NaCl	= 5.52/58.5	= 0.0945
Lb-moles H_2O	= 94.48/18	= 5.26
Mole fraction NaCl	= 0.0945/5.35	= 0.0177
Mole fraction H_2O	= 0.9823	

From Fig. 146,

$$\gamma_{NaCl} = 0.593 \qquad \gamma_{H_2O} = 0.9832$$
$$a_{NaCl} = (0.0177)(0.593) = 0.0105 \qquad a_{H_2O} = (0.9823)(0.9832) = 0.964$$

From Equation (97),

$$\Delta G = (1.99)(672)(2.303)[(0.0945)\log 0.0105 + (5.26)\log 0.964]$$
$$= -833 \text{ Btu in the formation of 100 lb of solution}$$

Therefore the minimum useful work done *on* the solution in separating it into its components is +833 Btu or, for the separation process, $w_f = -833$.

Entropy of Dissolution. The entropy changes which accompany dissolution may be calculated from the enthalpy and free-energy changes by Equation (XI-5). Thus,

$$\Delta S = \frac{\Delta H - \Delta G}{T} \tag{99}$$

In Equation (99), ΔH is the integral heat of solution and ΔG the free energy of solution calculated from Equation (98). For ideal solutions $\Delta H = 0$ and the activity coefficients of Equation (98) become unity.

Illustration 11. Calculate the minimum useful work and entropy change of separation of 1 lb-mole of air at 65°F and 1.0 atm pressure into oxygen and nitrogen each at the temperature and pressure of the mixture.

Solution: For air at the given conditions mole fractions may be substituted for activities referred to the pure components at the temperature and pressure of the solution. From Equation (98),

$$\Delta G = (1.99)(525)(2.303)[(0.21)\log 0.21 + (0.79)\log 0.79] = -540 \text{ Btu}$$

Therefore, the free-energy change of separation is 540 Btu per lb-mole of air, which is the minimum work of separation done on the system, or $w_f = -540$ for the separation process.

From Equation (99), for the separation process, assuming $\Delta H = 0$,

$$\Delta S = -\frac{\Delta G}{T} = -\frac{540}{525} = -1.03 \text{ entropy units}$$

PROBLEMS

1. The system toluene–acetic acid forms an azeotrope containing 62.7 mole per cent toluene and having a minimum boiling point of 105.4°C at 760 mm Hg. The vapor-pressure data from the International Critical Tables are as follows:

t°C	*Toluene*	*Acetic Acid*
70	202.4	136.0
80	289.7	202.3
90	404.6	293.7
100	557.2	417.1
110	—	580.8
120	—	794.0
Normal boiling point, °C	110.7	118.5

Calculate the van Laar constants A and B for this system, and plot γ_1 and γ_2 as ordinates on a logarithmic scale against the mole fraction of toluene (component 1) on a uniform scale, neglecting the association of acetic acid in the vapor phase.

2. The following data on the vapor pressures in millimeters of Hg of carbon-tetrachloride–ethyl-alcohol solutions are from the International Critical Tables:

Weight Per Cent CCl_4	34.8°C	66°C	*Weight Per Cent* CCl_4	34.8°C	66°C
0	103	462	58.25	206	752
10.23	122	520	71.68	220	782
20.02	142	576	76.69	223	789
27.13	156	614	84.25	226	788
39.06	179	677	92.98	225	780
43.87	187	700	97.43	221	741
48.86	193	716	*98.6*	*213*	*717*
			99.9	*187*	*593*
			100.0	173	544

Numbers in italics are graphical interpolations.

From these evaluate the van Laar constants for the system at 34.8°C and 66°C.

3. From the results of problem 2 calculate vapor compositions corresponding to liquid compositions of 10, 20, 30, 40, 50, 60, 70, 80 and 90 mole per cent CCl_4 at pressures of 760 and 200 mm of Hg. Plot these results as y-x diagrams similar to Fig. 148. Assume that the activity coefficients are independent of temperature over the limited ranges involved. The vapor pressures of the pure component are as follows:

VAPOR PRESSURES, MM OF HG

t°C	CCl_4	C_2H_5OH	t°C	CCl_4	C_2H_5OH
20	91	43.9	55	379.3	280.6
25	114.5	59.0	60	450.8	352.7
30	143.0	78.8	65	530.9	448.8
35	176.2	103.7	70	622.3	542.5
40	215.8	135.3	75	—	666.1
45	262.5	174.0	80	843	812.6
50	317.1	222.2	Normal boiling point, °C	76.75	78.32

4. The composition of the azeotrope formed by carbon tetrachloride and ethyl alcohol at 760 mm of Hg is given in the International Critical Tables as 61.3 mole per cent CCl_4 and the boiling point as 64.95°C. Using the vapor-pressure data of problem 3, calculate the composition of the azeotrope at a temperature of 34.8°C and its total vapor pressure, using Equations (46) and (47). Evaluate the van Laar constants for the system at 34.8°C, and compare these results with those of problem 2.

5. Methyl cyclohexane and *n*-heptane are each partially miscible with aniline at 25°C. The equilibrium compositions in mole per cent are as follows:[20]

	Methyl cyclohexane–Aniline	*Heptane–Aniline*
Hydrocarbon layer	11.38% aniline	7.55% aniline
Aniline layer	82.35% aniline	93.91% aniline

Calculate the van Laar constants for the methyl cyclohexane–aniline system and for the *n*-heptane–aniline system at 25°C. Plot curves relating the activity coefficients of each of the respective hydrocarbons to its mole fraction in aniline.

6. (*a*) Using the results of problem 5 and Equation (44), calculate the activity coefficients of methyl cyclohexane and *n*-heptane in a solution of the following composition at 25°C, using the indicated liquid densities of the pure components:

	Mole Per Cent	*Liquid Density (Pure)* (g per cc)
n-Heptane	3.6	0.68
Methyl cyclohexane	5.4	0.76
Aniline	91.0	1.02

It may be assumed that the *n*-heptane and methyl cyclohexane form ideal binary solutions with each other.

(*b*) The vapor pressures at 25°C of *n*-heptane and methyl cyclohexane may be taken as 46 and 37 mm of Hg, respectively. Neglecting the vapor pressure of aniline, calculate the composition of the vapors and the total vapor pressure in equilibrium with the solution of part (*a*). Compare the effective relative volatilities ($\alpha = P_1\gamma_1/P_2\gamma_2$) of the heptane and methyl cyclohexane in the presence and absence of the aniline.

7. Calculate the vaporization equilibrium constant of acetone at a temperature of 300°F and an absolute pressure of 250 lb per sq in. The density of acetone at 20°C is 0.7915 g per cc. The vapor-pressure equation is given in Table IIa, page 73.

8. A stream of gas in a natural gasoline plant has the following composition in mole per cent:

Ethane	10
Propane	14
Isobutane	19
n-Butane	54
Isopentane	3
	100

(*a*) Calculate the pressure necessary to condense this gas completely at a temperature of 100°F.

(*b*) For a condenser operating at the pressure of part (*a*), calculate the temperature at which condensation will start and the temperature of 50 mole per cent condensation. Also calculate the composition of the first liquid to condense and the compositions of the liquid and vapor phases at 50 per cent condensation.

[20] K. A. Varteressian and M. R. Fenske, *Ind. Eng. Chem.*, **29**, 270 (1937).

9. A fractionating column is to produce overhead and bottoms products having the following compositions in mole per cent:

	Overhead	*Bottoms*
Propane	23	—
Isobutane	67	2
n-Butane	10	46
Isopentane	—	15
n-Pentane	—	37
	100	100

(*a*) Calculate the pressure at which the column must operate in order to condense completely the overhead product at 120°F.

(*b*) Assuming that the overhead product vapors are in equilibrium with the liquid on the top plate of the column, calculate the temperature of the overhead vapors and the composition of the liquid on the top plate when operating at the pressure of part (*a*).

(*c*) Calculate the temperature of the liquid leaving the reboiler of the column under the pressure of part (*a*), assuming equilibrium conditions to exist in the reboiler. Also calculate the composition of the vapors leaving the reboiler.

10. The heat of fusion of α-naphthylamine ($C_{10}H_9N$) at its melting point of 49°C is 3,150 cal per g-mole. Calculate the solubility at 30°C, assuming ideal behavior.

11. Calculate the ideal solubility of hydrogen cyanide gas in acetonitrile at a temperature of 100°C and a partial pressure of 5 atm. The density of liquid hydrogen cyanide is 0.699 at 20°C. The vapor pressures at 100°C are 9.2 and 1.89 atm. respectively.

12. Calculate the ideal solubility of chlorine in carbon tetrachloride at a partial pressure of 10 atm and a temperature of 100°C. The vapor pressures at 100°C are 37.6 and 1.92 atm. respectively.

13. The solubility of carbon monoxide in benzene at 20°C under a partial pressure of 1 atm is reported[18] as 0.1533 cc of gas measured at 0°C and 1 atm, per cc of liquid benzene. Assuming that the activity coefficients are unity, calculate the solubility in mole per cent at 20°C and 50 atm. Neglect the vapor pressure of benzene.

14. The following data for the solubility of carbon dioxide at a partial pressure of 1 atm in methyl alcohol are reported[18] in cubic centimeters of gas measured at 0°C and 1 atm per cc of liquid.

t°C	*Solubility*
15	4.366
20	3.918
25	3.515

(*a*) Assuming that ideal solutions are formed, calculate the average differential heats of dissolution of CO_2 in methyl alcohol in the temperature ranges of 15°–20° and 20°–25°C. Neglect the vapor pressure of alcohol.

(*b*) From the results of part (*a*), calculate the solubility at a temperature of 35°C and a pressure of 100 atm.

15. Calculate the minimum total work in Btu in separating 100 lb of a solution at 80°F containing 30 mole per cent benzene, 25 per cent toluene and 45 per cent xylene into the three pure components. Also calculate the entropy change per pound of solution.

16. Calculate the minimum work and the entropy change in separating 100 lb of a 2-molal aqueous solution of sucrose into its components at 25°C. Compare this result with the heat required to vaporize the water at the same temperature.

CHAPTER XVI

CHEMICAL EQUILIBRIUM

As explained in Chapter XI the criterion of equilibrium in a chemically reacting system at constant temperature and pressure is that the change in free energy of any possible reaction shall be zero. A negative free-energy change may be looked upon as the driving potential which is directing the reaction towards a state of rest and as a direct measure of the departure of the reacting system from its equilibrium state. From a kinetic viewpoint the reaction at equilibrium may still be considered as proceeding reversibly, but with equal rates in opposite directions and with no net change in composition.

By evaluation of the free-energy changes of the reactions of a chemical process it is possible to calculate the composition of the equilibrium mixture and to determine the extent of conversion of the initial reactants. Such considerations are essential in determining the most favorable conditions of temperature, pressure, composition, and ratios of reactants to obtain the greatest conversion of reactants and the highest yield of products.

THE EQUILIBRIUM CONSTANT

For developing the equations of chemical equilibrium the general reaction represented by the following stoichiometric equation is considered:

$$bB + cC + \cdots = rR + sS + \cdots \tag{1}$$

where b, c, r, s are the number of moles of reactants and products B, C, R, and S, respectively.

When this reaction proceeds isothermally at any temperature T, starting with each reactant in its standard state of unit activity and ending with products each at unit activity, the accompanying change of free energy is represented by the symbol $\Delta G°$. The corresponding activity a_B°, a_C°, a_R°, a_S° of each component in the standard state is unity.

When this reaction proceeds isothermally until equilibrium is established all the activities adjust themselves to new values at which the change in free energy is zero. At equilibrium the activities of the separate components are represented by symbols, a_B, a_C, a_R, and a_S.

Since the free-energy change is the difference between the free energies of the products and reactants,

$$\Delta G° = r\bar{G}_R^\circ + s\bar{G}_S^\circ + \cdots - b\bar{G}_B^\circ - c\bar{G}_C^\circ - \cdots \tag{2}$$

$$\Delta G = r\bar{G}_R + s\bar{G}_S + \cdots - b\bar{G}_B - c\bar{G}_C - \cdots \tag{3}$$

where $\bar{G}$ = partial molal free energy

Combining (2) and (3) gives

$$\Delta G - \Delta G^\circ = r(\bar{G}_R - \bar{G}_R^\circ) + s(\bar{G}_S - \bar{G}_S^\circ) + \cdots - b(\bar{G}_B - \bar{G}_B^\circ) - c(\bar{G}_C - \bar{G}_C^\circ) \quad (4)$$

Equation (4) may be written in terms of activities by being combined with Equation (XIV-53). Thus,

$$\Delta G - \Delta G^\circ = rRT \ln a_R + sRT \ln a_S + \cdots - bRT \ln a_B - cRT \ln a_C - \cdots$$

or

$$\Delta G - \Delta G^\circ = RT \ln \left(\frac{a_R^r\, a_S^s \cdots}{a_B^b\, a_C^c \cdots} \right) \quad (5)$$

At equilibrium, $\Delta G = 0$, and

$$-\frac{\Delta G^\circ}{T} = R \ln \left(\frac{a_R^r\, a_S^s \cdots}{a_B^b\, a_C^c \cdots} \right) = R \ln K \quad (6)$$

where K is the *equilibrium constant* at the temperature corresponding to the temperature of ΔG°. Thus,

$$K = \frac{a_R^r\, a_S^s \cdots}{a_B^b\, a_C^c \cdots} = e^{\frac{-\Delta G^\circ}{RT}} = e^{-\frac{\Delta H^\circ}{RT} + \frac{\Delta S^\circ}{R}} \quad (7)$$

where ΔG°, ΔH°, and ΔS° are the changes in free energy, enthalpy, and entropy, respectively, which accompany the stoichiometric reaction with all reactants and products in their standard states.

From Equation (7) it is evident that the equilibrium constant is entirely determined by the temperature and the free-energy change which would accompany the reaction of the indicated numbers of moles if each reactant were initially in its standard state and each product finally in its standard state of unit activity at the temperature of the system. This free-energy change ΔG° is termed the *standard free-energy change* of the reaction. The standard free-energy change depends upon the temperature, the definition of the standard state of each component, and the number of moles entering into the stoichiometric equation under consideration. Accordingly, the numerical value of an equilibrium constant is without significance unless it is accompanied by specification of these three factors.

The effect on the equilibrium constant of the form of the stoichiometric equation is illustrated by consideration of the synthesis of ammonia. This reaction may be designated as

$$N_2 + 3H_2 = 2NH_3$$

The corresponding equilibrium constant is then

$$K' = \frac{a_{NH_3}^2}{a_{N_2}\, a_{H_2}^3}$$

The same reaction may be written

$$\tfrac{1}{2}\,N_2 + \tfrac{3}{2}\,H_2 = NH_3$$

and

$$K = \frac{a_{NH_3}}{a_{N_2}^{\frac{1}{2}}\,a_{H_2}^{\frac{3}{2}}}$$

It is evident that, in this case, $K' = K^2$ and specification of the number of moles involved in the stoichiometric equation is essential to proper interpretation.

Effect of Pressure on the Equilibrium Constant. The effect which the pressure of the system has on the reaction equilibrium constant is dependent on the definition of the standard states of the components. As pointed out in Chapter XIV, the choice of the standard states is entirely arbitrary and may be on any convenient basis. When working with reacting systems which involve only pure solids or liquids and gases which may be assumed to approximate ideal solutions it is convenient to refer the activities of the gaseous components to the ideal-gaseous state at one atmosphere, making activities equal to fugacities in atmospheres. The activities of the pure solids and liquids are referred to the fugacities of the pure components under their own vapor pressures or under a pressure of 1 atm at the temperature of the system. With standard state defined in this manner the *reaction equilibrium constant is independent of the pressure of the system* which by virtue of the definitions can have no effect on the standard free-energy change.

When working with reacting systems which involve nonideal solutions, either gaseous or liquid, it is more convenient to select as standard states the pure components at the temperature and pressure of the system. As is developed in Chapter XIV, this basis has the advantage of permitting useful correlations between mole fractions and activity coefficients. Similarly, if it is desired to refer the standard state to infinite dilution, it is convenient to choose the dilute solution at the pressure of the system. It should be noted that standard states defined in this manner are dependent upon the pressure of the system and that the corresponding *equilibrium constants vary as a function of pressure.*

It is of primary importance that the exact definition of the standard state be clearly recognized in all equilibrium calculations.

STANDARD FREE-ENERGY CHANGES

It is pointed out in Chapter XI that free energy is an extensive property and that the free-energy change in any process is determined by the final and initial conditions and not by the intermediate path. Thus, free-energy changes may be treated in the same manner as enthalpy changes and are functions of temperature, pressure, and composition at the beginning and end of the change.

The free-energy change accompanying a reaction as it actually proceeds under the conditions of the reaction process is of little interest except in the consideration of electrochemical reactions where the decrease in free energy of the system represents the electric energy released under reversible conditions at constant pressure. Also, it has been pointed out in Chapter XIII that the decrease in free energy of a flow process proceeding reversibly at constant temperature without generation of electric energy is equal to the mechanical work done. Usually chemical reactions proceed in a highly irreversible manner without generation of either electric energy or other useful work, and the loss in free energy is not accompanied by a corresponding release of useful work.

The greatest value of the free-energy concept is in the calculation of compositions of reaction systems at equilibrium under which conditions the actual free-energy changes are zero. These compositions are, however, related by Equation (6) to the standard free-energy changes which are not equal to zero at the standard states. In considering any changes which take place in a system actual and standard free-energy changes must not be confused. Calculation of the standard free-energy change is important, because from it the equilibrium constant is obtained.

Standard Free-Energy Change at 25°C. The standard free-energy changes and the corresponding standard enthalpy and entropy changes which accompany a reaction at an arbitrary reference temperature may be calculated by the summation principles demonstrated in Chapter VIII for heats of reaction. The accepted reference temperature which has been extensively used for thermodynamic data is 25°C (298.1°K) instead of the 18°C standard frequently employed for heats of reaction.

It has been common practice in the past to present standard free-energy data in the form of tables of standard free energies of formation at 25°C. From such tables a standard free energy of reaction is calculated as the algebraic sum of the standard free energies of formation of the products less the algebraic sum of the standard free energies of formation of the reactants. When an element enters into a reaction, its standard free energy of formation is zero if its state of aggregation is that selected as the basis for the standard free energy of its compounds.

Illustration 1. Calculate the standard change of free energy at 298.1°K in the gas-phase alkylation of isobutane with ethylene to form neohexane. Free energies of formation at a fugacity of 1 atm and 298.1°K (cal per g-mole)

$$C_4H_{10}\text{, isobutane (gas); } \Delta G_f^\circ = -4{,}160 = G_a$$
$$C_2H_4\text{, ethylene (gas); } \Delta G_f^\circ = +16{,}300 = G_b$$
$$C_6H_{14}\text{, neohexane (gas); } \Delta G_f^\circ = -2{,}300 = G_r$$

For the reaction,

$$C_4H_{10}\ (g) + C_2H_4\ (g) = C_6H_{14}\ (g)$$

$\Delta G^\circ_{298.1} = G_r - G_a - G_b = -2{,}300 - (-4{,}160) - 16{,}300 = -14{,}400$ cal per g-mole in the ideal state at 1 atm.

Effect of Temperature on $\Delta G°$ and K. A general differential relationship between $\Delta G°$ and temperature is expressed by Equation (XI-104). Combination of this with Equation (6) yields a similar expression for the effect of temperature on the equilibrium constant. Thus

$$R\left(\frac{\partial \ln K}{\partial T}\right)_p = \frac{\Delta H^\circ}{T^2}$$

or

$$R\left(\frac{\partial \ln K}{\partial (1/T)}\right)_p = -\Delta H^\circ \tag{8}$$

Equation (8) may be used to predict the effect of temperature on the equilibrium constant or conversely to determine the standard heat of reaction from equilibrium constants known at different temperatures. Such a determination is carried out graphically by determining the slope of a curve relating $R \ln K$ to $1/T$.

A more generally useful integrated relationship between the standard free-energy change and temperature is obtained by expressing the standard heat of reaction and entropy change in terms of empirical heat-capacity equations. If heat capacities are represented by the general empirical expression $c_p^\circ = a + bT + cT^2 + d/T^2$, Kirchoff's Equation (VIII-65), page 305, may be written:

$$\Delta H^\circ_T = I_H + \Delta aT + \tfrac{1}{2}\Delta bT^2 + \tfrac{1}{3}\Delta cT^3 - \Delta d/T \tag{9}$$

The integration constant I_H is determined from a single value of $\Delta H°$ at a temperature within the range of applicability of the heat-capacity equations.

Since at constant pressure $dS = c_p\, dT/T$, it follows that

$$d(\Delta S^\circ) = \Delta C^\circ dT/T \quad \text{or} \quad \Delta S^\circ_T = I_S + \Delta a \ln T + \Delta bT + \tfrac{1}{2}\Delta cT^2 - \frac{\Delta d}{2T^2} \tag{10}$$

The integration constant I_S is determined from a single value of $\Delta S°$ at any temperature within the range of the accuracy of the heat-capacity equations.

Since $\Delta G° = \Delta H° - T\Delta S°$, from Equations (9) and (10),

$$\frac{\Delta G^\circ}{T} = \frac{I_H}{T} + (\Delta a - I_S) - \Delta a \ln T - \frac{\Delta bT}{2} - \frac{\Delta cT^2}{6} - \frac{\Delta d}{2T^2} \tag{11}$$

The integration constants I_H and I_S are determined from a single value of $\Delta H°$ together with a single value of either $\Delta S°$ or $\Delta G°$. These reference values may be at different temperatures.

The forms of Equations (9), (10), and (11) depend upon the empirical equations used for expressing heat capacities. Other equations may be used with corresponding changes in the form. The standard free energy of formation of a compound from its elements at any temperature level is a special modification of Equation (11) as illustrated in the following illustration:

Illustration 2. Derive a general equation for the standard free energy of formation of $NH_3(g)$ at temperature T°K with the standard state of each component being the ideal gas at 1 atm. The absolute entropies in the ideal-gaseous state at 298.1°K and 1 atm are as follows:

$$N_2(g);\ s_a^\circ = 45.79 \text{ cal/(g-mole)(°K)}$$
$$H_2(g);\ s_b^\circ = 31.23 \text{ cal/(g-mole)(°K)}$$
$$NH_3(g);\ s_r^\circ = 46.03 \text{ cal/(g-mole)(°K)}$$

From Table V, page 214,

$$N_2;\ c_p = 6.30 + 1.819(10^{-3})T - (0.345)(10^{-6})T^2$$
$$H_2;\ c_p = 6.88 + (0.066)(10^{-3})T + (0.279)(10^{-6})T^2$$
$$NH_3;\ c_p = 5.92 + (8.963)(10^{-3})T - (1.764)(10^{-6})T^2$$

For the reaction,

$$\tfrac{1}{2}N_2(g) + \tfrac{3}{2}H_2(g) \rightarrow NH_3(g)$$
$$\Delta H_{291}^\circ = -11{,}000 \text{ cal per g-mole } NH_3$$
$$\Delta a = 5.92 - \tfrac{1}{2}(6.30) - \tfrac{3}{2}(6.88) = -7.55$$
$$\Delta b = [8.963 - \tfrac{1}{2}(1.819) - \tfrac{3}{2}(0.066)]10^{-3} = (7.955)(10^{-3})$$
$$\Delta c = [-1.764 + \tfrac{1}{2}(0.345) - \tfrac{3}{2}(0.279)]10^{-6} = (-2.010)(10^{-6})$$
$$\Delta d = 0$$

Substitution in Equation (9) gives

$$I_H = -11{,}000 + 7.55(291) - 3.977 \times 10^{-3}(291)^2 + 0.670(10^{-6})(291)^3 = -9130$$
$$\Delta S_{298.1}^\circ = s_r^\circ - \tfrac{1}{2}s_a^\circ - \tfrac{3}{2}s_b^\circ = 46.03 - \tfrac{1}{2}(45.79) - \tfrac{3}{2}(31.23) = -23.72$$

From Equation (10),

$$\Delta S_{298.1}^\circ = I_S + \Delta a \ln T + \Delta b T + \tfrac{1}{2}\Delta c T^2$$
$$-23.72 = I_S - 7.55(5.6971) + 7.955 \times 10^{-3}(298.1) - \tfrac{1}{2}(2.01)(10^{-6})(298.1)^2$$

or $I_S = 17.00$

$$(\Delta a - I_S) = -7.55 - 17.00 = -24.55$$

Substitution in Equation (11) gives

$$\frac{\Delta G^\circ}{T} = \frac{-9{,}130}{T} + 7.55 \ln T - (3.977)(10^{-3})T + (0.335)(10^{-6})T^2 - 24.55$$

The standard free-energy change of any reaction at any temperature level may be obtained by the method of Illustration 2 in which Equation (11) is applied directly to the reaction combined with a knowledge of heat-capacity equations of the components and single values of ΔH° and ΔG° or ΔS° for the reaction, each at some known temperature. An alternate method is to evaluate the free energies of formation as functions of temperature and calculate the free-energy change of the reaction by combining the standard free energies of formation of the separate components. There is little point in this latter procedure except where numerous different reactions which involve the same components are under consideration.

For reactions in which $\Delta C_p = 0$ both ΔH° and ΔS° are independent of temperature. Equation (11) then reduces to

$$\Delta G^\circ = \Delta H_1^\circ - T\,\Delta S_1^\circ \tag{12}$$

From Equation (12) it is evident that the standard free-energy change of a reaction may vary widely with temperature, even though the standard heat of reaction and entropy change are constant.

Where the heat-capacity data for a reaction are not conveniently expressed by empirical equations the standard free-energy change at any temperature may be calculated by correcting separately the standard enthalpy and entropy changes by graphical integrations. Thus, if ΔC_p is the difference between the heat capacity of the products and the reactants, and it is desired to obtain the standard free-energy change at a temperature T_3 from values of standard enthalpy and entropy changes ΔH_1° and ΔS_2° at temperature T_1 and T_2, respectively:

$$\Delta G_3^\circ = \Delta H_1^\circ + \int_{T_1}^{T_3} \Delta C_p^\circ\, dT - T_3 \Delta S_2^\circ - T_3 \int_{T_2}^{T_3} \frac{\Delta C_p^\circ}{T}\, dT \tag{13}$$

The integrations of Equation (13) may be carried out graphically by plotting ΔC_p° and $\Delta C_p^\circ/T$ against T.

Standard Free Energies of Formation of Hydrocarbons. Standard free energies of formation of a few hydrocarbons are shown in Fig. 155, modified from a similar figure of Parks and Huffman.[1] In this figure the molal free energies of formation are expressed per carbon atom in order that the figure shall show visually the relative stabilities of the different compounds in their standard states with respect to the elements. For example, the fact that the acetylene curve (C_2H_2) is positive indicates that over the entire temperature range this compound is thermodynamically unstable in its standard state and tends to decompose spontaneously to an equilibrium mixture containing large proportions of carbon and hydrogen. Similarly, since the curve for propane (C_3H_8) is higher than that for ethane (C_2H_6), the latter compound is the more stable. Propane becomes relatively unstable at temperatures above 450°K.

The standard free energies of formation of the hydrocarbons are obtained by multiplying the values from Fig. 155 by the number of carbon atoms. The standard free-energy changes of reactions involving these hydrocarbons may then be calculated by summations of the standard free energies of formation.

[1] Parks and Huffman, "The Free Energies of Some Organic Compounds," Reinhold Publishing Corporation, New York (1932).

Illustration 3. Calculate the standard free-energy change of the following reaction at 1000°K from the data of Fig. 155:

$$2C_2H_4 = iC_4H_8$$

$$iC_4H_8;\ \Delta G_f^\circ = (4)(15{,}300) = 61{,}200 \text{ cal per g-mole}$$

$$C_2H_4;\ \Delta G_f^\circ = (2)(14{,}000) = 28{,}000 \text{ cal per g-mole}$$

For the reaction at 1000°K,

$$\Delta G^\circ = 61{,}200 - 56{,}000 = 5{,}200 \text{ cal}$$

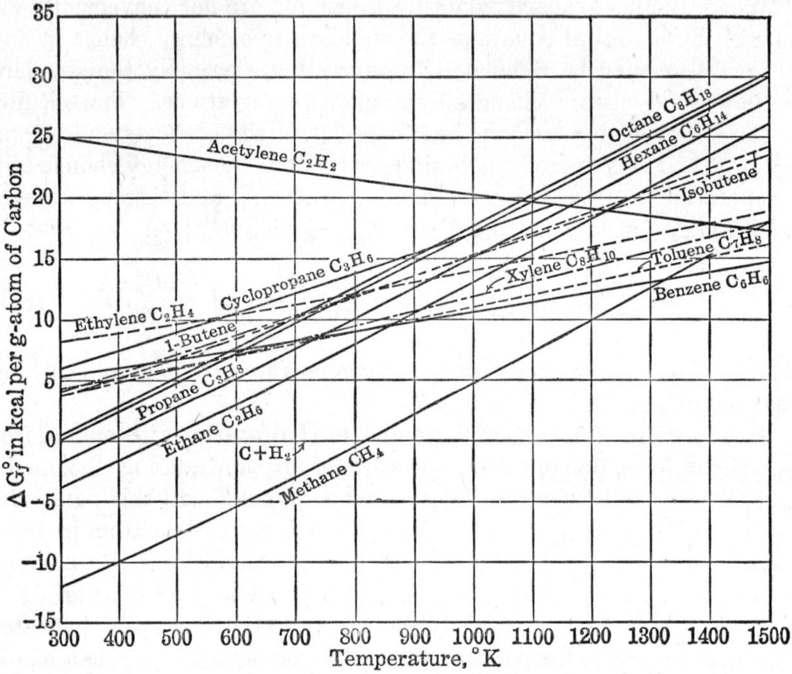

(Revision of chart from Parks and Huffman, "Free Energies of Some Organic Compounds," with permission of the Reinhold Publishing Corporation.)

FIG. 155. Standard Free Energies of Formation of Hydrocarbons.

(Reproduced in " C.P.P. Charts ")

Experimental Free-Energy Data. Four common methods are available for determining standard free energies of formation and the standard free-energy change of a chemical reaction.

1. For reactions which proceed reversibly in a galvanic cell at constant pressure and temperature, the free-energy change can be obtained directly from measurements of electric energy generated.

2. The standard free-energy change may be calculated from measurements of temperature, pressure, and composition of each phase in a system at equilibrium by means of Equation (6).

3. The standard change in free energy can be obtained directly from measurements of ΔH° and ΔS° by the equation $\Delta G^\circ = \Delta H^\circ - T\Delta S^\circ$.

Values of $\Delta H°$ are determined from thermochemical data, as described in Chapter VIII. Values of $\Delta S°$ are obtained from the absolute entropies of the products and reactants. These absolute entropies are calculated either from low-temperature heat-capacity measurements and the third law of thermodynamics, as discussed in Chapter XI, or from spectroscopic data and statistical calculations, as outlined in Chapter XVI.

4. Free energies of formation and reaction may be obtained through combination of free energy data for other series of reactions by the procedure demonstrated in Chapter VIII for the indirect evaluation of heats of reaction.

Presentation of Free-Energy Data. Three methods are in common use for the presentation of standard free-energy data. The earliest method was the tabulation of standard free energies of formation at an arbitrary base temperature of 25°C together with the constants for the empirical heat-capacity equation of each compound. From these data standard free-energy changes may be calculated at any desired temperature by the methods demonstrated in Illustration 1.

An alternate method is to present tables of heats of formation and absolute entropies at a selected base temperature. By combining these values the standard free energies of reaction at the base temperature are obtained. Free energies at other temperatures are then calculated from the heat-capacity equations. This method has the advantage of separating the heat of formation and entropy data on which the free-energy calculations are based. Changes are constantly being made in the accepted values, and tabulation of separate heats of formation and entropies minimize the confusion resulting from such changes. Furthermore, the entropy values are frequently useful for other calculations.

In Table XXXV are values of absolute entropies at 25°C. Similar data for many other compounds are to be found in the literature. The use of such data in conjunction with the heats of formation and reaction of Tables XIV and XV, pages 253 and 262, is demonstrated in Illustration 2.

With the extension of calorimetric measurements into the region of low temperatures and the development of statistical methods based on spectroscopic data it has become possible to express values of enthalpy and free energy of many substances relative to their values at the absolute zero of temperature. A further improvement has been introduced by reporting values of $(H_T^\circ - H_0^\circ)$ and $(G_T^\circ - G_0^\circ)/T$ as direct tabular or graphical functions of temperature rather than formulating the results in terms of the constants of empirical heat-capacity equations. Consistent with this practice heats of formation $(\Delta H_f^\circ)_0$ at 0°K are now being reported relative to the elements at 0°K instead of at some arbitrary temperature level. By this scheme the agreement of different values can be judged directly by numerical values rather than by the comparison of different equations which appear in as many forms as there are empirical heat-capacity equations for the separate components. The relative agreement of several forms of equations can be judged only by repetition of calculations and usually not by inspection.

The standard heat of formation of a compound relative to the elements, all in their standard state of unit activities at any temperature level, is expressed by the following equation:

$$(\Delta H_f^\circ)_T = [(H_T^\circ - H_0^\circ) + (\Delta H_f^\circ)_0]_C - \Sigma(H_T^\circ - H_0^\circ)_E \tag{14}$$

where

$(\Delta H_f^\circ)_T$ = standard heat of formation at temperature T
H_T° = enthalpy of the compound or element at temperature T
H_0° = enthalpy of the compound or element at 0°K
$(\Delta H_f^\circ)_0$ = standard heat of formation at 0°K

The subscripts C and E refer to the compound and to the element, respectively. At the absolute zero where the temperature, entropy, and vapor pressure all become zero the four energy functions of a substance in a specified state become equal. Thus

$$U_0^\circ = H_0^\circ = G_0^\circ = A_0^\circ \tag{15}$$

As a result it is common practice to report data in terms of the exactly equivalent functions $(G_T^\circ - G_0^\circ)/T$, $(G_T^\circ - H_0^\circ)/T$, or $(G_T^\circ - U_0^\circ)/T$, and $(\Delta H_f^\circ)_0$ or $(\Delta G_f^\circ)_0$.

The standard free energy of formation of a compound at a temperature T from the elements at the same temperature is expressed by an equation similar to (14).

$$\left(\frac{\Delta G_f^\circ}{T}\right)_T = \left[\left(\frac{G_T^\circ - H_0^\circ}{T}\right) + \frac{(\Delta H_f^\circ)_0}{T}\right]_C - \Sigma\left(\frac{G_T^\circ - H_0^\circ}{T}\right)_E \tag{16}$$

Similarly, the standard free-energy change and standard heat of reaction of any reaction at any temperature may be expressed in terms of heats of formation at the absolute zero. Thus,

$$\Delta H_T^\circ = \Sigma[(H_T^\circ - H_0^\circ) + (\Delta H_f^\circ)_0]_P - \Sigma[(H_T^\circ - H_0^\circ) + (\Delta H_f^\circ)_0]_R \tag{17}$$

$$\left(\frac{\Delta G^\circ}{T}\right)_T = \Sigma\left[\frac{G_T^\circ - H_0^\circ}{T} + \frac{(\Delta H_f^\circ)_0}{T}\right]_P - \Sigma\left[\frac{G_T^\circ - H_0^\circ}{T} + \frac{(\Delta H_f^\circ)_0}{T}\right]_R \tag{18}$$

where the subscript P indicates the products and R the reactants.

Standard enthalpy and free-energy changes are readily calculated by Equations (17) and (18) from values of $(\Delta H_f^\circ)_0$ and tables or curves which express $(H_T^\circ - H_0^\circ)$ and $(G_T^\circ - H_0^\circ)/T$ as functions of temperature. It will be noted that for any compound the group

TABLE XXXV

STANDARD MOLAL ENTROPIES AT 298.1°K

(cal)/(g-mole)(°K)

Standard States:

Gases: The ideal-gaseous state at 1 atm
Liquids: The liquid state at 1 atm
Solids: The solid state at 1 atm
Aqueous Solutions: The hypothetical ideal 1.0 molal solution in which $a/m = 1.0$

INORGANIC COMPOUNDS

From K. K. Kelley, *U.S. Bur. Mines Bull.* 434 (1941).

		Aluminum	
Al (s)	6.75	Al_2O_3	12.5
Al (g)	39.31	Al_2SiO_5 (Sillimanite)	27.0
Al^{+++} (aq)	−76		
		Antimony	
Sb (s)	10.5	Sb_2O_4 (s)	30.4
Sb (g)	43.07	Sb_2O_5 (s)	29.9
Sb_2 (g)	60.9	Sb_2S_3 (s)	39.6
Sb_2O_3 (s)	29.4	$SbCl_3$ (g)	80.9
Sb_4O_6 (g)	102	$SbCl_3$ (s)	44.8
		Argon	
A (g)	36.99		
		Arsenic	
As (s)	8.4	As_2O_5 (s)	25.2
As (g)	41.62	$AsCl_3$ (g)	78.2
As_2 (g)	57.3	$AsCl_3$ (l)	55.8
As_4 (g)	69	AsF_3 (g)	69.2
As_2O_3 (s)	25.6	AsF_3 (l)	50.1
As_4O_6 (g)	101		
		Barium	
Ba (g)	40.67	BaF_2 (s)	23.0
Ba^{++} (aq)	2.3	$BaCO_3$ (s) (Witherite)	26.8
BaO (s)	16.8	$Ba(NO_3)_2$ (s)	51.1
BaO (g)	56.0	$BaSO_4$ (s)	31.6
$BaCl_2 \cdot 2H_2O$ (s)	48.5		
		Beryllium	
Be (s)	2.28	BeO (s)	3.37
Be (g)	32.56	BeO (g)	47.2
		Bismuth	
Bi (s)	13.6	Bi_2O_3 (s)	36.2
Bi (g)	44.68	$BiCl_3$ (g)	85.5
Bi_2 (g)	65.4	$BiCl_3$ (s)	45.8
		Boron	
B (s)	1.7	BCl_3 (g)	68.6
B (g)	36.66	BF_3 (g)	61.1
BBr_3 (g)	76.7	B_4C (s)	6.47

TABLE XXXV (*Continued*)

		Bromine	
Br_2 (l)	36.7	Br^- (aq)	19.7
Br_2 (g)	58.63	BrCl (g)	57.4
Br (g)	41.81	BrO_3^- (aq)	38.5
		Cadmium	
Cd (s)	12.3	$CdCl_2$ (s)	31.2
Cd (g)	40.07	$Cd(OH)_2$ (s)	21
Cd^{++} (aq)	−15.6	$CdCO_3$ (s)	25.2
CdO (s)	13.1	$CdSO_4$ (s)	31.3
CdO (g)	46.9	$CdSO_4 \cdot H_2O$ (s)	39.7
CdS (s)	17	$CdSO_4 \cdot 8/3H_2O$ (s)	57.5
		Calcium	
Ca (s)	9.95	$Ca(OH)_2$ (s)	17.4
Ca (g)	37.0	$CaCO_3$ (s) (Calcite)	22.2
Ca^{++} (aq)	−11.4	$CaCO_3$ (s) (Aragonite)	21.2
CaO (s)	9.5	$CaC_2O_4 \cdot H_2O$ (s)	37.4
CaO (g)	52.3	$Ca_3(PO_4)_2$ (s) (α)	57.6
CaS (s)	13.5	$CaSiO_3$ (s) (Wollastonite)	19.6
CaF_2 (s)	16.4	$CaSO_4$ (s) (Natural Anhydrite)	25.5
CaH_2 (s)	9.9		
		Carbon	
C (Graphite)	1.36	CF_4 (g)	62.7
C (Diamond)	0.585	$COCl_2$	67.2
C (g)	37.77	CH_4	44.5
C_2 (g)	47.89	CN (g)	48.42
CO (g)	47.32	C_2N_2 (g)	57.9
CO_2 (g)	51.08	CNBr (g)	59.1
CS (g)	50.3	CNCl (g)	56.4
CS_2 (l)	36.2	CNI (g)	61.3
CS_2 (g)	56.84	$CO_3^{=}$ (aq)	−13.0
COS (g)	55.37	HCO_3^- (aq)	22.2
CBr_4 (g)	85.5	H_2CO_3 (aq)	45.1
CCl_4 (g)	74.2	CN^- (aq)	25
CCl_4 (l)	52.2	$C_2O_4^{=}$ (aq)	9.6
		Chlorine	
Cl_2 (g)	53.31	ClO^- (aq)	10.0
Cl (g)	39.47	ClO_2^- (aq)	24.1
Cl^- (aq)	13.5	ClO_3^- (aq)	39.4
Cl_2O (g)	63.7	ClO_4^- (aq)	43.6
ClO_2 (g)	59.6		
		Chromium	
Cr (s)	5.68	$CrCl_2$ (s)	27.7
Cr (g)	41.64	$CrCl_3$ (s)	31.0
Cr_2O_3 (s)	19.4	$CrO_4^{=}$ (aq)	10.5

TABLE XXXV (*Continued*)

		Cobalt	
Co (s)	6.8	$CoCl_2$ (s)	25.4
Co (g)	42.89		
		Copper	
Cu (s)	7.97	CuBr (s)	22.8
Cu (g)	39.75	CuBr (g)	59.9
Cu_2 (g)	58.9	CuCl (s)	20.8
Cu^{++} (aq)	−26.5	CuCl (g)	57.3
Cu_2O (s)	24.1	CuF (g)	54.9
CuO (s)	10.4	CuI (s)	23.1
Cu_2S (s)	28.9	CuI (g)	61.6
Cu_2S (s) (Covellite)	15.9	CuH (g)	46.9
$CuCO_3$ (s)	17.7	$CuSO_4$ (s)	25.3
		Fluorine	
F_2 (g)	48.58	F^- (aq)	−2.3
F (g)	37.93	F_2O (g)	58.95
		Helium	
He	30.13		
		Hydrogen	
H_2	31.23	H_2S (g)	49.15
D_2	34.62	HBr (g)	47.48
H (g)	27.40	HCl (g)	44.66
H^+ (aq)	0	HF (g)	41.53
H_2O (l)	16.75	HI (g)	49.36
H_2O (g)	45.13	HCN (l)	26.96
D_2O (l)	18.08	HCN (g)	48.23
D_2O (g)	47.38		
		Iodine	
I_2 (s)	27.9	IBr (s)	33.0
I_2 (g)	62.29	ICl (g)	59.11
I (g)	43.19	ICl_3 (s)	41.1
I^- (aq)	25.3	IO_3^- (aq)	28.0
IBr (g)	61.84		
		Iron	
Fe (s)	6.47	FeS (s)	16.1
Fe (g)	43.12	FeS_2 (s) (Pyrite)	12.7
Fe^{++} (aq)	−25.9	$FeCl_2$ (s)	29.4
Fe^{+++} (aq)	−61.0	Fe_3C (s)	24.2
FeO (s)	13.4	$FeCO_3$ (s) (Siderite)	22.2
Fe_2O_3 (s)	21.5	Fe_4N	37.4
Fe_3O_4 (s)	35.0		

TABLE XXXV (*Continued*)

		Lead	
Pb (s)	15.49	Pb_3O_4 (s)	50.5
Pb (g)	41.90	PbS (s)	21.8
Pb^{++} (aq)	3.9	PbS (g)	61.2
PbO (s) (Yellow)	16.6	$PbCl_2$ (s)	32.6
PbO (g)	57.4	$PbCl_2$ (g)	75.9
PbO_2 (s)	18.3	$PbSO_4$ (s)	35.2
		Lithium	
Li (s)	6.70	LiF (s)	9.7
Li (g)	33.15	LiI (g)	56.5
Li_2 (g)	47.0	LiH (s)	5.9
Li^+ (aq)	4.7	LiH (g)	40.8
LiCl (g)	49.9	LiOH (s)	12.8
Li_2CO_3	21.5		
		Magnesium	
Mg (s)	7.77	$Mg(OH)_2$ (s)	15.09
Mg (g)	35.51	$MgCO_3$ (s) (Magnesite)	15.7
Mg^{++} (aq)	−31.6	$MgSiO_3$ (s)	15.4
MgO (s)	6.55	$MgSO_4$ (s)	20
MgO (g)	50.7		
		Manganese	
Mn (s)	7.61	Mn_2O_3 (s)	22.9
Mn (g)	41.50	Mn_3O_4 (s)	35.5
MnO (s)	14.4	MnS (s)	18.7
MnO (g)	54.0	$MnCl_2$ (s)	25.9
MnO_2 (s)	13.9	$MnCO_3$ (Rhodochrosite)	20.5
MnO_4^- (aq)	46.7	MnC	33.6
		Mercury	
Hg (l)	18.5	$HgBr_2$ (g)	74.7
Hg (g)	41.8	$HgBr_2$ (s)	38.9
Hg_2^{++} (aq)	17.7	$HgCl_2$ (g)	70.4
HgO (s) (Red)	16.6	$HgCl_2$ (s)	34.6
HgO (s) (Yellow)	17.6	HgI_2 (g)	78.7
HgBr (s)	26.4	HgI_2 (s)	42.2
HgCl (s)	23.5	Hg_2SO_4 (s)	48.0
		Molybdenum	
Mo (s)	6.83	MoS_2 (s)	15.0
Mo (g)	43.47	MoS_3 (s)	15.9
		Neon	
Ne (g)	34.96		
		Nickel	
Ni (s)	7.12	$NiCl_2$ (s)	25.6
Ni (g)	43.53	$Ni(CO)_4$ (g)	97
NiO (s)	9.2		

TABLE XXXV (*Continued*)

		Nitrogen	
N_2 (g)	45.79	NOCl (g)	63.0
N (g)	36.62	NH_3 (g)	46.03
N_2O (g)	52.58	NH_4Cl (s)	31.8
NO (g)	50.34	NH_4HCO_3 (s)	28.3
NO_2 (g)	57.47	NH_4HS (s)	27.1
N_2O_4 (g)	72.7	NH_4OH (aq)	42.8
N_2O_5 (s)	36.6	NH_4^+ (aq)	26.4
N_2O_5 (g)	82	NO_2^- (aq)	29.9
NOBr (g)	65.2	NO_3^- (aq)	35.0
		HNO_3 (liq)	37.19
		Oxygen	
O_2 (g)	49.03	OH (g)	43.9
O (g)	38.48	OH^- (aq)	−2.49
O_3 (g)	57.1		
		Phosphorus	
P_2 (g)	52.1	PCl_5 (s)	40.8
P_4 (g)	66.8	PF_3 (g)	64.2
P (s) (White)	10.6	PH_3 (g)	50.35
P (g)	38.99	PN (g)	50.44
PBr_3 (g)	83.4	H_3PO_4 (aq)	44.0
PCl_3 (g)	74.7	$H_2PO_4^-$ (aq)	28.0
PCl_3 (l)	52.2	$HPO_4^{=}$ (aq)	−2.3
PCl_5 (g)	87.7	$PO_4^{\equiv}$ (aq)	−45
		Potassium	
K (s)	15.2	KCl (g)	57.7
K (g)	38.30	KF (g)	56.7
K_2 (g)	59.5	KI (s)	24.1
K^+ (aq)	24.2	KH (g)	47.3
KBr (s)	22.6	$KClO_3$ (s)	34 2
KBr (g)	60.4	KNO_3 (s)	31.8
KCl (s)	19.76	K_2SO_4 (s)	44.8
		Silicon	
Si (s)	4.50	SiO_2 (Cristobalite)	10.35
Si (g)	40.13	SiO_2 (Tridymite)	10.5
SiO_2 (Quartz)	10.1	SiO_2 (Glass)	11.2
$SiCl_4$ (g)	79.2	SiH_4 (g)	48.7
$SiCl_4$ (l)	57.3	SiC (s)	3.95
SiF_4 (g)	68.0	Si_3N_4 (s)	22.8
		Silver	
Ag (s)	10.20	AgBr (g)	62.1
Ag (g)	41.33	AgCl (g)	59.8
Ag^+ (aq)	17.54	AgCl (s)	23.0
Ag_2O (s)	29.1	$AgNO_3$ (s)	33.7
Ag_2S (α)	35.0	Ag_2SO_4 (s)	47.9
AgBr (s)	25.60	$Ag(NH_3)_2^+$ (aq)	57.8

TABLE XXXV (*Continued*)

		Sodium	
Na (s)	12.2	NaH (g)	45.0
Na (g)	36.72	NaOH (s)	14.2
Na_2 (g)	54.9	Na_2CO_3 (s)	32.5
Na^+ (aq)	14.0	$NaHCO_3$ (s)	24.4
Na_2O (g)	17	$NaNO_3$ (s)	27.8
NaBr (g)	58.1	Na_4SiO_4 (s)	46.8
NaCl (s)	17.3	Na_2SiO_3 (s)	27.2
NaCl (g)	55.5	$Na_2Si_2O_5$ (s)	39.4
NaF (s)	13.1	Na_2SO_4 (s)	35.7
NaF (g)	53.8	$Na_2SO_4 \cdot 10H_2O$ (s)	140.5
NaI (g)	60.0	NaK	58.6
		Sulfur	
S (rh)	7.62	SO_3 (l)	31.7
S (mono)	7.78	SF_6 (g)	69.6
S (g)	40.1	H_2S (aq)	29.4
S_2 (g)	54.41	HS^- (aq)	14.9
S_6 (g)	92	H_2SO_3 (aq)	54.7
S_8 (g)	109	HSO_3^- (aq)	32.6
SO (g)	53.07	$SO_3^{=}$ (aq)	3
SO_2 (g)	59.24	HSO_4^- (aq)	30.6
SO_3 (g)	63.8	$SO_4^{=}$ (aq)	4.4
		Tin	
Sn (white)	12.3	Sn (g)	40.24
Sn (gray)	10.7	Sn^{++} (aq)	−4.9
SnO (s)	13.5	$SnCl_4$ (g)	87.2
SnO_2 (s)	12.5	$SnCl_4$ (l)	62.1
		Titanium	
Ti (s)	6.6	$TiCl_4$ (g)	84.4
Ti (g)	43.07	$TiCl_4$ (l)	60.4
TiO_2 (Rutile)	12.4		
		Vanadium	
V (s)	7.0	V_2O_4 (s)	24.5
V (g)	43.55	V_2O_5 (s)	31.3
V_2O_3 (s)	23.5		
		Zinc	
Zn (s)	9.95	$ZnBr_2$ (s)	33.0
Zn (g)	38.46	$ZnCl_2$ (s)	25.9
Zn^{++} (aq)	−25.7	ZnI_2 (s)	38.5
ZnO (s)	10.4	$ZnCO_3$ (Smithsonite)	19.7
ZnO (g)	54.1	$ZnSO_4$ (s)	30.7
ZnS (s)	13.8		
		Zirconium	
Zr (s)	9.5	Zr (g)	43.33

TABLE XXXV (*Continued*)

ORGANIC COMPOUNDS

Ideal-gaseous state at 1 atm unless otherwise noted

Hydrocarbons

	$S°$	*Ref.*		$S°$	*Ref.*
Methane	44.46	28	Styrene (l)	56.78	11
Acetylene	48.08	28	Ethyl benzene (l)	61.3	26
Ethylene	52.75	28	Nonane (l)	94.0	26
Ethane	54.86	28	Mesitylene	92.34	29
Methyl acetylene (allylene)	56.83	20	*n*-Butyl benzene (l)	76.9	26
Propadiene (allene)	56.04	22	Neopentane	73.23	28
Propylene	65.06	28	Isopentane	82.03	28
Propane	64.69	28	Cyclopentane	70.70	4
Cyclopropane	56.79	21	Cyclopentane (l)	48.87	4
Dimethyl acetylene	67.93	33	*n*-Hexane	92.45	28
Dimethyl acetylene (l)	46.63	33	2-Methyl pentane	90.15	28
Isobutene	71.28	28	2-Methyl pentane (l)	69.9	28
cis-2-Butene	73.58	28	2, 2-Dimethyl butane	85.75	28
trans-2-Butene	72.98	28	2, 2-Dimethyl butane	64.4	28
1-Butene	75.38	28	2, 3-Dimethyl butane	86.49	28
Butane	74.21	28	3-Methyl pentane	89.96	28
Butane (l)	55.2	28	Benzene	64.39	29
Isobutane	70.51	28	Benzene (l)	41.49	29
Isobutane (l)	52.09	28	Cyclohexane	71.41	6
Trimethylethylene	79.7	18	Cyclohexane (l)	49.3	6
Pentane	83.23	28	Methyl cyclopentane (l)	59.3	26
2-Methyl hexane	99.48	28	*n*-Heptane	101.57	28
Methyl cyclohexane (l)	59.4	26	Naphthalene (s)	39.9	26
2, 2, 3-Trimethyl butane	92.31	28	Decane (l)	102.7	26
Toluene	76.44	29	*β*-Methyl naphthalene (s)	48.8	26
Toluene (l)	52.40	29	Duodecane	147	16
Ethyl cyclopentane (l)	67.1	26	Diphenyl (s)	49.2	26
n-Octane	110.69	28	*n*-Heptyl cyclohexane (l)	106.8	26
2, 2, 4-Trimethyl pentane	101.35	28	Anthracene (s)	49.6	26
2, 2, 3, 3-Tetramethyl butane	94.05	28	Phenanthrene (s)	50.6	26
o-Xylene	84.50	29	Pyrene (s)	51.4	26
o-Xylene (l)	58.80	29	*n*-Dodecyclohexane (l)	147.5	26
m-Xylene	85.60	29	1, 3, 5-Triphenyl benzene (s)	87.9	26
m-Xylene (l)	60.42	29	2-Methyl 2-butene	82.8	10
p-Xylene	84.27	29	3-Methyl 1-butene	82.0	10
p-Xylene (l)	59.20	29	2-Methyl 1-butene	81.0	10
Styrene	82.07	11			

Alcohols			*CHO Compounds*		
Methyl alcohol (g)	57.72	13	Formaldehyde (g)	52.42	32
Ethyl aclohol (g)	66.39	2	Acetone (g)	72.7	31
Ethyl alcohol (l)	38.4	31	Acetone (l)	47.9	31
Isopropyl alcohol (l)	43.0	31	Dimethyl ether (g)	63.72	19
Isopropyl alcohol (g)	73.4	31	Dimethyl ether (l)	44.98	19

TABLE XXXV (*Continued*)

Halogen Compounds	$S°$	*Ref.*	*Nitrogen Compounds*	$S°$	*Ref.*
Chloroform (g)	70.82	32	Methylamine (g)	57.73	5
Methylene chloride (g)	64.68	32	Methylamine (l)	35.90	5
Methyl chloride (g)	55.94	23	Dimethylamine (g)	65.24	3
Methyl chloride (l)	36.74	23	Dimethylamine (l)	43.58	3
Fluorotrichloromethane (g)	74.07	25	Methyl cyanide (g)	68.02	9
Fluorotrichloromethane (l)	53.92	25	Methyl isocyanide (g)	68.70	9
Ethylene dichloride (l)	49.84	27	Bromoform (g)	79.14	32
Ethylene dichloride (g)	72.74	27	Methylene bromide (g)	70.16	7
Phosgene (g)	67.24	32	Methyl bromide (g)	58.74	8
Acids			Ethylene dibromide (l)	53.37	27
			Ethylene dibromide (g)	79.37	27
Formic (g) (monomer)	60.0	12	Methyl fluoride (g)	53.30	7
Formic (g) (dimer)	83.1	12	Methyl iodide (g)	60.85	7
Acetic acid (g)	70.1	12	*Sulfur Compounds*		
Lactic acid (s)	34.30	15			
Hippuric acid (s)	57.2	14	Methyl mercaptan (g)	61.02	30
			Dimethyl sulfide (g)	68.28	24
Carbohydrates			Dimethyl sulfide (l)	46.94	24
1-Sorbose (s)	52.8	17			
β-Maltose monohydrate (s)	99.8	1			
α-Lactose monohydrate (s)	99.1	1			
β-Lactose (s)	92.3	1			
α-*d*-Galactose (g)	49.1	17			

REFERENCES

1. ANDERSON, A. G., and G. STEGEMAN, *J. Am. Chem. Soc.*, **63**, 2120 (1941).
2. ASTON, J. G., *Ind. Eng. Chem.*, **34**, 514–21 (1942).
3. ASTON, J. G., M. L. EIDINOFF, and W. S. FORSTER, *J. Am. Chem. Soc.*, **61**, 1539–43 (1939).
4. ASTON, J. G., H. L. FINK, and S. C. SCHUMANN, *J. Am. Chem. Soc.*, **65**, 341–6 (1943).
5. ASTON, J. G., C. W. SILLER, and G. H. MESSERLY, *J. Am. Chem. Soc.*, **59**, 1743–51 (1937).
6. ASTON, J. G., G. J. SZASZ, and H. L. FINK, *J. Am. Chem. Soc.*, **65**, 1135–9 (1943).
7. EDGELL, W. F., and G. GLOCKLER, *J. Chem. Phys.*, **9**, 484–5 (1941).
8. EGAN, C. J., and J. D. KEMP, *J. Am. Chem. Soc.*, **60**, 2097–2101 (1938).
9. EWELL, R. H., and J. F. BOURLAND, *J. Chem. Phys.*, **8**, 635–6 (1940).
10. EWELL, R. H., and P. E. HARDY, *J. Am. Chem. Soc.*, **63**, 3460–5 (1941).
11. GUTTMAN, L., E. F. WESTRUM, JR., and K. S. PITZER, *J. Am. Chem. Soc.*, **65**, 1246–7 (1943).
12. HALFORD, J. O., *J. Chem. Phys.*, **10**, 582–4 (1942).
13. HALFORD, J. O., and B. PECHERER, *J. Chem. Phys.*, **6**, 571–5 (1938).
14. HUFFMAN, H. M., *J. Am. Chem. Soc.*, **63**, 688–9 (1941).

15. HUFFMAN, H. M., E. L. ELLIS, and H. BORSOOK, *J. Am. Chem. Soc.*, **62**, 297–9 (1940).
16. HUGGINS, M. L., *J. Chem. Phys.*, **8**, 181 (1940).
17. JACK, G. W., and G. STEGEMAN, *J. Am. Chem. Soc.*, **63**, 2120 (1941).
18. KASSEL, L. S., *J. Chem. Phys.*, **4**, 435–41 (1936).
19. KENNEDY, R. M., M. SAGENKAHN, and J. G. ASTON, *J. Am. Chem. Soc.*, **63**, 2268 (1941).
20. KISTIAKOWSKY, G. B., and W. W. RICE, *J. Chem. Phys.*, **8**, 610–18 (1940).
21. LINNETT, J. W., *J. Chem. Phys.*, **6**, 700 (1938).
22. LINNETT, J. W., and W. H. AVERY, *J. Chem. Phys.*, **6**, 690 (1938).
23. MESSERLY, G. H., and J. G. ASTON, *J. Am. Chem. Soc.*, **62**, 886–90 (1940).
24. OSBORNE, D. W., R. N. DOESCHER, and D. M. YOST, *J. Am. Chem. Soc.*, **64**, 165 (1942).
25. OSBORNE, D. W., C. S. GARNER, R. N. DOESCHER, and D. M. YOST, *J. Am. Chem. Soc.*, **63**, 3496–9 (1941).
26. PARKS, G. S., *Chem. Rev.*, **27**, 75–83 (1940).
27. PITZER, K. S., *J. Am. Chem. Soc.*, **62**, 331–5 (1940).
28. PITZER, K. S., *Chem. Rev.*, **27**, 39–57 (1940).
29. PITZER, K. S., and D. W. SCOTT, *J. Am. Chem. Soc.*, **65**, 803–29 (1943).
30. RUSSELL, H., JR., D. W. OSBORNE, and D. M. YOST, *J. Am. Chem. Soc.*, **64**, 165–9 (1942).
31. SCHUMANN, S. C., and J. G. ASTON, *J. Chem. Phys.*, **6**, 485–8 (1938).
32. STEVENSON, D. P., and J. Y. BEACH, *J. Chem. Phys.*, **6**, 25–9, 108 (1938).
33. YOST, D. M., D. W. OSBORNE, and C. S. GARNER, *J. Am. Chem. Soc.*, **63**, 3492–6 (1941).

$[(H_T^\circ - H_0^\circ) + (\Delta H_f^\circ)_0]$ represents its enthalpy in the standard state at temperature T relative to the elements in their standard states at 0°K. Similarly, $[(G_T^\circ - H_0^\circ) + (\Delta H_f^\circ)_0]$ is the standard free energy of the compound at temperature T relative to the elements at 0°K. It is suggested by Aston[2] that these quantities be termed "single thermodynamic functions" and tabulated as such. Thus,

$$H_{fT}^\circ = (H_T^\circ - H_0^\circ) + (\Delta H_f^\circ)_0 \tag{19}$$

$$G_{fT}^\circ = (G_T^\circ - H_0^\circ) + (\Delta H_f^\circ)_0 \tag{20}$$

where

H_{fT}° and G_{fT}° = the *relative enthalpy* and *relative free energy*, respectively, of the compound at temperature T referred to the elements at 0°K

[2] J. G. Aston, *Chem. Rev.*, **27**, 63 (1940).

TABLE XXXVI

$-\left(\frac{G_T^\circ - H_0^\circ}{T}\right)$ FOR THE IDEAL-GASEOUS STATE AT 1 ATM

cal/(g-mole)(°K)

T, °K	298.1	400	500	600	800	1000	1500	$(\Delta H_f^\circ)_0$ kcal per g-mole
Methane	36.42	38.82	40.72	42.36	45.18	47.62	52.81	−15.96
Ethane	45.25	48.20	50.72	53.06	57.28	61.12	69.49	−16.48
Ethylene	44.05	46.70	48.8	50.8	54.4	57.5	64.2	14.51
Propane	52.83	56.62	59.98	63.13	69.00	74.44	86.3	−19.44
Propylene	54.3	57.6	60.6	63.3	68.3	72.9	82.8	8.58
n-Butane	58.54	63.56	68.02	72.16	79.93	87.12	102.65	−23.25
Butene-1	62.0	66.3	70.2	73.9	80.6	86.8	100.3	5.49
cis-2-butene	60.0	64.3	68.2	71.7	78.4	84.4	97.5	3.65
trans-2-butene	59.4	63.7	67.6	71.1	77.8	83.8	96.9	2.70
*iso*butane	56.14	60.85	65.13	69.21	76.90	83.99	99.54	−24.52
*iso*butene	57.0	61.6	65.6	69.4	76.3	82.5	95.9	1.64
n-Pentane	64.19	70.32	75.80	80.89	90.51	99.37	118.49	−27.03
2-Methyl butane	64.70	70.40	75.64	80.63	90.12	98.87	118.01	−28.45
Tetramethylmethane	56.28	61.87	67.13	72.18	81.71	90.55	109.96	−31.07
n-Hexane	69.86	77.12	83.63	89.69	101.15	111.68	134.40	−31.05
2-Methyl pentane	69.7			88.7		110.4		−32.2
3-Methyl pentane	69.7			88.6		110.4		−32.0
2,2-Dimethyl butane	66.0			84.6		106.4		−34.6
2,3-Dimethyl butane	66.3			85.4		107.3		−34.0
Acetylene	40.01	42.49	44.56	46.38	49.50	52.14	57.43	54.34
Graphite	0.545	0.854	1.180	1.510	2.164	2.798	4.206	0
Hydrogen H_2 (g)	24.436	26.438	27.965	29.218	31.204	32.752	35.605	0
H_2O (g)	37.191	39.529	41.316	42.789	45.153	47.039	50.647	−57.108
CO	40.364	42.408	43.963	45.238	47.271	48.876	51.880	−27.18
CO_2	43.578	45.848	47.681	49.261	51.921	54.137	58.513	−93.949
O_2	42.081	44.127	45.691	46.984	49.062	50.715	53.826	0
N_2	38.834	40.877	42.431	43.705	45.729	47.322	50.301	0
NH_3		43.338	45.205	46.782	49.402	51.580	55.968	
NO	42.985	45.140	46.769	48.100	50.214	51.878	54.979	21.579

TABLE XXXVII

$H_T^\circ - H_0^\circ$ FOR THE IDEAL-GASEOUS STATE AT 1 ATM

kcal per g-mole

T, °K	298.1	400	500	600	800	1000	1500
Methane	2.397	3.31	4.35	5.55	8.3	11.4	21.4
Ethane	2.865	4.27	6.02	8.03	12.78	18.37	34.56
Ethylene	2.52*	3.71*	5.12*	6.73*	10.48*	14.76*	27.10*
Acetylene	2.41						
Propane	3.535	5.59	8.08	11.06	17.91	25.92	48.77
Propylene	3.24*	4.99*	7.08*	9.49*	15.15*	21.69*	40.57*
n-Butane	4.66	7.43	10.77	14.63	23.68	34.13	63.74
*iso*butane	4.29	7.08	10.46	14.39	23.50	34.00	64.00
1-Butene	3.99						
cis-2-butene	4.06						
trans-2-butene	4.06						
*iso*butene	4.25						
n-Pentane	5.68	9.12	13.26	18.04	29.21	42.08	78.42
2-Methyl butane	5.17	8.63	12.84	17.70	28.98	41.91	78.69
Tetramethyl methane	5.05	8.56	12.84	17.81	29.37	42.54	79.50
n-Hexane	6.71	10.83	15.77	21.46	34.76	50.04	93.10
2-Methyl pentane	6.07			21.1		49.9	
3-Methyl pentane	6.04			21.1		50.1	
2, 2-Dimethyl butane	5.87			20.9		50.2	
2, 3-Dimethyl butane	6.020			21.3		50.3	
Graphite	0.251	0.51	0.83	1.20	2.07	3.07	6.0
Hydrogen, H_2 (g)	2.023	2.731	3.430	4.128	5.537	6.966	10.696
Water, H_2O (g)	2.365	3.190	4.019	4.874	6.669	8.583	13.89
CO	2.073	2.784	3.490	4.209	5.701	7.258	11.363
CO_2	2.240	3.197	4.227	5.328	7.697	10.233	17.02
O_2	2.069	2.789	3.524	4.280	5.855	7.499	11.77
N_2	2.074	2.772	3.485	4.188	5.672	7.230	11.220
NO	2.193	2.900	3.670	4.344	5.872	7.460	11.610

*Calculated from data of Table XXI, page 336.

Because of its smaller variation with temperature it is convenient to tabulate values of G_{fT}°/T. In these terms Equations (17) and (18) reduce to

$$\Delta H_T^\circ = \Sigma(H_{fT}^\circ)_P - \Sigma(H_{fT}^\circ)_R \tag{21}$$

$$\frac{\Delta G_T^\circ}{T} = \Sigma\left(\frac{G_{fT}^\circ}{T}\right)_P - \Sigma\left(\frac{G_{fT}^\circ}{T}\right)_R \tag{22}$$

It is evident that values of the single functions relative to the elements at 0°K are the most convenient form for the presentation and use of thermodynamic data. The disadvantage of this method at present is that the term $(\Delta H_f^\circ)_0$ is frequently based on relatively uncertain calorimetric determinations of heats of reaction or combustion whereas the groups $(H_T^\circ - H_0^\circ)$ and $(G_T^\circ - H_0^\circ)$ may be derived with considerable accuracy from spectroscopic and low-temperature heat-capacity data.

For this reason it is considered desirable to maintain separate tabulations of the component groups in order that improved calorimetric data may be more readily utilized.

In Table XXXVI are values of $(G_T - H_0^\circ)/T$ and $(\Delta H_f^\circ)_0$ for a number of compounds[3, 4] and elements in the ideal-gaseous state at 1 atm for seven temperatures. In Table XXXVII is a corresponding summary of values of $(H_T^\circ - H_0^\circ)$. By interpolation between the temperatures of the tables standard enthalpy and free-energy changes are obtained by Equations (17) to (22) with no uncertainties regarding the accuracy of empirical heat-capacity equations. Values of $(\Delta G^\circ/T)_T$ calculated from

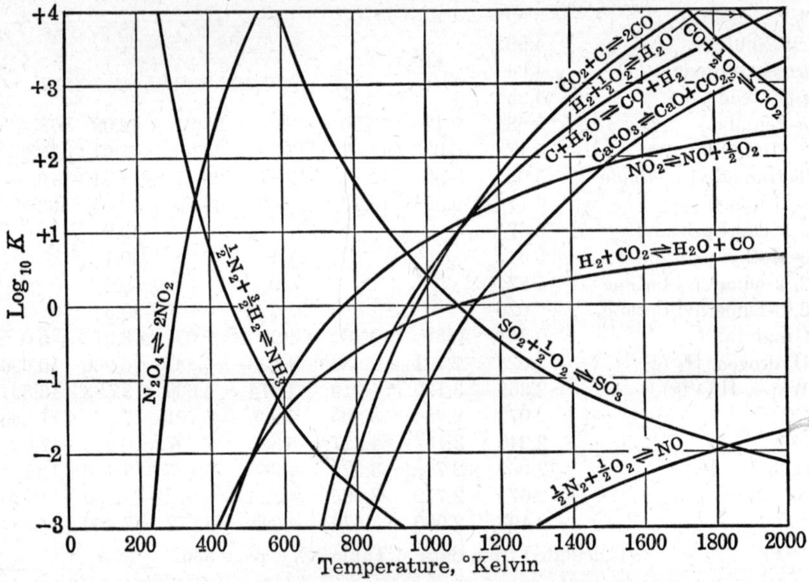

FIG. 156. Equilibrium Constants of Chemical Reactions.

(Reproduced in "C.P.P. Charts")

Equations (18) or (22) are directly related to the equilibrium constant. Thus rearranging Equation (6) gives

$$4.576 \log K = -\left(\frac{\Delta G^\circ}{T}\right)_T \tag{23}$$

where $(\Delta G^\circ/T)_T$ is expressed in cal/(g-mole)(°K) or Btu/(lb-mole)(°R).

In Fig. 156 are values of log K for several reactions plotted as functions of temperature. These values were calculated by Equation (23) using the data of Tables XIV, XV, XXXV, and XXXVII.

[3] K. S. Pitzer, *Chem. Rev.*, **27**, 39 (1940).

[4] F. D. Rossini, E. J. R. Prosen, and K. S. Pitzer, *J. Research Natl. Bur. Standards*, **27**, 529–41 (1941).

Illustration 4. From the data of Tables XXXVI and XXXVII calculate the values of ΔH_f° and $\Delta G_f^\circ/T$ for the formation of isobutene from the elements at 298.1°K

$$4C(s) + 4H_2(g) \longrightarrow C_4H_8(g)$$

From Tables XXXVI and XXXVII

	$iC_4H_8(g)$	$H_2(g)$	$C(s)$
$\left(\frac{G_T^\circ - H_0^\circ}{T}\right)_{298.1}$	−57.0	−24.436	−0.545
$(\Delta H_f^\circ)_0$	1.64	0	0
$(H_T^\circ - H_0^\circ)_{298.1}$	4.25	2.023	0.251

From Equation (14),

$$(\Delta H_f^\circ)_{298.1} = [4.25 + 1.64] - [4(0.251) + 4(2.023)] = -3.206 \text{ kcal per g-mole}$$

From Equation (16),

$$\left(\frac{\Delta G_f^\circ}{T}\right)_{298.1} = \left[-57.0 + \frac{1000 \times 1.64}{298.1}\right] - [4(-24.436) + 4(-0.545)]$$
$$= 48.43 \text{ cal/(g-mole)(°K)}$$

Illustration 5. From the data of Tables XXXVI and XXXVII calculate the values of ΔH°, $\frac{\Delta G^\circ}{T}$ and the equilibrium constant K for the dehydrogenation of propane to propylene at 800°K.

From Table XXXVI at 800°K,

	$C_3H_8(g)$	$H_2(g)$	$C_3H_6(g)$
$\left(\frac{G_T^\circ - H_0^\circ}{T}\right)$	−69.00	−31.204	−68.3
$(\Delta H_f^\circ)_0$	−19.440	0	8.580

From Table XXXVII,

$(H_T^\circ - H_0^\circ)$	17.91	5.537	15.15

For the reaction,

$$C_3H_8(g) = C_3H_6(g) + H_2(g)$$

From Equation (17),

$$\Delta H_{800}^\circ = [15.15 + 8.580 + 5.537] - [17.910 - 19.44] = 30.797 \text{ kcal per g-mole}$$

From Equation (18),

$$\left(\frac{\Delta G^\circ}{T}\right)_{800} = \left[-68.3 + \frac{8{,}580}{800} - 31.204\right] - \left[-69.00 - \frac{19{,}440}{800}\right]$$

$$= -68.3 - 31.204 + 69.00 + \frac{8{,}580 + 19{,}440}{800} = 4.521 \text{ cal/(g-mole)(°K)}$$

From Equation (23),

$$\log K = -4.521/4.576 = -0.9880 = -0.988$$
$$K = 0.103$$

log 1/K = −1 K = 0,1

1/K = 10

EQUILIBRIUM COMPOSITIONS

B. F. Dodge[5] has aptly discussed the feasibility of a chemical reaction thus:

"The statement is sometimes made that a given reaction is thermodynamically impossible. This is a loose statement which has no meaning in the absence of qualifying statements. For example, any reaction, starting with pure reactants uncontaminated by any of the products, will have a tendency to proceed to some extent even though this may be only infinitesimal. Thus, the reaction

$$H_2O(g) = H_2(g) + \tfrac{1}{2}O_2(g)$$

proceeds to some extent at 25°C, and we can even calculate with considerable assurance the percentage of water vapor that would be decomposed. From the accurately known value of $\Delta G°$ of this reaction at 25°C the equilibrium constant is about 1×10^{-40}, and the extent of decomposition is infinitesimally small but definite."

"From the value of the standard free-energy change for any reaction we can form an opinion about the feasibility of the reaction without further calculation. Thus, if $\Delta G° = 0$ at a given temperature, then $K = 1$, and it is obvious that the reaction must proceed to a considerable extent before equilibrium is reached. The situation becomes less favorable as $\Delta G°$ increases in the positive direction but there is no definite value that one can choose as clearly indicating that the reaction is not feasible from the standpoint of industrial operation. At 600°K the $\Delta G°$ for the methanol synthesis reaction is +11,000 calories per gram-mole, and yet the reaction is certainly feasible at this temperature. In this case the unfavorable free energy change for the standard state is partially overcome by utilizing high pressure to displace the equilibrium. Other means can also be used such as changing the ratio of reactants or removing one of the reaction products."

"For the purpose of ascertaining quickly and only approximately if any given reaction is promising at a given temperature, the following rough classification may be useful:

$\Delta G° < 0$	Reaction is promising
$\Delta G° > 0$ but $< +10{,}000$	Reaction is of doubtful promise but warrants further study
$\Delta G° > +10{,}000$	Very unfavorable, would be feasible only under unusual circumstances

It should be understood that these are only approximate criteria that are useful in preliminary exploratory work."

[5] B. F. Dodge, *Trans. Am. Inst. Chem. Engrs.*, **34**, 540 (1938).

From Equation (7) compositions of systems at equilibrium can be calculated from a knowledge of the standard free-energy change and the relations of activities to stoichiometric compositions. These relationships involve the activity coefficients discussed in Chapter XIV. If these coefficients are included, Equation (7) becomes, for the general case,

$$K = \frac{(\gamma_R N_R)^r (\gamma_S N_S)^s \cdots}{(\gamma_B N_B)^b (\gamma_C N_C)^c \cdots}$$

$$\frac{N_R^r N_S^s \cdots}{N_B^b N_C^c \cdots} = K \frac{\gamma_B^b \gamma_C^c \cdots}{\gamma_R^r \gamma_S^s \cdots} \tag{24}$$

Gaseous Systems. As previously mentioned, in a system which involves only gaseous components which behave as approximately ideal solutions, it is convenient to choose as the standard state the component gases each at unit fugacity expressed in atmospheres. This state is hypothetical at temperatures where saturation pressures are below unit fugacity. With this definition, the activity of component B in ideal solutions of gases is

$$a_B = f_B = N_B f_{B\pi} = N_B \nu_{B\pi} \pi \tag{25}$$

where

$f_{B\pi}$, $\nu_{B\pi}$ = the fugacity and fugacity coefficient, respectively, of pure component B at the temperature and total pressure π of the system

Combining (25) and (7) gives

$$K = \left(\frac{N_R^r N_S^s \cdots}{N_B^b N_C^c \cdots}\right)\left(\frac{\nu_R^r \nu_S^s \cdots}{\nu_B^b \nu_C^c \cdots}\right)\pi^{r+s+\cdots-b-c\cdots} \tag{26}$$

In a system containing n_I moles of inert gases not entering into the reaction,

$$N_R = \frac{n_R}{n_B + n_C + \cdots + n_R + n_S + \cdots + n_I} \tag{27}$$

where n_R = moles of R in the equilibrium mixture.

The ratio of the fugacity coefficients in Equation (26) is constant for a given temperature and pressure and is designated by K_ν.

$$K_\nu = \frac{\nu_R^r \nu_S^s \cdots}{\nu_B^r \nu_C^c \cdots} \tag{28}$$

However, unlike the equilibrium constant K, the term K_ν is affected by changes in pressure as well as in temperature.

For reactions at low pressures, in the neighborhood of atmospheric or below, the fugacity coefficients may be taken as unity and the term K_ν neglected. At higher pressures this term may have a marked effect on the equilibrium calculations and may be determined from Fig. 142, page 622.

Combining (26), (27), and (28), gives

$$\left[\frac{n_R^r n_S^s \cdots}{n_B^b n_C^c \cdots}\right] K_\nu \left[\frac{\pi}{n_B + n_C + \cdots + n_R + n_S + \cdots + n_I}\right]^{(r+s+\cdots)-(b+c+\cdots)} = K \quad (29)$$

Equation (29) is the most useful form of the equilibrium equation from which, if the equilibrium constant K is known, the composition of a reacting system at equilibrium may be directly calculated. This calculation is best carried out by expressing the number of moles of each active material present at equilibrium in terms of the equilibrium conversion x of a given reactant and the numbers of moles of components in the original unreacted mixture. These values are substituted in Equation (29), which is then solved for the equilibrium conversion. If the final equation to be solved is of a complicated form, graphical methods may offer the most convenient solution.

Illustration 6. The gases from the pyrites burner of a contact sulfuric acid plant have the following composition by volume:

SO_2	7.8%
O_2	10.8%
N_2	81.4%
	100.0%

This gaseous mixture is passed into a converter where in the presence of a catalyst the SO_2 is oxidized to SO_3. The temperature in the converter is maintained at 500°C and the pressure at 760 mm of Hg. Calculate the composition of the gases leaving the converter, assuming that equilibrium conditions are reached. It may be assumed that the fugacity coefficients and hence K_ν are equal to unity.

Solution:

Basis: 100 lb-moles of the original mixture.

$$SO_2\ (g) + \tfrac{1}{2}O_2\ (g) = SO_3\ (g)$$

From Fig. 156, at 500°C (773°K), $\log K = 1.93$; $K = 85$

$\pi = 1.0$ atm

Limiting reactant = SO_2

Let x = equilibrium conversion of SO_2. Composition of equilibrium mixture:

$$SO_2 = 7.8 - 7.8x \text{ lb-moles} = n_{SO_2}$$

$$SO_3 = 7.8x = n_{SO_3}$$

$$O_2 = 10.8 - 7.8\left(\frac{x}{2}\right) \text{ lb-moles} = n_{O_2}$$

$$N_2 = 81.4 \text{ lb-moles} = n_{N_2}$$

From Equation (29):

$$\frac{n_{SO_3}}{(n_{SO_2})(n_{O_2})^{\frac{1}{2}}}\left[\frac{\pi}{n_{SO_2} + n_{SO_3} + n_{O_2} + n_{N_2}}\right]^{-\frac{1}{2}} = K \quad \text{(a)}$$

Substituting in this equation gives

$$\frac{7.8x}{(7.8 - 7.8x)(10.8 - 3.9x)^{\frac{1}{2}}}\left[\frac{\pi}{100 - 3.9x}\right]^{-\frac{1}{2}} = K \qquad \text{(b)}$$

Since $\pi = 1.0$ and $K = 85$, for graphical solution:

$$\frac{x}{1.0 - x}\left(\frac{100 - 3.9x}{10.8 - 3.9x}\right)^{\frac{1}{2}} - 85 = \Delta$$

The solution is obtained by assuming a series of values of x and calculating the corresponding values of Δ. These results may be plotted, the value of x corresponding to $\Delta = 0$ being the correct solution. Since the quantity in parentheses varies but little when x is close to 1.0, the correct solution also may be readily obtained by trial without a plot. The use of this method gives

$$x = 0.9585 \text{ or } 95.85 \text{ per cent}$$

Composition of gases leaving the converter:

$SO_2 = 7.8 - (7.8)(0.9585)$	=	0.32 lb-mole	0.3%
$SO_3 = (7.8)(0.9585)$	=	7.49 lb-moles	7.8%
$O_2 = 10.8 - (3.9)(0.9585)$	=	7.07 lb-moles	7.3%
$N_2 =$		81.40 lb-moles	84.6%
Total	=	96.28 lb-moles	100.0%

This same procedure may be followed to calculate the degree of completion in any system at conditions for which the necessary data are available. By repeating the calculations of Illustration 6 to correspond to other conditions of temperature and pressure, the effects of varying these conditions may be quantitatively predicted.

Under conditions of high pressure the fugacity coefficients cannot be neglected, but the general method of solution is the same. Values of K_ν may be calculated directly from fugacity coefficients of the individual components obtained from Fig. 142. Where a number of calculations are to be made for a particular system of reactants and products a chart may be prepared relating K_ν to temperature and total pressure. Figure 156a is such a chart for the ammonia-synthesis system where

$$K_\nu = \frac{\nu_{NH_3}}{\nu_{N_2}^{\frac{1}{2}}\,\nu_{H_2}^{\frac{3}{2}}}$$

Illustration 7. Calculate the equilibrium percentage conversion of nitrogen to ammonia at 700°K and a pressure of 300 atm if the gas enters the converter with a composition of 75 mole per cent H_2 and 25 mole per cent N_2.

$$\tfrac{1}{2}N_2 + \tfrac{3}{2}H_2 = NH_3$$

At 300 atm, 700°K, $K = 0.0091$ (Fig. 156), and $K_\nu = 0.72$ (Fig. 156a).

Let x represent the number of moles of ammonia formed at equilibrium starting with $\frac{1}{2}$ mole N_2 and $\frac{3}{2}$ moles H_2. Then

$$\frac{n_{NH_3}}{n_{N_2}^{\frac{1}{2}} n_{H_2}^{\frac{3}{2}}} = \frac{x}{(\frac{1}{2} - \frac{1}{2}x)^{\frac{1}{2}}(\frac{3}{2} - \frac{3}{2}x)^{\frac{3}{2}}} = \frac{0.77x}{(1 - x)^2}$$

and

$$2 - x = \text{total number of moles at equilibrium}$$

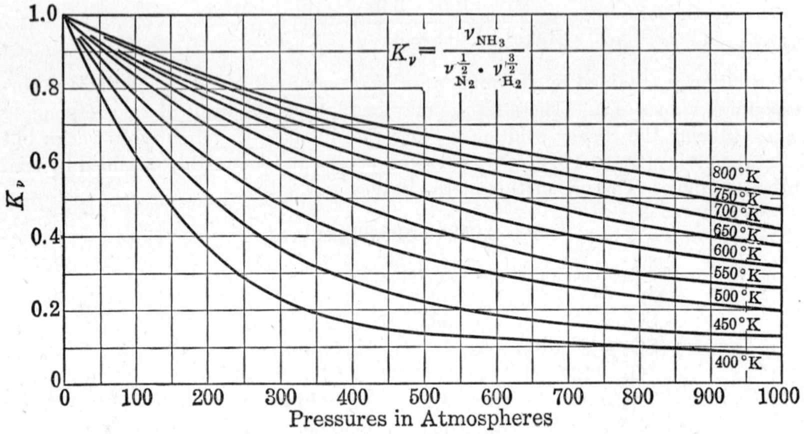

FIG. 156a. Ratio of Fugacity Coefficients for Ammonia Synthesis.

Substitution of the preceding values in Equation (29) gives

$$\left[\frac{0.77x}{(1 - x)^2}\right] 0.72 \left(\frac{300}{(2 - x)}\right)^{1-2} = 0.0091$$

or

$$x = 0.589$$

The percentage conversion of N_2 to NH_3 is hence 58.9 per cent, and the composition of the equilibrium mixture in mole per cent is

$$N_2 = 14.6 \text{ per cent}$$
$$H_2 = 43.8 \text{ per cent}$$
$$NH_3 = 41.6 \text{ per cent}$$

Effect of Reaction Conditions on Equilibrium Conversion. Equation (29) may be rearranged:

$$\frac{(n_R)^r (n_S)^s \cdots}{(n_B)^b (n_C)^c \cdots} = \frac{K}{K_\nu} \left[\frac{n_B + n_C + \cdots + n_R + n_S + \cdots + n_I}{\pi}\right]^{(r+s+\cdots) - (b+c+\cdots)} \quad (30)$$

From inspection of this equation the effects which are produced by changes in the conditions of the reaction on the equilibrium conversion can be predicted. Any change which will increase the right-hand side of Equation (30) will tend to increase the ratio of products to reactants in the equilibrium mixture and correspond to an increased conversion.

Effect of Temperature. From Equation (30) it is apparent that an increased value of the equilibrium constant K must correspond to an increased conversion. The value of the equilibrium constant for the commonly chosen standard state depends only upon temperature, as already discussed. It follows that the *equilibrium conversion is increased by a rise in temperature in the case of an endothermic reaction and decreased in the case of an exothermic reaction.*

Effect of Pressure. It has been pointed out that the equilibrium constant K is independent of pressure with the standard states for which Equation (30) was derived. The value of K_ν is affected by pressure in a manner for which generalized predictions cannot be made. When the compressibility of the products is greater than that of the reactants, an increase in pressure will decrease K_ν and hence increase the conversion. In addition, from inspection of Equation (30) it is apparent that the pressure under which a reaction proceeds will affect its equilibrium conversion in case the reaction produces a change in the total number of moles of gaseous components present in the system. If there is no change in the number of moles of gases, the exponent $(r + s + \cdots) - (b + c + \cdots)$ will equal zero, and the magnitude of the pressure π will have no effect on the extent of the reaction except as it affects K_ν.

However, *if a reaction produces a decrease in the total number of moles of gaseous components the equilibrium degree of completion is increased by an increase in pressure.* If the total number of moles of gases is increased as a result of the reaction, an increase in pressure reduces the equilibrium degree of completion. This is in accord with the classical Le Chatelier–Braun principle.

Effect of Dilution with an Inert Gas. Dilution of a reacting system with an inert gas corresponds to an increase of n_I of Equation (30). The effect produced is similar to that of a decrease in pressure. Hence, *if a reaction produces an increase in the number of moles of gaseous components the equilibrium degree of completion is increased by dilution with an inert gas.* If no change in the total number of moles of gases accompanies a reaction, the presence of inert gases has no effect on the equilibrium conversion.

Effect of Excess Reactants. If component B of Equation (30) is the limiting reactant, an increase in the number of moles of the other reactants $C \cdots$ increases the number of moles of products, R, S, $\cdots$, and also the degree of conversion of reactant B at equilibrium. Therefore, *the presence of excess of one reactant tends to increase the equilibrium conversion of the other reactant.*

Effect of Presence of Products in Initial Reacting System. It is apparent from Equation (30) that the presence in the original unreacted system of any of the compounds which are products of a reaction reduces the amounts of these compounds which are formed by the reaction in proceeding to equilibrium conditions. Therefore, the *addition of reaction products to the initial reaction system reduces the equilibrium conversion of any reactant.*

Equilibrium-Conversion Charts. Where numerous calculations are to be made of equilibrium composition corresponding to various conditions of temperature and pressure for a reacting system of *constant initial composition* it is frequently convenient to prepare charts expressing the equilibrium degree of completion as a function of temperature and pressure. Such a chart obviates the necessity of frequent repetitions of tedious graphical solutions of the type of Illustration 6.

Even in a complicated system the data for an equilibrium-conversion chart may be readily calculated by arbitrarily selecting a series of values of conversion. From Equation (29) the equilibrium constant is calculated corresponding to each of these selected degrees of completion and to a selected constant pressure. The temperature corresponding to each equilibrium constant may then be obtained from Fig. 156. These results are plotted, temperatures being used as abscissas and fractional conversion as ordinates. By repeating these calculations to correspond to other selected pressures a complete chart may be constructed. These calculations are best tabulated. In the following illustration this method is applied to the system discussed in Illustration 6.

Illustration 8. For the reacting system discussed in Illustration 6, calculate and plot curves relating the equilibrium conversion to temperature at pressures of 1.0 and 2.0 atm, respectively, assuming $K_v = 1.0$.

TABLE A

1	2	3	4	5	6	7	8
x	$1.0 - x$	$3.9x$	$10.8 - 3.9x$	$100 - 3.9x$	$[(5)/(4)]^{\frac{1}{2}}$	$(1)/(2)$	$K(\pi)^{\frac{1}{2}}$
0.50	0.5	1.95	8.85	98.0	3.33	1.0	3.33
0.70	0.30	2.73	8.07	97.3	3.47	2.33	8.09
0.90	0.10	3.51	7.29	96.5	3.64	9.0	32.8
0.95	0.05	3.70	7.10	96.3	3.68	19.0	70.1
0.98	0.02	3.82	6.98	96.2	3.71	49.0	181

Basis: 100 lb-moles of the original mixture.

Let x = fractional conversion of SO_2 at equilibrium

From Equation (b) of Illustration 6:

$$K(\pi)^{\frac{1}{2}} = \frac{(x)}{(1.0 - x)}\left(\frac{100 - 3.9x}{10.8 - 3.9x}\right)^{\frac{1}{2}}$$

A series of values of x are selected and the corresponding values of K calculated, first with π equal to 1.0 and then to 2.0 atm. The corresponding temperatures T°K are then obtained from Fig. 156. The calculations are summarized in Tables A and B. In the column headings the numbers in parentheses represent the results contained in the columns bearing these numbers.

TABLE B

	$\pi = 1.0$			$\pi = 2.0$		
x	K	$\log K$	T°K	K	$\log K$	T°K
0.5	3.33	0.523	978	2.35	0.371	1005
0.7	8.09	0.908	910	5.71	0.757	935
0.9	32.8	1.516	822	23.2	1.366	841
0.95	70.1	1.846	782	49.5	1.695	800
0.98	181	2.258	737	128	2.107	752

These calculations may be continued to correspond to other pressures. The results are plotted in curves E of Fig. 157 relating the equilibrium degree conversion as ordi-

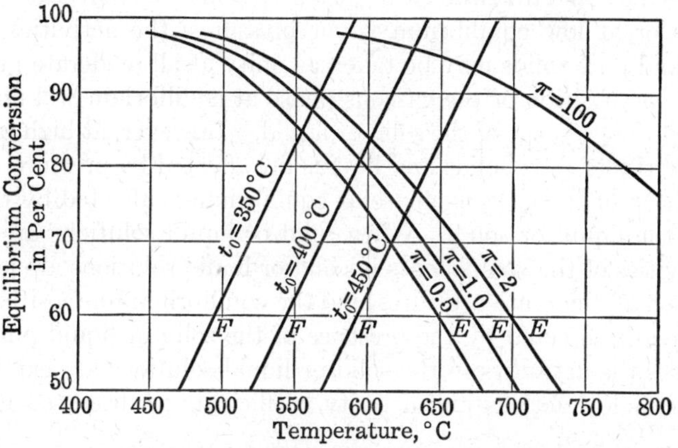

(Composition of Entering Gas: 7.8% SO_2, 10.8% O_2, 81.4% N_2)

FIG. 157. Equilibrium Conversion and Adiabatic Temperature Relations for the Oxidation of SO_2.

nates to temperature in degrees centigrade. This chart is applicable only to systems of the particular initial composition here considered. Change in relative proportions of any of the original reactants renders the chart inapplicable.

A chart of the type of curves E of Fig. 157 may be prepared for any reacting system and from it equilibrium conversion may be readily estimated by interpolation. For example, at a temperature of 600°C and a pressure of 1.5 atm it is estimated from the E curves of Fig. 157 that the equilibrium conversion is 82 per cent. If the initial composition of the reacting system undergoes change a new chart must be prepared to correspond to each different composition.

Heterogeneous Reactions. Equation (29) may be applied to any gaseous system if the gas mixture behaves as an ideal solution. When a component of a reaction is involved in a heterogeneous reaction as a pure liquid or pure solid, its activity may be taken as unity provided the pressure on the system does not differ much from the chosen standard state. The effect of pressure upon the activity of a solid or liquid may be calculated from Equation (XIV-30); this effect is negligible at moderate pressures.

If in the reaction represented by Equations (1) and (26) there are, in addition to gaseous components B, C, R and S, d moles of a solid or liquid reactant D and t moles of a solid or liquid product T, Equation (26) becomes

$$K = \frac{a_T^t}{a_D^d}\left(\frac{N_R^r N_S^s}{N_B^b N^c}\right)\left(\frac{\nu_R^r \nu_S^s}{\nu_B^b \nu_C^c}\right)\pi^{r+s-b-c} \qquad (31)$$

Where the standard state for solids and liquids is taken at atmospheric pressure or at low equilibrium vapor pressures, the activities of pure liquids and pure solids may be taken as unity at all moderate pressures, and the composition of the gaseous phase at equilibrium will not be affected by the presence of the solid or liquid. However, at high pressures the activities of pure solids and liquids are affected by pressure, and the composition of the gaseous phase at equilibrium is affected by the presence of the liquid or solid. When solid or liquid solutions are formed the activities of the components in solid or liquid solutions are no longer unity even at moderate pressures, and the equilibrium composition of the gas is greatly affected by the presence of the solid or liquid phase. In such cases the activities in the solid or liquid solutions are expressed in terms of mole fractions and activity coefficients as discussed in Chapter XIV.

Illustration 9. Ferrous oxide, FeO, is reduced to metallic iron by passing a mixture of 20 per cent CO and 80 per cent N_2 over it at a temperature of 1000°C under a pressure of 1 atm. Assuming that equilibrium is reached, calculate the weight of metallic iron produced per 1000 cu ft of gas admitted measured at 1 atm and 1000°C. Fugacity coefficients may be assumed equal to unity. The reaction taking place is as follows:

$$FeO\ (s) + CO\ (g) \rightleftarrows Fe\ (s) + CO_2\ (g)$$

At 1000°C the value of K for this reaction is 0.403.

Basis: 1.0 lb-mole of entering gas.

Since the activities of FeO and Fe equal 1.0, Equation (31) may be arranged in the form of (29). Then

$$\left(\frac{n_{CO_2}}{n_{CO}}\right)\left[\frac{\pi}{n_{CO} + n_{CO_2} + n_{N_2}}\right]^{1-1} = K \quad \text{or} \quad \frac{n_{CO_2}}{n_{CO}} = K$$

Let x = fractional conversion of CO at equilibrium. At equilibrium:

$$n_{CO_2} = 0.20x$$

$$n_{CO} = 0.20(1 - x)$$

$$\frac{0.20x}{0.20(1 - x)} = 0.403$$

or

$$x = 0.287$$

CO_2 produced = (0.287)(0.20) = 0.0574 lb-mole
Fe produced = 0.0574 lb-atom = 3.2 lb
Volume of entering gas at 1000°C = $(359)(\frac{1273}{273})$ = 1673 cu ft

Fe produced per 1000 cu ft of gas = $\frac{(3.2)(1000)}{1673} = 1.9$ lb

Pressures of Decomposition and Vaporization. Many solid compounds decompose to yield another solid and a gas as in the calcination of calcium carbonate to form lime and carbon dioxide. Such decompositions will proceed only when the activity of the gaseous product in contact with the solid is less than the equilibrium value determined by the temperature and the nature of the reaction.

A solid decomposition reaction may be represented by the following equation:

$$bB\text{ (s)} \rightleftarrows rR\text{(s)} + sS\text{ (g)} \tag{32}$$

The activities of the solids are approximately unity at moderate pressures so that as long as any of compounds B and R are present Equation (7) may be applied to this reaction as follows:

$$K = a_S^s = e^{-\frac{\Delta G^\circ}{RT}} = e^{-\frac{\Delta H^\circ}{RT} + \frac{\Delta S^\circ}{R}} \tag{33}$$

At low pressures, where the activity of a gas may be taken as equal to its partial pressure, $a_S^* = p_S^*$ and

$$\ln p_S^* = -\frac{\Delta G^\circ}{sRT} = -\frac{\Delta H^\circ}{sRT} + \frac{\Delta S^\circ}{sR} \tag{34}$$

It may be noted that if ΔH° and ΔS° are independent of temperature the form of Equation (34) is similar to that of the Clausius–Clapeyron vapor-pressure equation. In general both ΔH° and ΔS° vary somewhat with temperature but to a much smaller extent than ΔG°.

Illustration 10. Calculate the decomposition pressure of limestone at 1000°K and the temperature necessary to produce a decomposition pressure of 1.0 atm.

$$CaCO_3\text{ (s)} \rightleftarrows CaO\text{ (s)} + CO_2\text{ (g)}$$

From Table XXXV,

$$\Delta S^\circ_{298.1} = 9.5 + 51.1 - 22.2 = 38.4$$

From Table XIV,

$$\Delta H^\circ_{291} = (-151.7) + (-94.03) - (-289.3) = 43.57 \text{ kcal per g-mole}$$

From Table V, for CO_2:

$$c_p = 6.85 + (8.533)(10^{-3})T - (2.475)(10^{-6})T^2$$

From the data of Kelley,[6]

for CaO: $$c_p = 10.00 + (4.84)(10^{-3})T - \frac{108{,}000}{T^2}$$

for $CaCO_3$: $$c_p = 19.68 + (11.89)(10^{-3})T - \frac{307{,}600}{T^2}$$

Using the symbols of Equation (9), yields

$$\Delta a = 6.85 + 10.00 - 19.68 = -2.83$$
$$\Delta b = (8.533 + 4.84 - 11.89)10^{-3} = (1.483)(10^{-3})$$
$$\Delta c = -(2.475)(10^{-6})$$
$$\Delta d = (-108{,}000) - (-307{,}600) = 199{,}600$$

Substitution in Equation (9) gives

$$\Delta H_{291} = I_H + \Delta aT + \tfrac{1}{2}\Delta bT^2 + \tfrac{1}{3}\Delta cT^3 - \frac{\Delta d}{T}$$

$$43{,}570 = I_H + (-2.83)(291) + \tfrac{1}{2}(1.483)(10^{-3})(291)^2 + \tfrac{1}{3}(-2.475)(10^{-6})(291)^3 - \frac{199{,}600}{291}$$

or $$I_H = 45{,}037 \quad \text{and} \quad \Delta H^\circ_{1000} = 41{,}924$$

Substitution in Equation (10) gives

$$\Delta S^\circ_{298.1} = I_S + \Delta a \ln T + \Delta bT + \tfrac{1}{2}\Delta cT^2 - \frac{\Delta d}{2\,T^2}$$

$$38.4 = I_S + (-2.83)(5.697) + (1.483)(10^{-3})(298.1) + \tfrac{1}{2}(-2.475)(10^{-6})(298.1)^2 - \frac{1}{2}\left(\frac{199{,}600}{(298.1)^2}\right)$$

or $$I_S = 55.31 \quad \text{and} \quad \Delta S^\circ_{1000} = 35.90$$

From Equation (34),

$$\log p^*_S = \frac{1}{2.303}\left[\frac{-41{,}924}{(1.987)(1000)} + \frac{35.90}{1.987}\right] = -1.316$$

or $$p_S = 0.0483 \text{ atm at } 1000°\text{K}$$

At a decomposition pressure of 1 atm, from Equation (34) $\Delta G^\circ_T = 0$. The corresponding temperature T is calculated from Equation (11):

$$\Delta G^\circ_T = I_H + (\Delta a - I_s)T - \Delta aT \ln T - \tfrac{1}{2}\Delta bT^2 - \tfrac{1}{6}\Delta cT^3 - \frac{\Delta d}{2T}$$

or, substitution of known values gives

$$\Delta G^\circ_T = 45{,}037 - 58.14T + 6.517T \log T - 0.7415(10^{-3})T^2 + (0.4125)(10^{-6})T^3 - \frac{99{,}800}{T} = 0$$

This equation is solved graphically or by trial and error to give

$$T = 1180°\text{K, the decomposition temperature of } CaCO_3 \text{ at } 1.0 \text{ atm}$$

Equations (23) and (34) are directly applicable to the equilibrium between a pure solid or liquid and its vapor. In this case ΔG°, ΔH°, ΔS°

[6] K. K. Kelley, "High-Temperature Specific Heat Equations for Inorganic Substances," *U.S. Bur. Mines Bull.*, 371 (1934).

represent the changes in free energy, enthalpy, and entropy, respectively, in the isothermal transformation of the solid or liquid to the ideal vapor at a pressure of 1.0 atm. In vaporization at low pressures $\Delta H°$ becomes equal to the heat of vaporization at the given temperature.

Illustration 11. Calculate the vapor pressure and latent heat of vaporization of molten copper at 3,000°K from the following data:

For solid copper at 298.1°K: Cu(s) ⇆ Cu(g)

$$\mathrm{H}° - \mathrm{H}_0° = 1199 \text{ cal per g-atom}$$
$$\mathrm{S}° = 7.92$$

For copper vapor in the ideal state at 298.1°K and 1.0 atm relative to solid copper at 0°K:

$$\mathrm{H}° - \mathrm{H}_0° = 82{,}720$$
$$\mathrm{S}° = 39.75$$

Atomic heat capacities (above 298°K):

$$\text{Solid copper, } c_p = 5.44 + 0.001462T$$
$$\text{Liquid copper, } c_p = 7.50$$
$$\text{Gaseous copper, } c_p = 4.97$$

Heat of fusion at 1357°K (melting point) = 3,110 cal per g-atom

For liquid copper at 3,000°K:

$$\mathrm{H}° - \mathrm{H}_0° = 1199 + \int_{298}^{1357} (5.44 + 0.001462T)\,dT + 3110 + 7.50(3000 - 1357) = 23{,}670$$

$$\mathrm{H}° - \mathrm{H}_0° \text{ for copper vapor at 3000°K} = 82{,}720 + 4.97(3000 - 298)$$
$$= 96{,}120 \text{ cal per g-atom}$$

For liquid copper at 3,000°K:

$$\mathrm{S}° = 7.92 + \int_{298}^{1357} \frac{(5.44 + 0.001462T)}{T}\,dT + \frac{3110}{1357} + 7.50 \ln \frac{3000}{1357} = 26.02$$

$$\mathrm{S}° \text{ for copper vapor at 3,000°K} = 39.75 + 4.97 \ln \tfrac{3000}{298} = 39.75 + 11.45$$
$$= 51.20 \text{ cal/(g-atom)(°K)}$$

In vaporization at 3,000°K:

$$\Delta\mathrm{H}° = 96{,}120 - 23{,}670 = 72{,}450 \text{ cal per g-atom}$$
$$\Delta\mathrm{S}° = 51.20 - 26.02 = 25.18 \text{ cal/(g-atom)(°K)}$$
$$\Delta\mathrm{G}° = \Delta\mathrm{H}° - T\Delta\mathrm{S}° = 72{,}450 - 3000(25.18) = -3090 \text{ cal per g-atom}$$

From Equation (23)

$$\log K = \frac{3090}{(4.576)(3000)} = 0.225$$

or

$$K = P\nu = 1.68$$

Since at this moderate pressure ν is approximately unity the vapor pressure P is 1.68 atm.

ADIABATIC REACTION EQUILIBRIUM TEMPERATURES

In Chapter VIII methods were demonstrated whereby the temperature attained by a reacting system may be calculated if the reaction goes to completion without loss of heat from the system. In a reaction reaching equilibrium conditions which do not approach 100 per cent conversion, the temperature attained depends upon the degree of conversion actually produced, which in turn depends on temperature. The equilibrium temperature attained by such a reaction when proceeding adiabatically may best be determined by a graphical calculation, making use of an equilibrium-conversion chart for the system.

On an equilibrium-conversion chart of the type of curves E of Fig. 157, page 721, a curve is plotted relating calculated adiabatic reaction temperature to conversion. This curve may be established by selecting a series of values of degrees of conversion and calculating the reaction temperature corresponding to each by the method discussed in Chapter VIII,

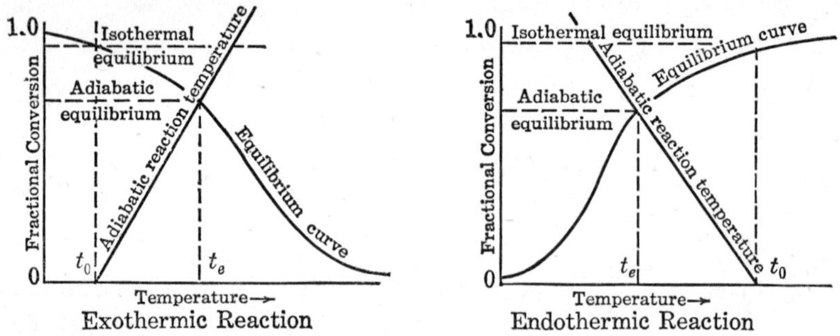

FIG. 158. Equilibrium Conversion in Exothermic and Endothermic Reactions for Isothermal and Adiabatic Operations.

page 308. The values on this curve are independent of pressure if it is assumed that enthalpies are independent of pressure. However, a new curve must be plotted to correspond to each change in the initial temperatures of any of the reactants.

In reactions proceeding adiabatically the characteristic behaviors of exothermic and endothermic reactions are shown diagrammatically in Fig. 158. In exothermic reactions the equilibrium conversion decreases with increase in temperature whereas the temperature steadily rises as the reaction proceeds. The adiabatic reaction obviously cannot proceed beyond the intersection of these two lines. In endothermic reactions the reverse is true: the equilibrium degree of conversion is favored by a rise in temperature whereas the temperature falls as the reaction proceeds. It may be noted that isothermal operation leads to a higher equilibrium conversion in either an endothermic or exothermic reaction.

Illustration 12. The mixture of 7.8 per cent SO_2, 10.8 per cent O_2, and 81.4 per cent N_2 discussed in Illustrations 6 and 8 enters the converter at a temperature of 400°C. Calculate the equilibrium temperature attained in the converter, assuming that it is thermally insulated so that the heat loss is negligible.

Solution:

Basis: 100 g-moles of the original gaseous mixture.

Reference temperature: 18°C.

Let x = degree of completion

From the heat-of-formation data, Table XIV, page 257,

$$\Delta H = (-93{,}900) - (-70{,}920) = -22{,}980 \text{ cal per g-mole}$$

$$SO_2 \text{ converted} = 7.8x \text{ g-moles}$$

$$\text{Total standard heat of reaction} = (7.8x)(22{,}980) = 179{,}240x \text{ cal}$$

Enthalpy of Reactants: Mean heat capacity between 18° and 400°C is obtained from Table VI, page 216.

$$SO_2 = (10.91)(7.8)(400 - 18) = 32{,}510$$
$$O_2 = (7.40)(10.8)(400 - 18) = 30{,}530$$
$$N_2 = (7.09)(81.4)(400 - 18) = 220{,}460$$
$$\text{Enthalpy of reactants} = 283{,}500$$

Enthalpy of Products Relative to 18°C:

$$\Delta H = \int_{291}^{T} nc_p\, dT; \qquad T \text{ is in °K}$$

$$SO_2 = (7.8 - 7.8x)\int_{291}^{T} [8.12 + (6.825)(10^{-3})T - (2.103)(10^{-6})T^2]\, dT$$

$$SO_3 = 7.8x\int_{291}^{T} [8.20 + (10.236)(10^{-3})T - (3.156)(10^{-6})T^2]\, dT$$

$$O_2 = (10.8 - 3.9x)\int_{291}^{T} [6.13 + (2.99)(10^{-3})T - (0.806)(10^{-6})T^2]\, dT$$

$$N_2 = 81.4\int_{291}^{T} [6.30 + (1.819)(10^{-3})T - (0.345)(10^{-6})T^2]\, dT$$

By integration and addition the total enthalpy of products

$$= [643.4(T - 291) + (116.7)(10^{-3})(T^2 - 291^2) - (17.74)(10^{-6})(T^3 - 291^3)]$$
$$- x[23.3(T - 291) - (7.5)(10^{-3})(T^2 - 291^2) + (1.69)(10^{-6})(T^3 - 291^3)]$$

From an energy balance, solving for x, there results

$$x = \frac{-480{,}170 + 643.4T + (116.7)(10^{-3})T^2 - (17.74)(10^{-6})T^3}{173{,}060 + 23.3T - 7.5(10^{-3})T^2 + (1.69)(10^{-6})T^3}$$

This equation permits direct calculation of values of x corresponding to selected values of T.

t°C	x
519	0.5
570	0.7
627	0.95

These values are plotted in the F curve of Fig. 157, page 721, corresponding to an initial temperature t_0, of 400°C. This curve crosses the E curve corresponding to a pressure of 1.0 atm at the point corresponding to a temperature of 598°C and a degree of completion of 82.5 per cent. These are the equilibrium conditions of the reaction.

The F curves of Fig. 157 relate calculated reaction temperatures to conversion, each corresponding to a different initial temperature of the reactants entering the converter. These curves were established by the method demonstrated in Illustration 12. Such a chart may be constructed in a similar manner to apply to any system for which the necessary data are available. From Fig. 157 the equilibrium temperature and degree of completion corresponding to any selected adiabatic operating conditions may be readily estimated. For example, if the reacting gases enter the converter at a temperature of 375°C and a pressure of 1.5 atm, an equilibrium temperature of 585°C and a degree of completion of 87 per cent will be attained.

If a reaction cannot be assumed to proceed under adiabatic conditions, a chart of the type of Fig. 157 can be constructed which takes into account the loss of heat from the reacting system. Such losses in no way affect the E curves. However, the heat losses, expressed as a function of the temperature of the system, must be included in the energy-balance equation from which the data for the F curves are calculated.

Equilibrium-Reaction Temperatures at High Pressures. When reactions proceed at high pressures, the method described for obtaining the equilibrium-reaction temperature should be modified to take into consideration the effect of pressure upon enthalpies, heat capacities, and activities of the component gases. For example, if a 1:3 mixture of nitrogen and hydrogen enters a catalytic reaction chamber at 400°C and reacts adiabatically to equilibrium at 300 atm, the temperature leaving will be 590°C with a corresponding percentage conversion of nitrogen to ammonia of 25 per cent. If the results were calculated with the effect of pressure upon heat capacities and activities neglected, the calculated results would be 600°C with a corresponding conversion of 21 per cent. The smaller temperature rise under actual conditions is due to the increased heat capacity of ammonia at 300 atm, and the increased percentage conversion is due to the decrease in activities of ammonia as compared to nitrogen and hydrogen.

Equilibrium-reaction temperatures are readily calculated by the general methods already demonstrated if ideal solutions are formed. This condition is generally satisfactorily approximated where all reactants and products are well above their critical temperatures. For other cases involving reactions at conditions in the vicinity of the critical point of the mixture, the precise methods for estimating activities described in Chapter XIV must be used.

Illustration 13. It is desired to calculate the equilibrium percentage conversion of SO_2 to SO_3 and the corresponding temperature when the system described in Illustration 12 reacts adiabatically at 100 atm and the entering temperature is 400°C.

The equilibrium-temperature line E, corresponding to 100 atm, is constructed as in Illustration 8 except that the deviation of K_v from unity is calculated from Fig. 142 as in Illustration 7. The adiabatic heating line F is constructed by the method employed in Illustration 12, except that allowance is made for the effect of the pressure of 100 atm upon the enthalpies of the entering and reacting mixtures. These corrections are obtained from Fig. 106. The procedure is as follows:

The critical constants of the component gases are:

	T_c°K	p_c atm
SO_2	430.2	77.7
N_2	125.9	33.5
O_2	154.2	49.7
SO_3	491.3	83.6

For the entering gas, which contains 7.8 per cent SO_2, 81.4 per cent N_2, 10.8 per cent O_2:

Pseudocritical temperature = 0.078(430.2) + 0.814(125.9) + 0.108(154.2) = 152°K

Pseudocritical pressure = 0.078(77.7) + 0.814(33.5) + 0.108(49.7) = 38.6 atm

$$\text{Pseudoreduced temperature, } T_r' = \frac{673}{152} = 4.40$$

$$\text{Pseudoreduced pressure, } p_r' = \frac{100}{38.6} = 2.59$$

From Fig. 106,
$$\frac{\text{H}^* - \text{H}}{T_c} = 0$$

The effect of the elevated pressure on the enthalpy at 400°C is negligible. When the gas has been heated adiabatically by reaction to equilibrium, the approximate conversion is 96% at 631°C.

The corresponding composition is

$SO_2 = 7.8(1 - 0.96)$	=	0.31 moles	0.3%
N_2	=	81.40 moles	84.6%
$O_2 = 10.8 - \dfrac{(7.8 - 0.31)}{2}$	=	7.05 moles	7.3%
$SO_3 = 7.8 - 0.31$	=	7.49 moles	7.8%
		96.25 moles	100.0%

The correction of enthalpy for pressure may be shown by Fig. 106 to be negligible and the adiabatic line at 100 atm coincides with the adiabatic line for 1 atm.

EQUILIBRIUM OF SIMULTANEOUS REACTIONS

Where several reactions occur among a given group of reactants, the composition of the resultant products at equilibrium depends upon the simultaneous equilibria of all the separate reactions. If large numbers of products are formed as in the pyrolysis of heavy hydrocarbons, or of cellulose, the resultant composition cannot ordinarily be calculated. However, even in such cases where many reactions are possible, restriction to a single course may be accomplished by the choice of a specific catalyst which promotes one selected reaction. This selection is dependent on kinetic rather than on thermodynamic considerations, and it is common for equilibrium to be approached with respect to one or two reactions while many other reactions which are thermodynamically possible do not occur to any appreciable extent. Under such circumstances thermodynamic methods are only partly successful and must be supplemented by knowledge of the actual course of the reaction and nature of the products formed before equilibrium calculations can be carried out.

For example, in the reaction of carbon monoxide with hydrogen the number of products theoretically possible is almost unlimited and the free energy in the formation of many of these products may be extremely favorable; yet it is possible by the selection of a special catalyst to exclude consideration of nearly all possible products save methanol. An apparent equilibrium yield can be established by considering only the reaction $CO + 2H_2 \rightleftharpoons CH_3OH$, and ignoring all other possible reactions. However, such an equilibrium is valid only with respect to the one reaction and is not a true equilibrium for all possible reactions. If the reacting system is permitted to remain at reaction conditions for an indefinite time the other relatively slow reactions will take place with resultant changes in composition and ultimately a true equilibrium involving all possible products will be reached. The calculation of such true-equilibrium compositions for complex systems is of little practical significance, and consideration is ordinarily limited to the relatively few rapid reactions.

Isomerization. The simplest case of simultaneous reactions is encountered in the isomerization of organic compounds, as, for example, the isomerization of normal butane to isobutane. Similarly, normal pentane may be isomerized to either isopentane (2-2-methyl butane) or neopentane (2-dimethyl propane). Isomers in equilibrium proportions are produced in many reactions, particularly in the presence of catalysts.

Calculation of equilibrium compositions in isomerization is simplified by the monomolecular nature of the reactions and by the fact that the physical properties of the isomers are so similar that deviations from ideal behavior cancel. The composition of the equilibrium mixture of isomers is calculated on the basis of a unit quantity of the starting compound in the equilibrium mixture. The corresponding quantity of each isomer is then equal to the equilibrium constant of the reaction by which it is formed.

Illustration 14. From the data of Table XXXVI calculate the composition of the mixture obtained by isomerizing *n*-pentane to equilibrium at 400°K.

	n-Pentane	*Isopentane*	*Neopentane*
$\left(\frac{G^\circ - H_0^\circ}{T}\right)_{400}$	−70.32	−70.40	−61.87
$(\Delta H_f^\circ)_0$	−27.030	−28.450	−31.070
$(\Delta G^\circ/T)_{400}$	0	−3.63	−1.65
K	1	6.21	2.30

The standard free-energy change $(\Delta G°/T)_{400}$ accompanying the isomerization of the normal compound is calculated by the method of Illustration 5. The composition of an equilibrium mixture containing 1 mole of n-pentane is obtained directly from Equation (29) to give the following:

	Moles	*Mole per Cent*
n-Pentane	1.0	10.5
Isopentane	6.21	65.4
Neopentane	2.30	24.1
Total	9.51	100.0

By repeating the calculations of Illustration 14 at other temperatures the complete isomerization equilibrium diagram is developed as shown in Fig. 159. From this diagram it is evident that low temperatures are required for maximum production of neopentane while isopentane is at a maximum at approximately 500°K. Similar diagrams for more complex systems are readily developed once the basic thermodynamic constants are evaluated.

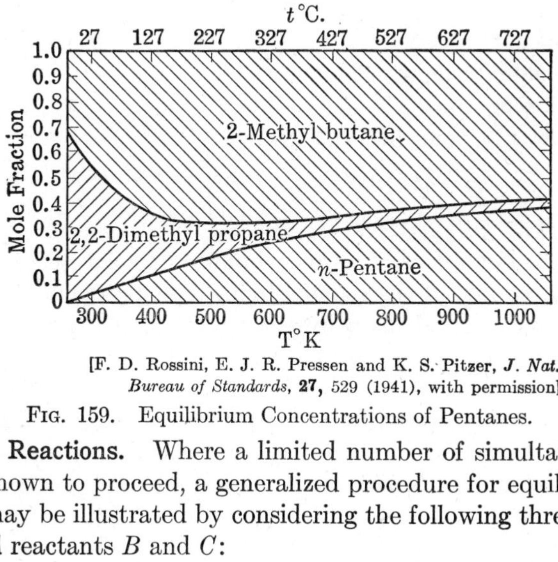

[F. D. Rossini, E. J. R. Pressen and K. S. Pitzer, *J. Nat. Bureau of Standards*, **27**, 529 (1941), with permission]

FIG. 159. Equilibrium Concentrations of Pentanes.

Complex Reactions. Where a limited number of simultaneous reactions are known to proceed, a generalized procedure for equilibrium calculations may be illustrated by considering the following three reactions from initial reactants B and C:

(1) $$bB + cC \rightleftarrows rR + sS$$

(2) $$bB + rR \rightleftarrows tT + uU$$

(3) $$cC + sS \rightleftarrows vV$$

If each reaction proceeded to completion, the over-all equation would be

(4) $$2bB + 2cC = tT + uU + vV$$

If the foregoing are all gaseous reactions the respective equilibrium compositions are expressed by Equation (29):

$$K_1 = \left(\frac{n_R^r n_S^s}{n_B^b n_C^c}\right) K_{\nu 1} \left(\frac{\pi}{\Sigma n}\right)^{(r+s)-(b+c)} \tag{35}$$

$$K_2 = \left(\frac{n_T^t n_U^u}{n_B^b n_R^r}\right) K_{\nu 2} \left(\frac{\pi}{\Sigma n}\right)^{(t+u)-(b+r)} \tag{36}$$

$$K_3 = \left(\frac{n_V^v}{n_C^c n_S^s}\right) K_{\nu 3} \left(\frac{\pi}{\Sigma n}\right)^{v-(c+s)} \tag{37}$$

where $\Sigma n = n_B + n_C + n_R + n_S + n_T + n_U + n_V + n_I$

From Equation (6):

$$\Delta G_1^\circ = -RT \ln K_1 \tag{38}$$

$$\Delta G_2^\circ = -RT \ln K_2 \tag{39}$$

$$\Delta G_3^\circ = -RT \ln K_3 \tag{40}$$

Combining (38), (39), and (40) gives

$$\Delta G_4^\circ = \Delta G_1^\circ + \Delta G_2^\circ + \Delta G_3^\circ = -RT \ln K_1K_2K_3 = -RT \ln K_4 \tag{41}$$

or

$$K_4 = K_1K_2K_3 \tag{42}$$

Thus, the equilibrium constant of the over-all reaction is the product of the equilibrium constants of the intermediate reactions. By substitution of (35), (36), and (37) in (42)

$$K_4 = \left(\frac{n_T^t n_U^u n_V^v}{n_B^{2b} n_C^{2c}}\right) K_{\nu 4} \left(\frac{\pi}{\Sigma n}\right)^{t+u+v-2b-2c} \tag{43}$$

Equation (43) is applicable to any equilibrium mixture of the system under consideration, though the intermediate products R and S are present but are not indicated by the equation. However, by itself it does not completely define the composition of the equilibrium mixture and fixes only the relative proportions of components B, C, T, U, and V. Thus, an over-all equilibrium equation representing the result of a sequence of reactions is useful for the calculation of complete equilibrium compositions *only when intermediate products are not present in significant quantities.*

In the general case where all intermediate and final products must be considered, it is necessary that the equilibrium equations of all reactions be satisfied by the composition of the system at equilibrium. Determination of the equilibrium composition therefore requires simultaneous solution of the independent equilibrium equations. The number of equations to be solved is equal to the number of independent equations involved. For example, in the system just considered there are three independent reactions, (1), (2), and (3). Reaction (4) is not independent since it results from combination of the others. Thus, simultaneous solution of the equilibrium Equations (35), (36), and (37) completely establishes the equilibrium composition, and this composition will of necessity satisfy Equation (43). It is a general rule that *the number of independent reactions which must be considered is the least number which includes every reactant and product present to an appreciable extent in all phases of the equilibrium and accounts for the formation of each product from the original reactants.*

For simultaneous solution of the equilibrium equations a definite quantity of the initial system is selected as a basis, and the number of moles converted by each reaction in proceeding to equilibrium is designated as an algebraic unknown. For example, in the system considered above an equilibrium calculation may be based on 100 moles of the initial reactants. The number of moles of B converted by reaction (1) in going to equilibrium may be designated as x. Similarly, y may represent the moles of B converted by reaction (2), and z the moles of C converted by reaction (3). The number of moles of all components in Equations (35), (36), and (37) may be expressed in terms of these three variables x, y, and z, which are then evaluated by simultaneous solution. The solution may require a graphical or trial-and-error procedure where the equations are complex.

These principles are well illustrated by consideration of the catalytic process for the production of hydrogen and carbon monoxide by the reaction of steam and methane at high temperatures. This process is extensively used for obtaining both pure hydrogen and mixtures of hydrogen and carbon monoxide for use in synthesis. A thermodynamic analysis of the operation involved was presented by Dodge.[5] The kinetic mechanism of a catalytic reaction of this type may involve several intermediate steps which are of importance in determining the rate of reaction, as discussed in Chapter XVIII. However, a sound thermodynamic treatment is possible without consideration or knowledge of the true mechanism. It may be assumed that the principal reaction proceeds in two stages with intermediate formation and removal of carbon. Thus, these reactions with the corresponding equilibrium constants at 600°C from the data of Tables XV and XXXV are as follows:

$$(1) \qquad CH_4 = C + 2H_2; \qquad K_1 = \frac{a^2_{H_2}}{a_{CH_4}} = 2.13$$

$$(2) \qquad C + H_2O = CO + H_2; \qquad K_2 = \frac{(a_{CO})(a_{H_2})}{a_{H_2O}} = 0.269$$

The over-all reaction is

$$(3) \qquad CH_4 + H_2O = CO + 3H_2; \qquad K_3 = \frac{(a^3_{H_2})(a_{CO})}{(a_{CH_4})(a_{H_2O})} = 0.574$$

In order that no carbon may appear in the equilibrium mixture represented by these three reactions it is necessary that sufficient steam be added so that the ratio $a^2_{H_2}/a_{CH_4}$ may be equal to or greater than K_1 and that the ratio $(a_{CO})(a_{H_2})/a_{H_2O}$ be equal to or less than K_2. If the first ratio is greater than K_1, then carbon added to such a system can react with hydrogen and form methane until the ratio is reduced to K_1. If the second ratio is less than K_2, then carbon added to such a system can react with steam until the ratio rises to K_2.

Many other reactions are possible in this system. A few, including the more likely, follow, together with the corresponding equilibrium constants at 600°C.

		At 600°C
(4)	$CO + H_2O = CO_2 + H_2$;	$K_4 = 2.21$
(5)	$2CO = C + CO_2$;	$K_5 = 8.14$
(6)	$CO_2 = CO + \frac{1}{2}O_2$;	$K_6 = 4.95(10^{-13})$
(7)	$H_2O = H_2 + \frac{1}{2}O_2$;	$K_7 = 1.12(10^{-12})$
(8)	$2CH_4 = C_2H_6 + H_2$;	$K_8 = 5.51(10^{-5})$

Consideration of the values of these equilibrium constants indicates that at 600°C reactions (6), (7), and (8) can proceed only to negligible extents and hence that O_2 and C_2H_6 cannot be appreciably present at equilibrium. When the ratio of steam to methane in the feed is sufficiently high so that carbon cannot be present at equilibrium, the equilibrium composition may be calculated from consideration of only reactions (3) and (4) which involve all the significant reactants in the absence of carbon. In order to determine the minimum steam ratio required for freedom from carbon, the equilibrium compositions corresponding to a series of steam ratios are calculated on the assumption of no carbon formation. The resultant ratios $a^2_{H_2}/a_{CH_4}$ and $(a_{CO})(a_{H_2})/(a_{H_2O})$ are plotted against the steam–methane ratio. The minimum steam ratio is that where $a^2_{H_2}/a_{CH_4} = K_1$ and $(a_{CO})(a_{H_2})/(a_{H_2O}) = K_2$.

This procedure is demonstrated in the following illustration:

Illustration 15. Calculate the composition of the equilibrium mixture obtained when 5 moles of steam react with 1 mole of methane at 600°C and 1.0 atm, assuming that no carbon is present. Also determine the minimum ratio of steam required for freedom from carbon.

Basis: 1 mole CH_4, 5 moles H_2O.

Let x = moles CH_4 converted by reaction (3)

y = moles CO converted by reaction (4)

At equilibrium,

$$\text{Moles } CH_4 = 1 - x$$
$$\text{Moles } H_2O = 5 - x - y$$
$$\text{Moles } CO = x - y$$
$$\text{Moles } H_2 = 3x + y$$
$$\text{Moles } CO_2 = y$$
$$\text{Total moles} = 6 + 2x$$

Since the pressure is 1.0 atm, K_ν may be taken as 1.0. Substitution of the foregoing values in Equation (29) gives

Reaction (3):

$$\left[\frac{(x-y)(3x+y)^3}{(1-x)(5-x-y)}\right]\left[\frac{1.0}{6+2x}\right]^2 = K_3 = 0.574$$

Reaction (4):

$$\left[\frac{(y)(3x+y)}{(x-y)(5-x-y)}\right] = K_4 = 2.21$$

These two equations are solved by assuming several values of x and calculating the corresponding values of y from each equation. Because of the complicated forms of the individual equations the solutions are best carried out graphically. The values of y from each equation are plotted against the assumed values of x. The intersection of these two curves gives the correct solution of the two equations. Thus

$$x = 0.9124$$
$$y = 0.633$$

Equilibrium Mixture:

	Moles	*Mole per Cent*	*Mole per cent (Moisture Free)*
CH_4	$1 - x = 0.0876$	1.12	2.01
H_2O	$5 - x - y = 3.4546$	44.15	
CO	$x - y = 0.2794$	3.57	6.39
H_2	$3x + y = 3.3702$	43.07	77.11
CO_2	$y = 0.6330$	8.09	14.49
Total	$6 + 2x = 7.8248$	100.00	100.00

The corresponding activity ratios for reactions (1), (2), and (5) are

$$\frac{a_{H_2}^2}{a_{CH_4}} = 16.5 > (K_1 = 2.13)$$

$$\frac{a_{CO}a_{H_2}}{a_{H_2O}} = 0.0348 < (K_2 = 0.269)$$

$$\frac{a_{CO_2}}{a_{CO}^2} = 63.4 > (K_5 = 8.14)$$

It follows from these ratios that with 5 moles of steam per mole of methane no carbon is present at equilibrium. This calculation is repeated for steam ratios of 2 and 1.25. The corresponding values of $a_{CO}a_{H_2}/a_{H_2O}$ are plotted in Fig. 160, from which the minimum steam ratio is determined as 1.38. The limiting steam ratio can be calculated directly from reactions (1) (2) and (4), which embrace all components including carbon and shows the formation of CO.

It must be emphasized that the solution of Illustration 15 is based on the assumption that equilibrium is attained. In actual operations which do not reach equilibrium quite different results might be obtained, depending on the rates of the various reactions. Thus, if reaction (1) were fast and (2) were slow, carbon might form even with high steam ratios. Conversely, if reaction (3) were primarily a catalytic reaction and reactions (1) and (5) were relatively slow, operation might be possible at low steam ratios without carbon formation. These effects are determined by the kinetics of the reactions and the types of catalysts present.

Where three significant and independent reactions are involved, the solution of three simultaneous equations is required with the evaluation of three unknowns, x, y, and z. Where algebraic methods of solution fail or become too complex, a more general graphical solution involves finding the simultaneous intersection of three lines in space; this is equivalent to finding the point of intersection of two three-sided pyramids which meet at a common apex. This can be done by assuming various values of z and plotting the three equations corresponding to reactions I, II, and III in terms of x and y as shown in Fig. 161. In this figure the intersection of the three lines gives a triangular area A at $z = 5.0$; at another value of $z = 4.0$ the area of intersection is reduced to B; and by further decrease in z, the area of intersection diminishes to zero at $z = 2$ and increases again as values of z are further decreased.

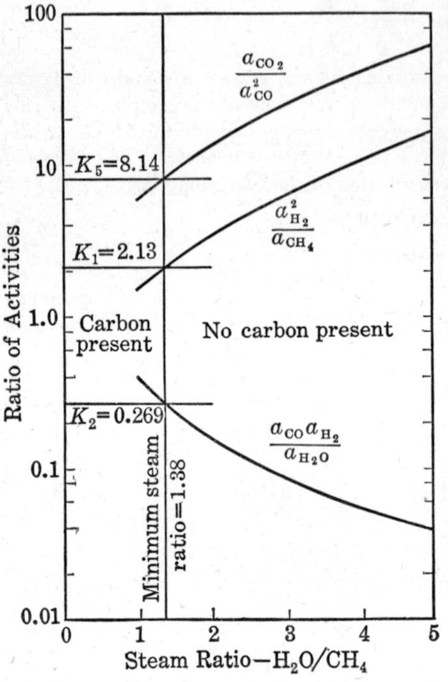

FIG. 160. Minimum Steam Ratio for Absence of Carbon.

It is fortuitous that in industrial processes rarely more than three independent and significant reactions are involved under equilibrium conditions at a given temperature level.

Metallurgical Reactions. The methods previously developed are directly applicable to the calculation of equilibriums in metallurgical processes such as calcining, roasting, and smelting. The majority of these reactions are heterogeneous and complex. Such problems may be handled by extension of the procedure demonstrated in Illustration 15.

An interesting and unique case is encountered where the primary reactants are all solids while the products include substances in the gaseous and liquid phases. In such cases a reaction pressure is developed, analogous to the decomposition pressures previously discussed. The reaction pressure is a result of the nature of the reactants and the temperature, and it is not possible to vary independently temperature and pressure as is the case where a gaseous reactant is present.

When a reaction starts out in a heterogeneous system any or all of the solid or liquid phases may be absent when equilibrium is reached.

A typical metallurgical reaction in which all reactants are solids is the reduction of zinc oxide by carbon. This problem has been studied in detail by Maier.[7]

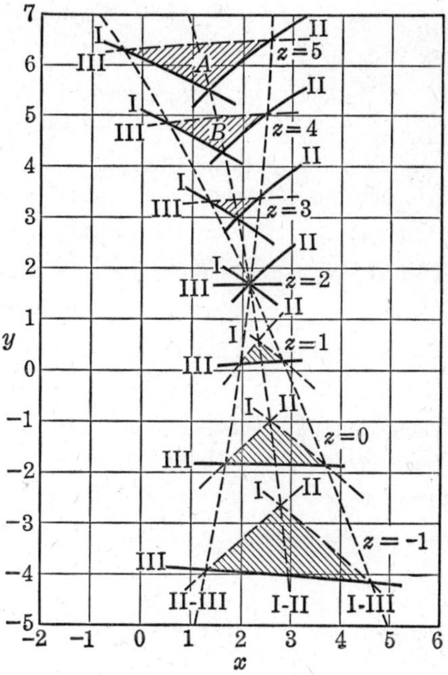

FIG. 161. Solution of Three Complex Simultaneous Equations.

Illustration 16. Zinc oxide (ZnO) is reduced by roasting it with carbon in a closed retort from which the gaseous and liquid products of reaction may be continuously removed. The atmosphere is excluded from the retort so that the reaction is free to proceed under its own equilibrium pressure.

(*a*) Calculate the equilibrium reaction pressure as a function of the roasting temperature.

(*b*) Determine the temperature at which the operation must be conducted in order that the products may be withdrawn at a pressure of 1.0 atm.

(*c*) Calculate the minimum temperature and the corresponding pressure at which zinc is produced as a liquid.

(*d*) Calculate the temperature and pressure at which the operation must be conducted in order that 50 per cent of the zinc may be produced directly in the liquid state.

Solution: Although the over-all effect is the reaction of the two solids, it may be assumed, in order to account for all the products, that the actual reaction proceeds in two stages:

(1) $$\mathrm{ZnO\ (s) + CO\ (g) = Zn\ (g) + CO_2\ (g)}$$

(2) $$\mathrm{C\ (s) + CO_2\ (g) = 2CO\ (g)}$$

The over-all reaction becomes

(3) $$\mathrm{ZnO\ (s) + C\ (s) = Zn\ (g) + CO\ (g)}$$

In order to define completely the equilibrium composition, including the CO_2 present, reactions (1) and (2) must be considered simultaneously.

The standard free-energy changes of the two reactions may be expressed as func-

[7] C. G. Maier, "Zinc Smelting from a Chemical and Thermodynamic Viewpoint," *U.S. Bur. Mines Bull.*, 324 (1930).

tions of temperature by means of the data of Tables V, XIV, and XXXV, together with the data of Kelley[6, 8] on the heat capacities and heats of vaporization:

For reaction (1),

$$\Delta G_1^\circ = -RT \ln K_1 = 49265 - 70.50T + 5.83T \ln T - 2.496(10^{-3})T^2 + 0.336(10^{-6})T^3 - \frac{0.912(10^5)}{T} \quad \text{(a)}$$

For reaction (2),

$$\Delta G_2^\circ = -RT \ln K_2 = 40608 - 24.96T - 2.977T \ln T + 3.484(10^{-3})T^2 - 0.259(10^{-6})T^3 - 58450/T \quad \text{(b)}$$

Basis of Calculations. 1 lb-mole of ZnO reduced which liberates 1 lb-atom of oxygen appearing as either CO or CO_2 in the products of the over-all reaction.

(*a*) Let x = lb-moles of CO produced

Then $(1 - x)$ lb-atoms of oxygen is present in CO_2 or

$$\tfrac{1}{2}(1 - x) = \text{lb-moles of } CO_2 \text{ formed}$$

The total gas formed is the sum of the CO, Zn, and the CO_2, or

$$\tfrac{1}{2}(3 + x) = \text{total gaseous moles formed}$$

These values are substituted in Equation (29) with the assumption that all zinc is vaporized and that activities are equal to partial pressures.

$$K_1 = \frac{\pi(1 - x)}{x(3 + x)} \quad \text{(c)}$$

$$K_2 = \frac{4x^2\pi}{(3 + x)(1 - x)} \quad \text{(d)}$$

Combining (c) and (d) by eliminating π gives

$$4K_1x^3 = K_2(1 - x)^2 \quad \text{(e)}$$

For any given temperature, values of K_1 and K_2 are obtained from Equations (a) and (b), and corresponding values of x are calculated from Equation (e). The resultant pressures are then obtained from Equations (c) or (d). These results are shown in the first four columns of Table A. In Fig. 162 the equilibrium pressure is plotted against temperature.

(*b*) From Fig. 162, where $\pi = 1.0$ atm, $T = 1170°$K. From Table A, at 1.0 atm, $x = 0.978$. The composition of the gaseous products is obtained directly from this value. Thus, the total number of moles of products is $3.978/2 = 1.989$.

	Moles	*Partial Pressure*	*Mole per Cent*
Zn	1.0	0.503 atm	50.3
CO	0.978	0.491	49.1
CO_2	0.011	0.006	0.6
	1.989	1.000	100.0

For each lb-atom of carbon consumed, 0.503/0.497 or 1.012 lb-atoms of zinc are produced, or 5.51 lb of zinc are formed per lb of carbon consumed.

(*c*) *Minimum Temperature for Producing Liquid Zinc.* If operating conditions are

[8] K. K. Kelley, "The Free Energies of Vaporization and Vapor Pressures of Inorganic Substances," *U.S. Bur. Mines Bull.*, **383** (1935).

such that liquid zinc may be produced directly at equilibrium conditions, the vaporization equilibrium of zinc must be considered as a simultaneous reaction. Thus

(4) $$Zn\ (l) \rightleftarrows Zn\ (g)$$

The equilibrium constant for this vaporization is equal to the vapor pressure of the liquid zinc if ideal behavior is assumed, or $K_4 = P_{Zn}$. In Fig. 162 are plotted vapor pressures as a function of temperature taken from the data of Maier.[7]

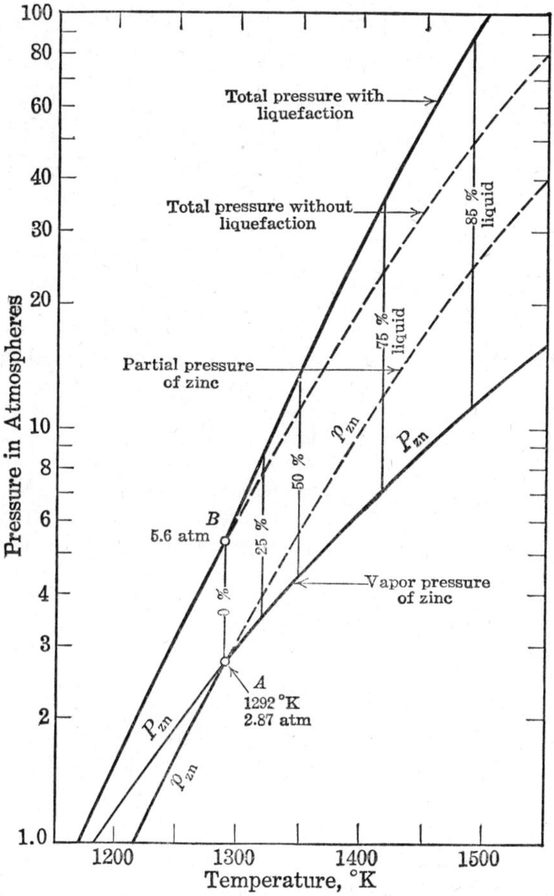

FIG. 162. Equilibrium Conditions in the Reduction of Zinc Oxide.

The minimum temperature at which liquid zinc can be produced occurs when the partial pressure of the zinc vapor is just equal to the vapor pressure of molten zinc. This corresponds to the dew point of zinc vapor. The partial pressure of zinc vapor p_{Zn} at different reduction temperatures is calculated by the method of part (*b*) and tabulated in column 6 of Table A. In column 7 are the vapor pressures P_{Zn} of liquid zinc. The partial pressure of zinc vapor p_{Zn} and the vapor pressure of molten zinc P_{Zn} are plotted against temperature in Fig. 162. The intersection point A of these two curves indicates that the minimum temperature at which molten zinc can occur is 1292°K, where the partial pressure of zinc vapor is 2.87 atm and the total pressure is 5.6 atm. However, under these conditions a negligible amount of zinc is formed, and higher temperatures and pressures must be employed to obtain appreciable yields of the liquid.

TABLE A

COMPLETE VAPORIZATION OF ZINC ASSUMED

T°K	K_1	K_2	x	π *Atm*	p_{Zn} *Atm*	P_{Zn} *Atm* $= K_4$
1100	0.001666	15.35	0.980	0.322	0.162	0.43
1200	0.009014	70.8	0.978	1.607	0.808	1.20
1300	0.03853	255.3	0.976	6.31	3.174	3.06
1400	0.1300	760.4	0.975	20.01	10.07	6.33
1500	0.3740	1940	0.973	54.25	27.31	12.2

By application of the phase rule, Equation (XIV-27), to this system point B is identified as an univariant point. At this point two solids, one liquid, and one vapor phase can exist simultaneously. At temperatures below B two solids and one gas phase exist. At temperatures above 1292°K the system is univariant; thus, fixing the temperature automatically fixes the pressure.

(*d*) *Yield of Liquid Zinc by Direct Reduction.* When liquid zinc is present at equilibrium conditions, the partial pressure of the zinc vapor must equal the vapor pressure of the liquid zinc at the equilibrium temperature.

Since Equation (29) is applicable only where each component is present in only one phase, Equation (6) must be used as the basis for developing equilibrium expressions. On the basis of 1 lb-mole of ZnO reduced, let x represent the lb-moles of CO formed and y the lb-moles of zinc vapor present at equilibrium. Then the lb-moles of CO_2 are $\frac{1}{2}(1 - x)$, and the total moles of gaseous products are $\frac{1}{2}(1 + x + 2y)$.

Equilibrium partial pressures, if ideal behavior of the gases is assumed, are as follows:

$$p_{CO} = \frac{2x\pi}{1 + x + 2y} \tag{f}$$

$$p_{CO_2} = \frac{(1 - x)\pi}{1 + x + 2y} \tag{g}$$

$$P_{Zn} = \frac{2y\pi}{1 + x + 2y} = K_4 \tag{h}$$

or, from (h),

$$\pi = \frac{P_{Zn}(1 + x + 2y)}{2y} \tag{i}$$

These values are substituted in Equation (6) with the assumption that partial pressures are equal to activities.

Reaction (1),

$$ZnO + CO \longrightarrow Zn + CO_2$$

$$\frac{P_{Zn}(1 - x)}{2x} = K_1 \tag{j}$$

Reaction (2),

$$C + CO_2 \longrightarrow 2CO$$

$$\frac{4\pi x^2}{(1 - x)(1 + x + 2y)} = K_2 \tag{k}$$

Substitution of (i) in (k) gives

$$\frac{4P_{Zn}x^2}{(1 - x)2y} = K_2 \tag{l}$$

Solving (j) for x gives

$$x = \frac{P_{Zn}}{P_{Zn} + 2K_1} \tag{m}$$

Substitution of (m) in (l) gives

$$\frac{P^3_{Zn}}{yK_1(P_{Zn} + 2K_1)} = K_2 \tag{n}$$

or

$$y = \frac{P^3_{Zn}}{(P_{Zn} + 2K_1)K_1K_2} \tag{o}$$

From Equation (o) y is evaluated directly for any selected temperature which fixes the values of P_{Zn}, K_1, and K_2. The corresponding compositions and total pressures are evaluated from Equations (m) and (i). These results are tabulated in Table B together with the percentages of the total zinc formed which is present in the liquid state.

TABLE B

THE PRESENCE OF LIQUID ZINC CONSIDERED

Temperature °K	K_1	K_2	P_{Zn}	y	x	π	*Per Cent Liquid* Zn $100(1 - y)$
1290	0.0339	226	2.80	1.00	0.976	5.555	0
1353	0.074	464	4.6	0.603	0.970	12.1	39.7
1400	0.1300	760.4	6.33	0.390	0.961	22.2	61.0
1500	0.3740	1940	12.2	0.192	0.941	83.9	80.8

The values of π in Table B are plotted in Fig. 162 for comparison with those of Table A which were based on the assumption of complete vaporization. From inspection of Fig. 162, 50 per cent liquefaction of zinc will occur at 1373°K, 16.3 atm pressure.

It is evident that the direct production of liquid zinc in yields approaching complete recovery would require operating conditions which cannot be attained with the structural materials at present available.

EQUILIBRIA IN LIQUID SOLUTIONS

The fundamental thermodynamic principles of equilibrium are universally applicable, and Equation (6) is rigorous for reactions involving liquid or solid solutions of all types. However, the application of thermodynamic methods to such systems has not been particularly fruitful because of the complex relationships frequently existing between activities and concentrations. This situation is illustrated by the activity coefficients developed in Chapter XIV for NaCl in aqueous solutions. When more than one solute is involved, as is commonly the case where reactions occur in solution, the complexity of the relationships is compounded to such an extent as to render the methods of little practical value except where ideal behavior is approximated and where activity coefficients can be assumed to be constant.

Constant activity coefficients can be satisfactorily assumed when one is dealing with very dilute solutions or in systems involving only closely related types of materials such as organic homologs. In such cases the calculation of equilibrium compositions is the same as for gaseous systems except that pressure is not an appreciable factor. Thus, from Equation (24),

$$K = \left(\frac{N_R^r N_S^s \cdots}{N_B^b N_C^c \cdots}\right)\left(\frac{\gamma_R^r \gamma_S^s \cdots}{\gamma_B^b \gamma_C^c \cdots}\right) = \left(\frac{N_R^r N_S^s \cdots}{N_B^b N_C^c \cdots}\right) K_\gamma \tag{44}$$

or, in a form similar to Equation (29),

$$K = K_\gamma\left(\frac{n_R^r n_S^s \cdots}{n_B^b n_C^c \cdots}\right)(n_B + n_C + \cdots + n_R + n_S + \cdots + n_I)^{(c+b+\cdots)-(r+s+\cdots)} \tag{45}$$

In working with liquid solutions it is necessary that the activities and standard free-energy data used in equilibrium calculations be based on the same standard states. The entropy of a substance in solution varies widely with change in concentration. Hence, the standard free energy of formation based on $a_1/N_1 = 1$ when $N_1 = 0$ will be very different from that based on $a_1/m_1 = 1$ when $m_1 = 0$, even though both are referred to the state of infinite dilution as explained in Chapter XIV. With this precaution the procedures in calculating equilibrium compositions are the same as those demonstrated for the gaseous systems.

Solubility of Carbonates. The application of thermodynamic principles to the solubility of the metal carbonates has been developed by Kelley and Anderson.[9] These principles are of importance in connection with studies of the recovery and purification of carbonate ores by leaching operations.

When a metal carbonate $MeCO_3$ is dissolved in water in the presence of carbon dioxide, the reactions shown in Table XXXVIII may take place.

TABLE XXXVIII

REACTIONS IN THE DISSOLUTION OF A METALLIC CARBONATE

(1)	$MeCO_3\ (s) = MeCO_3\ (aq)$
(2)	$MeCO_3\ (aq) = Me^{++} + CO_3^{--}$
(3)	$MeCO_3\ (s) + H_2O = Me^{++} + OH^- + HCO_3^-$
(4)	$MeCO_3\ (s) + H_2CO_3\ (aq) = Me^{++} + 2HCO_3^-$
(5)	$MeCO_3\ (s) + 2H_2O = Me(OH)_2\ (s) + H_2CO_3\ (aq)$

[9] K. K. Kelley and C. T. Anderson, "Contributions to the Data on Theoretical Metallurgy; IV Metal Carbonates–Correlation and Applications of Thermodynamic Properties," *U.S. Bur. Mines Bull.*, 384 (1935).

(6) $Me(OH)_2\,(s) = Me(OH)_2\,(aq)$

(7) $Me(OH)_2\,(aq) = Me^{++} + 2(OH)^-$

(8) $H_2O = H^+ + OH^-$

(9) $H_2CO_3\,(aq) = H^+ + HCO_3^-$

(10) $H_2CO_3\,(aq) = H_2O + CO_2\,(g)$

(11) $HCO_3^- + H_2O = OH^- + H_2CO_3\,(aq)$

(12) $HCO_3^- = H^+ + CO_3^{--}$

(13) $Me(HCO_3)_2(s) = Me(HCO_3)_2\,(aq)$

(14) $Me(HCO_3)_2\,(aq) = Me^{++} + 2(HCO_3)^-$

(15) $MeCO_3\,(s) + H_2CO_3\,(aq) = Me(HCO_3)_2\,(s)$

All of these possible reactions must be in equilibrium at saturation conditions. Since reaction (10) produces $CO_2(g)$ as a product, it follows that a definite equilibrium CO_2 pressure will result from dissolving a carbonate in pure water. Dissolution will at first take place by reactions (1–5), and carbon dioxide gas will tend to be formed as a result of reaction (10). If the solution is in contact with an atmosphere containing no partial pressure of carbon dioxide, the gas will be evolved with resultant formation of $Me(OH)_2(aq)$, Me^{++}, and OH^-. This reaction will proceed until the $Me(OH)_2(aq)$, Me^{++}, and OH^- concentration reaches saturation and $Me(OH)_2(s)$ is formed. The reaction then proceeds with the over-all formation of $Me(OH)_2(s)$ and $CO_2(g)$ by reactions (5) and (10) until the $MeCO_3(s)$ is completely consumed.

If dissolution were to take place in contact with a CO_2 free atmosphere as previously described, the equilibrium CO_2 pressure of the solution would progressively diminish until the first $Me(OH)_2(s)$ forms and then remain constant at the value fixed by the equilibrium of reactions (4) and (10). Thus, the partial pressure of CO_2 corresponding to equilibrium in reactions (5) and (10) is the minimum equilibrium CO_2 pressure exerted by a solution in equilibrium with $MeCO_3(s)$ and corresponds to an invariant point at which two solids, one liquid, and one gas phase are in equilibrium.

If a carbonate is dissolved in water in contact with an atmosphere containing a partial pressure of CO_2 greater than minimum value previously discussed, dissolution will proceed to an equilibrium fixed by reactions (1–4) and (7–15). If the CO_2 pressure in the atmosphere is less than the equilibrium CO_2 dissolution pressure of the pure carbonate, CO_2 will be evolved as equilibrium is approached, and the final solution will be basic. If the CO_2 pressure in the atmosphere is greater than the dissolution pressure, the solution will absorb CO_2 and be acidic when equilibrium is reached. As the CO_2 pressure in the atmosphere is further increased, the acidity of the solution increases until the concentration of $Me(HCO_3)_2(aq)$ reaches the saturated value and $Me(HCO_3)_2(s)$ begins to form. The CO_2 pressure at which this occurs is determined by the equilibrium of reactions (15) and (10) and is the maximum pressure which can be in equilibrium with a solution which is in equilibrium with $MeCO_3$(s). Higher CO_2 pressures will result in disappearance of the solid carbonate and formation of solutions whose compositions are fixed by the CO_2 pressure and the equilibria of reactions (2) and (8–14).

In dealing with the sparingly soluble carbonates the situation is simplified by the fact that the salts are completely dissociated in solution. Thus, reactions (1) and (2), (6) and (7), and (13) and (14) may be combined into single reactions and the dissolved salts, $MeCO_3$, $Me(OH)_2$, and $Me(HCO_3)_2$, disregarded. Furthermore, the solid forms of the acid carbonates of the metals which form sparingly soluble carbonates are not known to exist, and reactions (13) to (15) need not be considered. Additional simplification results from the fact that in dilute solutions the activity of water is constant and equal to 1.0.

The equilibrium constant of a reaction in dilute solution is determined from its standard enthalpy and entropy changes. In working with dilute aqueous solutions it is customary to express compositions in terms of molalities and to select the standard state such that $a/m = 1.0$ where $m = 0$. The standard enthalpy change in dilute solution is obtained from Tables XIV, XV, and XVI. The standard entropy change is determined from Table XXXV. It may be noted that values of zero are arbitrarily assigned to the heat of formation and the entropy of the hydrogen ion in dilute solution. The corresponding free-energy change of formation is then 4,655 cal per g-mole at 298.1°K. All other ionic heats of formation and entropies are thus expressed relative to the hydrogen ion since the individual absolute values are unknown.

Illustration 17. Calculate the equilibrium constant of the following reaction at 298°K:

$$CaCO_3\ (s)\ (ppt) \rightleftarrows Ca^{++}\ (aq) + CO_3^{--}\ (aq)$$

From Tables XIV and XVI, pages 253 and 272,

$$\Delta H_{291} = -160.3 - 129.74 - (-287.8) = -2.24 \text{ kcal per g-mole}$$

From Table XXXV,

$$\Delta S_{298} = -11.4 - 13.0 - 22.2 = -46.6 \text{ cal/(°K)(g-mole)}$$

Neglecting the difference between the heats of reaction at 291° and 298°K yields

$$\left(\frac{\Delta G}{T}\right)_{298} = \frac{-2240}{298} + 46.6 = 39.1$$

$$\log K_{298} = \frac{-39.1}{4.576} = -8.548$$

$$K_{298} = (2.83)(10^{-9})$$

In calculating equilibrium constants for ionic reactions great care must be exercised to use heat of formation and entropy data which are consistent with each other. The entropies of the ions are determined from equilibrium measurements and heat of reaction data. When such entropies are used for the reverse operation of calculating the equilibrium constant, exactly the same heat-of-reaction data must be used. This fact results in some confusion in the use of the heats of formation of $H_2O(l)$ and $CO_2(g)$ from Table XIV. The values in this table are recent determinations by Rossini which differ slightly from the values reported by Bichowsky and Rossini. However, the entropies of the ions and aqueous compounds in Table XXXV are all based on the original Bichowsky and Rossini values of −68.37 and −94.23 for the heats of formation of $H_2O(l)$ and $CO_2(g)$, respectively. Accordingly, these values should be used for the calculation of equilibrium constants in aqueous solutions rather than the more accurate values of Table XIV.

The equilibrium composition of a system involving ions is calculated by the procedure demonstrated in the preceding sections dealing with complex reactions. In order to define the equilibrium, it is desirable to consider the smallest number of reactions which include all reactants and products present at equilibrium. An algebraic equation relates the concentrations of the components of each reaction to its equilibrium constant. An additional algebraic equation expresses the electrical neutrality of the solution, that is, the equality of the positive and negative charges of all of the ions. A further equation which may prove useful expresses the material balance of the dissolution. The equilibrium composition is evaluated by simultaneous solution of these equations.

Illustration 18. Calculate the complete composition of the solution obtained when precipitated calcium carbonate is dissolved in water in contact with an atmosphere containing a partial pressure of CO_2 of 3×10^{-4} atm.

Solution: It may be assumed that the specified partial pressure of CO_2 is above that of the invariant point and that the system at equilibrium contains $CaCO_3$ (s), Ca^{++}, CO_3^{--}, HCO_3^-, H^+, OH^-, and H_2CO_3 (aq). There are therefore six unknown compositions to be evaluated, which requires five independent equilibrium equations in addition to the equation of electrical neutrality. The material-balance equation cannot be used in this case, because CO_2 may be gained or lost during the dissolution.

Although other selections are possible, it is convenient to consider the following reactions for the establishment of the equilibrium equations. The corresponding equilibrium constants are calculated by the method of Illustration 17.

	Reaction	K_{298}
(a)	$CaCO_3\ (s) = Ca^{++} + CO_3^{--}$	$(2.9)(10^{-9}) = (Ca^{++})(CO_3^{--})$
(b)	$H_2O\ (l) = H^+ + OH^-$	$(5.5)(10^{-15}) = (H^+)(OH^-)$
(c)	$H_2CO_3\ (aq) = H^+ + HCO_3^-$	$(9.3)(10^{-8}) = \dfrac{(H^+)(HCO_3^-)}{H_2CO_3\ (aq)}$
(d)	$H_2CO_3\ (aq) = H_2O + CO_2\ (g)$	$31.7 = \dfrac{CO_2\ (g)}{H_2CO_3\ (aq)}$
(e)	$HCO_3^- = H^+ + CO_3^{--}$	$(1.4)(10^{-11}) = \dfrac{(H^+)(CO_3^{--})}{(HCO_3^-)}$

The equation of electrical neutrality may be written as follows if the chemical formulas are used to denote molalities:

(f) $$2Ca^{++} + H^{+} = 2CO_3^{--} + HCO_3^{-} + OH^{-}$$

It should be noted that (f) is not a stoichiometric equation.

Since the CO_2 pressure is specified, it follows from (d) that the molality of H_2CO_3 (aq) is represented by

$$H_2CO_3\ (aq) = (3)(10^{-4})/31.7 = (0.95)(10^{-5}) \text{ g-moles per 1000 g } H_2O$$

If y is used to represent the molality of the HCO_3^- ion, the following expressions may be written from Equations (c), (b), (e), and (a) for the molalities of the other ions:

$$H^+ = (0.95)(10^{-5})(9.3)(10^{-8})/y = (8.8)10^{-13}/y$$
$$OH^- = (5.5)(10^{-15})y/(8.8)(10^{-13}) = (6.25)(10^{-3})y$$
$$CO_3^{--} = (1.4)(10^{-11})y^2/(8.8)(10^{-13}) = 15.9y^2$$
$$Ca^{++} = (2.9)(10^{-9})/(15.9)y^2 = (1.82)(10^{-10})/y^2$$

These values may be substituted in Equation (f), resulting in an equation which contains only y as an unknown. Thus,

$$\frac{(3.64)(10^{-10})}{y^2} + \frac{(8.8)(10^{-13})}{y} = 31.8y^2 + y[1 + 6.25(10^{-3})]$$

This equation is solved graphically or by trial and error.

$$y = HCO_3^- = (7.1)(10^{-4}) \text{ g-mole per 1000 g } H_2O$$

The concentrations of the other ions are evaluated from the equations relating them to y. Thus

$$H^+ = (1.2)(10^{-9}) \text{ g-mole per 1000 g } H_2O$$
$$OH^- = (4.4)(10^{-6}) \text{ g-mole per 1000 g } H_2O$$
$$CO_3^{--} = (7.9)(10^{-6}) \text{ g-mole per 1000 g } H_2O$$
$$Ca^{++} = (3.7)(10^{-4}) \text{ g-mole per 1000 g } H_2O$$

The procedure of Illustration 18 may be repeated for other CO_2 pressures and curves plotted relating the composition of the solution to the equilibrium partial pressure of CO_2 above it. Such curves for the solubility of $CaCO_3$ in water are shown in Fig. 163. It may be noted that the concentration of the calcium ion passes through a minimum and is increased by either increasing or decreasing the CO_2 pressure from this point. As the CO_2 pressure is increased, carbon dioxide is absorbed by the solution, but as the pressure is reduced CO_2 is evolved and the concentration of the OH^- ion increases until solid $Ca(OH)_2$ is formed. The appearance of the added phase results in an invariant point at which the composition of the solution and the partial pressure of the CO_2 are fixed as long as both solid phases are present. The conditions at the invariant point are established by the following equation in conjunction with those of Illustration 18:

(g) $$CaCO_3\ (s) + 2H_2O \rightleftarrows Ca(OH)_2\ (s) + H_2CO_3\ (aq);$$
$$K_{298} = (9.3)(10^{-15}) = H_2CO_3\ (aq)$$

If calcium carbonate is dissolved in pure water in a closed liquid system which permits no gain or loss of carbon dioxide, an equilibrium partial pressure of CO_2 is established which is determined by the equations of Illustration 18 together with the following material-balance equation in which the formulas indicate molalities:

$$\text{(h)} \qquad Ca^{++} = CO_3^{--} + HCO_3^{-} + H_2CO_3$$

This is not a stoichiometric equation and merely indicates that each atom of calcium entering the solution is accompanied by an atom of

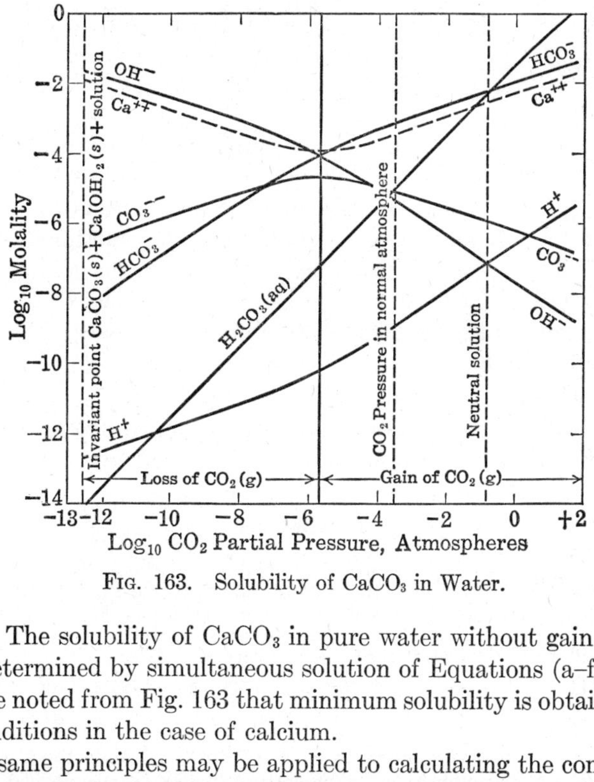

FIG. 163. Solubility of $CaCO_3$ in Water.

carbon. The solubility of $CaCO_3$ in pure water without gain or loss of CO_2 is determined by simultaneous solution of Equations (a–f) and (h). It may be noted from Fig. 163 that minimum solubility is obtained under these conditions in the case of calcium.

These same principles may be applied to calculating the compositions obtained in selective leaching and precipitation operations where several sparingly soluble metals or salts are involved. For each added constituent an additional equation is added to the group which must be solved simultaneously in order for the equilibrium composition to be evaluated. However, simplification is possible in many instances by neglect of components which are present in the solution in relatively small quantities. For example, from Fig. 163 it is evident that at CO_2 pressures above 10^{-4} atm the concentrations of OH^-, CO_3^{--}, and H^+

are negligible in comparison to the Ca^{++}, HCO_3^-, and H_2CO_3. In this range the dissolution may be considered as proceeding by the following stoichiometric equation:

$$CaCO_3\ (s) + H_2O\ (l) + CO_2\ (g) = Ca^{++} + 2HCO_3^-;$$

$$K_{298} = (6.0)(10^{-7}) = \frac{(Ca^{++})(HCO_3^-)^2}{(CO_2)\ (g)}$$

Since it is assumed that $HCO_3^- = 2\ Ca^{++}$ the foregoing equilibrium equation may be written

$$K_{298} = (6.0)(10^{-7}) = \frac{4(Ca^{++})^3}{CO_2\ (g)}$$

or

$$Ca^{++} = (5.3)(10^{-3})[CO_2\ (g)]^{\frac{1}{3}}$$

where the chemical formulas indicate activities in molalities and atmospheres, respectively. In this manner the molalities of the Ca^{++} and HCO_3^- ions are determined directly, and the molalities of the other components may be individually evaluated from the equations of Illustration 18. This procedure is generally applicable to the sparingly soluble carbonates, either singly or in complex mixtures if the CO_2 pressure is equal to or greater than that of the normal atmosphere.

The application of these methods to complex metallurgical problems is discussed in detail by Kelley and Anderson.[9] By assuming that the activity coefficients cancel out of the equilibrium equations it is possible to solve effectively many such problems by thermodynamic principles.

PROBLEMS

1. Normal butane is isomerized to isobutane by the action of a catalyst at moderate temperatures. It is found that equilibrium is reached at the following compositions:

Temperature, °C	*Mole per Cent n-Butane*	*Mole per Cent Isobutane*
44	31	69
118	43	57

Assuming that activities are equal to mole fractions, calculate the standard free-energy change of the reaction at each temperature and the average values for the heat-of-reaction and entropy change over this temperature range.

2. From the data of Tables V, XIV, and XXXV derive equations relating $\Delta G°/T$ to temperature for the following reactions:

(a) $$SO_2\ (g) + \tfrac{1}{2}O_2\ (g) = SO_3\ (g)$$

(b) $$\tfrac{1}{2}N_2\ (g) + \tfrac{3}{2}H_2\ (g) = NH_3\ (g)$$

(c) $$C\ (s) + H_2O = H_2 + CO$$

(d) $$CO_2 + C = 2\ CO$$

The atomic heat capacity of graphitic carbon,[10] related to T°K, is $2.673 + 0.002617T - 116{,}900/T^2$.

3. From the liquid-state values in Table XXXV estimate the entropies in the ideal-gaseous state at 298.1°K for the following compounds:

	T_B°C	*Density*, 20°C, g per cc
(a) *n*-Decane	174	0.747
(b) *n*-Butyl benzene	180	0.862
(c) *n*-Heptyl cyclohexane	223	0.801

In the absence of data on the physical properties of these compounds use Equations (III-16) and (III-1) for estimating vapor pressures at 298.1°K and heats of vaporization at the normal boiling points and Equation (VII-32) for the effect of temperature on heat of vaporization. Critical temperatures should be estimated by Equation (III-8), critical pressures by (III-18), and liquid densities by Equation (XII-38).

4. From the data of Tables XXXVI and XXXVII calculate the equilibrium constants and heats of reaction of the following reactions at temperatures of 298, 500, 800, and 1000°K:

(a) $$C_2H_6\ (g) = C_2H_4\ (g) + H_2\ (g)$$

(b) $$2C_2H_4\ (g) = i\text{-}C_4H_8\ (g)$$

(c) $$C_3H_8 + C_2H_4 = C_5H_{12}\ \text{(tetramethylmethane)}$$

5. The following values of $(H^\circ - H_0^\circ)$ for the ideal-gaseous state were calculated by Pitzer and Scott,[11] and the corresponding heats of vaporization at the normal boiling point and critical temperatures are from the International Critical Tables. From the data of Tables XVI and XXXVII calculate the values of $(\Delta H_f^\circ)_0$ the heat of formation at 0°K in the ideal-gaseous state for each of these compounds. The difference between the heats of combustion at 291°K and 298°K may be neglected as less than the probable errors of the measurements.

	$(H^\circ - H_0^\circ)_{298}$ kcal per g-mole	λ_b cal per g	$T_b^\circ C$	$T_c^\circ C$
(a) Benzene	3.415	94.4	80.1	289
(b) Toluene	4.314	86.5	111	321
(c) Ortho-xylene	5.604	82.8	144.5	358.3

6. Nitrogen tetroxide dissociates into nitrogen peroxide according to the following reaction:

$$N_2O_4\ (g) = 2NO_2\ (g)$$

The standard free-energy change in calories per gram-mole of this reaction in the ideal state at 1 atm is represented by the following equation from the International Critical Tables:

$$-\Delta G^\circ = -13{,}600 + 41.6T^\circ\text{K}$$

Calculate the equilibrium composition of the mixture formed from the dissociation of pure N_2O_4 under the following conditions, assuming $K_v = 1.0$:

(a) At a temperature of 273°K and 1 atm
(b) At a temperature of 400°K and 1 atm

[10] K. K. Kelley, *U.S. Bur. Mines Bull.*, 371 (1934).
[11] K. S. Pitzer and P. W. Scott, *J. Am. Chem. Soc.*, **65**, 803–29 (1943).

7. Isobutane is alkylated by ethylene to form neohexane, 2, 2-dimethyl butane, at elevated temperatures and pressures. From the data of Table XXXVI calculate the composition of the equilibrium mixture resulting from reacting a mixture of 4 moles of isobutane and 1.0 mole of ethylene at 700°K and 100 atm.

8. From the data of Table XXXVII calculate the heats of reaction in the ideal-gaseous state of the following reactions at 600°K and 1000°K:

(a) $$C_3H_8 \rightarrow CH_4 + C_2H_4$$

(b) $$C_3H_8 \rightarrow C_3H_6 + H_2$$

(c) $$CO + \tfrac{1}{2}O_2 \rightarrow CO_2$$

(d) $$CH_4 + H_2O \rightarrow CO + 3H_2$$

9. Benzene may be produced by the catalytic dehydrogenation of *n*-hexane. From the results of problem 5 and the data of Tables XV and XXXV calculate the composition of the equilibrium mixture and the percentage conversion of hexane at a temperature of 1050°F and 1 atm. The following heat-capacity equations may be used where T is in °K.

Benzene[12] $$c_p^* = 0.23 + 77.83 \times 10^{-3}T - (27.16)(10^{-6})T^2$$

n-Hexane[13] $$c_p^* = 4.296 + 118.661 \times 10^{-3}T - (42.13)(10^{-6})T^2$$

10. Propane may be dehydrogenated catalytically to form propylene. From the data of Table XXXVI develop curves relating the percentage of propane dehydrogenated at equilibrium to temperature in the range of 800–1200°F and at pressures of 0.5, 1.0, and 2.0 atm.

11. Water and chlorine react at elevated temperatures according to the following equation:

$$\tfrac{1}{2}Cl_2\ (g) + \tfrac{1}{2}H_2O\ (g) = HCl\ (g) + \tfrac{1}{4}O_2\ (g)$$

Using the data of Tables V, XIV, and XXXV and the heat-capacity equation of problem XIII-1, page 592,

(*a*) Calculate an equation for $\Delta G°/T$ for this reaction.

(*b*) Calculate the composition of the equilibrium mixture at 500°C starting with equal volumes of chlorine and water and assuming that $K_\nu = 1.0$.

(*c*) Repeat the calculation of part (*b*) for a pressure of 100 atm, evaluating K_ν from Fig. 142 and the critical data of Table XI, page 234.

12. Sulfur dioxide is reduced by hydrogen according to the following equation:

$$3H_2\ (g) + SO_2\ (g) = H_2S\ (g) + 2H_2O\ (g)$$

(*a*) From the data of Tables V, XIV, and XXXV derive an equation for the standard free-energy change of this reaction.

(*b*) Calculate the equilibrium composition of the mixture obtained at 1200°C and atmospheric pressure, starting with three parts of hydrogen and one part of SO_2.

(*c*) Repeat the calculation of part (*b*) at 100 atm, using Fig. 142 and the critical data from Table XI, page 234.

13. Methyl alcohol is synthesized by passing a mixture of CO and H_2 over a catalyst according to the following equation:

$$CO\ (g) + 2H_2\ (g) = CH_3OH\ (g)$$

[12] J. W. Andersen, G. H. Beyer, and K. M. Watson, *Natl. Petroleum News*, Tech. Sec., **36**, *R*476 (July 5, 1944). Also "Process Engineering Data," National Petroleum Publishing Company, Cleveland (1944).

[13] H. M. Spencer and G. N. Flannagan, *J. Am. Chem. Soc.*, **64**, 2511 (1942).

From the data of Tables V, XV, and XXXV calculate the composition of the equilibrium mixture obtained at a temperature of 300°C and a pressure of 240 atm, starting with two parts of H_2 and one part of CO, assuming that only this reaction takes place. The critical temperature of CH_3OH is 240°C and the critical pressure 79 atm. The heat capacity is given by the following equation (Chapter XVII, page 801) with T in °K:

$$c_p = 5.72 + (24.85)(10^{-3})T - (8.167)(10^{-6})T^2$$

14. Methyl alcohol is oxidized by air to formaldehyde at 550°C in a catalyst chamber at atmospheric pressure. Calculate the percentage yield of formaldehyde, using the theoretical air supply and assuming no further oxidation.

$$CH_3OH\ (g) + \tfrac{1}{2}O_2\ (g) \rightarrow HCHO\ (g) + H_2O\ (g)$$

Entropy and heat-of-reaction data are given in Tables XXXV and XV. Heat capacities may be taken from Table V and problem 12 and the following equation for HCHO[12] with T in °K:

$$c_p = 4.50 + (13.95)(10^{-3})T - (3.73)(10^{-6})T^2$$

15. In the Birkeland–Eyde process the nitrogen of the atmosphere is oxidized in a long flaming electric arc:

$$\tfrac{1}{2}N_2 + \tfrac{1}{2}O_2 = NO$$

Assuming that only this reaction takes place, calculate the percentage conversion of nitrogen to NO in air of average atmospheric composition at a pressure of 1.0 atm and at temperatures of, respectively, 2000°K and 3000°K.

16. Carbon dioxide is reduced by graphite according to the equation:

$$C\ (\text{graphite}) + CO_2\ (g) = 2CO\ (g)$$

Assuming that equilibrium is attained, calculate the degree of completion of the reduction of pure CO_2 under the following conditions, using the data of Table XXXVI:

(*a*) A temperature of 1000°K and a pressure of 1.0 atm, assuming $K_\nu = 1.0$
(*b*) A temperature of 1500°K and a pressure of 1.0 atm, assuming $K_\nu = 1.0$
(*c*) A temperature of 1000°K and a pressure of 100 atm

17. A mixture of 79 per cent N_2 and 21 per cent CO_2 by volume is passed over hot carbon (graphite) at a temperature of 1000°K and a pressure of 1.0 atm. Using the data of Table XXXVI calculate the equilibrium composition of the gases, and compare this result with that of part (*a*) of problem 16.

18. Carbon monoxide is burned with pure oxygen in the theoretically required proportions. Calculate the degrees of completion of the oxidation if equilibrium is attained at temperatures of, respectively, 1000°K and 3000°K, under a pressure of 1.0 atm. Evaluate $\Delta G°/T$ at 3000°K by extending the data of Tables XXXVI and XXXVII from 1500°K through graphical integration of the heat capacities of Table IV, page 212.

19. Water gas leaves a generator containing 51.1 per cent H_2, 2.3 per cent CO_2, and 46.6 per cent CO by volume on the dry basis. Ten per cent of the steam which was introduced into the bottom of the generator passed through the bed of hot coke without decomposition and is present in the gases. This gaseous mixture is passed into a reaction chamber under a pressure of 1.0 atm in contact with a catalyst of chromium oxide and allowed to attain equilibrium at a temperature of 700°K (423°C). Calculate the equilibrium composition of the gaseous mixture, using the data of Fig. 156.

20. One volume of the initial wet water gas described in problem 19 is mixed with three volumes of additional water vapor. This mixture is passed into the reaction chamber operated at the conditions described in problem 19 and allowed to reach equilibrium.

(*a*) Calculate the equilibrium composition of the gaseous mixture.

(*b*) Calculate the composition of the residual gas if the CO_2 and H_2O are removed from the gaseous mixture of part (*a*) after equilibrium is attained.

21. The hydrate of sodium carbonate decomposes according to the following equation:

$$Na_2CO_3 \cdot H_2O\ (s) = Na_2CO_3\ (s) + H_2O\ (g)$$

The equilibrium pressure in atmospheres of water vapor in this reaction is given by the following equation:

$$\log p = 7.944 - 3000.0/T°K$$

Derive an expression for the standard free-energy change of this reaction.

22. Zinc oxide is reduced with carbon monoxide under a pressure of 1.0 atm according to the following reaction:

$$ZnO\ (s) + CO\ (g) = Zn\ (g) + CO_2\ (g)$$

Calculate the degree of completion of the oxidation of CO at atmospheric pressure under the following different conditions, assuming that equilibrium conditions are attained and that ZnO is always present:

(*a*) At a temperature of 1000°C, using pure CO

(*b*) At a temperature of 1500°C, using pure CO

(*c*) At a temperature of 1000°C, using a mixture of 27.5 per cent CO, 4.3 per cent CO_2, and 68.2 per cent N_2 by volume

(*d*) At a temperature of 1500°C, using a mixture of 27.5 per cent CO, 4.3 per cent CO_2, and 68.2 per cent N_2 by volume

23. Sodium bicarbonate is calcined according to the following equation:

$$2NaHCO_3\ (s) = Na_2CO_3\ (s) + H_2O\ (g) + CO_2\ (g)$$

Calculate the pressure of the equimolecular mixture of H_2O and CO_2 in equilibrium with $NaHCO_3$ at a temperature of 100°C.

24. (*a*) Calculate the boiling point of aluminum at atmospheric pressure from the following data:[9]

Temperature, °K	$\frac{G° - H°_{0g}}{T}$ *Gas*	$\frac{G° - H°_{0s}}{T}$ *Solid or Liquid*
298.1	−33.766	−3.13
1000	−40.217	−9.09
2000	−43.762	−14.64
2100	−44.009	−15.01
2200	−44.245	−15.37
2300	−44.470	−15.71
2400	−44.681	−16.04

The value of $\Delta H°_0$ for vaporization is 66.920 kcal per g-atom.

(*b*) Calculate the vapor pressure at 2000°K.

25. Considering each of the following reactions:

(1) $SO_2(g) + \frac{1}{2}O_2(g) = SO_3(g)$

(2) $\frac{1}{2}N_2(g) + \frac{1}{2}O_2(g) = NO(g)$

(3) $C(s) + CO_2(g) = 2CO(g)$

Tabulate the effects of the following changes upon (*a*) the velocity of reaction (moles transformed per unit time per unit volume), (*b*) the equilibrium degree of completion, and (*c*) the actual degree of completion obtained in a specified time interval:

(1) Increase of temperature
(2) Increase of pressure
(3) Provision of a positive catalyst
(4) Dilution with an inert gas
(5) Agitation of the reacting system

Tabulate each effect as an increase, decrease, no effect, or indeterminate.

26. In the American process of synthesizing ammonia a mixture of three volumes of hydrogen and one volume of nitrogen is passed into a reaction chamber in contact with a catalyst of granular iron oxide combined with oxides of potassium and aluminum. Using the data of Illustration 7 and Fig. 156:

(*a*) Plot curves relating the equilibrium degree of completion of this reaction to temperature at pressures of 1.0, 100, and 300 atm. The temperature range from 400°K to 800°K should be included in the calculations.

(*b*) Calculate the equilibrium degree of completion of the afore-mentioned reaction at a pressure of 200 atm and a temperature of 750°K.

27. In the synthesis of ammonia described in problem 26 the mixture of N_2 and H_2 is introduced into the reaction chamber under a pressure of 300 atm and a temperature of 400°C. Assuming that heat loss from the reaction chamber is negligible, calculate the equilibrium-reaction temperature, using the curves of problem 26.

28. From the results of problem 10 and the data of Table XXXVII calculate the equilibrium temperatures and conversions reached when propane is dehydrogenated in an adiabatic reactor with an inlet temperature of 1200°F and pressures of 1.0 and 2.0 atm, respectively.

29. From the data of Table XXXVI calculate the complete compositions of the equilibrium mixtures resulting from the polymerization of pure ethylene to the butenes if it is assumed that no other reactions occur at the following conditions:

	Temperature, °C	*Pressure*, lb per sq in.
(*a*)	150°C	300
(*b*)	150°C	600
(*c*)	200°C	300
(*d*)	200°C	600

30. From the data of Table XXXVII calculate the composition of the equilibrium mixture of the five isomeric hexanes at a temperature of 600°K. Calculate the heat of reaction in converting pure normal hexane to the equilibrium mixture in the ideal-gaseous state at this temperature.

31. When propane is pyrolyzed the following reactions take place:

$$C_3H_8 \rightarrow C_3H_6 + H_2$$
$$C_3H_8 \rightarrow C_2H_4 + CH_4$$

Assuming that only these two reactions occur, calculate from the data of Table XXXVI the composition of the equilibrium mixture formed by heating propane at 1400°F and 1 atm.

32. Propane also decomposes to form carbon and hydrogen at high temperatures. Calculate the composition of the equilibrium mixture of problem 29 if this reaction is considered.

33. When propane is pyrolyzed at elevated pressures, a portion of the ethylene formed polymerizes to form butenes according to the following reaction:

$$2C_2H_4 \rightarrow C_4H_8$$

Assuming that only this reaction and the two of problem 31 occur, calculate the complete composition of the equilibrium mixture formed by heating propane at 1100°F and 1000 lb per sq in. The butenes formed will be present in their equilibrium proportions.

34. Magnesium is produced in the carbothermic process by heating briquettes of a mixture of carbon and magnesia. The following reaction results:

$$MgO(s) + C(s) \rightarrow Mg(g) + CO(g)$$

Using the following data for the physical properties[6,8] of Mg and MgO, calculate the equilibrium temperatures required for the reaction at pressures of 1.0 atm and 0.1 atm. Also calculate the corresponding heats of reaction, at the conditions of operation, per pound of Mg produced. From these results consider the feasibility of the direct production of liquid magnesium.

Heat Capacities:

MgO(s), $c_p = 10.86 + (1.197)(10^{-3})T - (2.087)(10^5)/T^2$ cal/(°K)(g-mole)

C(s), $c_p = 2.673 + (2.617)(10^{-3})T - (1.169)(10^5)/T^2$ cal/(°K)(g-mole)

Mg(g), $c_p = 4.97$ cal/(g-mole)(°K)

Heat of sublimation, Mg (s) = Mg (g); $\Delta H_{291} = 36.12$ kcal per g-mole

Vapor pressure of Mg:

Temperature, °K	789 (s)	881 (s)	998 (l)	1,159	1,236	1,303	1,380
Vapor pressure, atmospheres	0.0001	0.001	0.01	0.1	0.25	0.5	1.0

35. Magnesium is produced in the ferrosilicon reduction process by heating a mixture of magnesium oxide and ferrosilicon. The principal reaction may be assumed to be as follows:

$$2MgO + Si(s) = 2Mg(g) + SiO_2(s)$$

The heat capacities of silicon and silica are expressed by the following equation:[5]

$$Si(s);\ c_p = 5.74 + (0.617)(10^{-3})T - (1.01)(10^5)/T^2 \text{ cal/(°K)(g-mole)}$$
$$SiO_2(s);\ c_p = 10.87 + (8.712)(10^{-3})T - (2.412)(10^5)/T^2 \text{ cal/(°K)(g-mole)}$$

(*a*) Using the data of problem 34 calculate the equilibrium temperature of the reaction at operating pressures of 1.0, 10^{-3}, and 10^{-6} atm, respectively.

(*b*) Calculate the heat of reaction, at the operating conditions, per pound of magnesium produced at the three pressures of part (*a*). Compare these results with those of problem 34.

36. Calculate the compositions of the solutions formed when $MgCO_3$ (s) is dissolved in water in equilibrium with CO_2 at the following partial pressures.

	p_{CO_2} atm		p_{CO_2} atm
(*a*)	10^{-8}	(*d*)	10^{-2}
(*b*)	10^{-6}	(*e*)	1
(*c*)	10^{-4}	(*f*)	50

Also calculate the composition of the solution at the invariant point where $MgCO_3$ (s) and $Mg(OH)_2$ (s) are both present. From these data plot the solubility relationships in the form of Fig. 163 and calculate the composition of the solution formed by dissolving $MgCO_3$ (s) in pure water without gain or loss of CO_2.

37. A low-grade manganese ore consists essentially of a mixture of 35 per cent $MnCO_3$, 25 per cent $CaCO_3$, and 40 per cent $FeCO_3$ by weight. It is desired to concentrate the manganese for recovery. Calculate the composition of the equilibrium solutions resulting if this ore is leached with water in contact with an atmosphere of CO_2 at pressures of (*a*) 1 atm, (*b*) 10 atm. The entropy of the Mn^{++} ion may be taken as -15.9.

38. The ore of problem 37 is leached with a calcium chloride solution having a molality of 1.0. Assuming that the metallic salts are completely dissociated in solution and that the activity coefficients are unity, calculate the calcium and manganese and iron contents of the solution formed at equilibrium. The relatively small concentrations of carbonate and the bicarbonate ions may be neglected as compared with the Cl^- ion concentration and the over-all reactions considered as

$$Ca^{++} + MnCO_3(s) = CaCO_3(s) + Mn^{++}$$
$$Ca^{++} + FeCO_3(s) = CaCO_3(s) + Fe^{++}$$

Calculate the pounds of leaching solution required to dissolve all the manganese from 1000 lb of ore and the weight of iron dissolved and calcium precipitated in the operation.

This reaction forms the basis of a proposed process[9] for treating such ores in which the extract solution is treated with $Ca(OH)_2$ and blown with air to precipitate the oxides of Mn and Fe and regenerate the $CaCl_2$. The complete over-all reaction thus is

$$2MnCO_3 + 2Ca(OH)_2 + \tfrac{1}{2}O_2 = 2CaCO_3 + Mn_2O_3 + 2H_2O$$

CHAPTER XVII

THERMODYNAMIC PROPERTIES FROM MOLECULAR STRUCTURE

The application of thermodynamic principles is in many cases hampered by lack of data which are sufficiently reliable to be useful for engineering design. The accuracy with which standard free energies must be known is evident from consideration of the error ΔK, in the equilibrium constant which corresponds to an error $\Delta(\Delta G°)$ in the standard free-energy change. Thus, from Equation (XVI-6) at constant temperature,

$$-\Delta(\Delta G°) = RT \ln\left(\frac{\Delta K}{K} + 1\right) \tag{1}$$

An error of 10 per cent in the equilibrium constant ($\Delta K/K = 0.10$) at a temperature of 300°K corresponds to an error in the standard free-energy change of only 0.057 kcal per g-mole. Such an error results from an equal error in the standard heat of reaction or an error of 0.2 entropy unit. Since the heats of combustion of even simple organic compounds generally exceed 200 kcal per g-mole it follows that for such a compound an error in the heat of combustion of only 0.03 per cent may result in an error of 10 per cent in the equilibrium constant based upon it. An error in heat of reaction of only 0.42 kcal per g-mole or 1.4 entropy units will result in a 100 per cent error in the equilibrium constant at 300°K.

It is evident that, where equilibrium is a limiting factor in a process, measurements of the highest accuracy are necessary in order to develop thermodynamic properties which are satisfactory for engineering calculations. Such measurements have been carried out on relatively few materials, and in the present state of the basic data it is commonly necessary to resort to every available device for extrapolation and interpolation from available measurements.

Developments in spectroscopy and statistical mechanics together with improved methods of low-temperature calorimetry have resulted in reliable heat-capacity and entropy values for a variety of compounds, as indicated by Table XXXV. The corresponding heats of formation are generally less reliable. The equilibrium constants calculated from much of the available heat-of-formation data are useful only for predicting the general feasibility of reactions and the ranges of operating conditions in which equilibrium effects may be neglected.

Where reasonably reliable entropy and heat-capacity data are available or can be estimated by the procedures outlined in this chapter, a complete expression for the equilibrium constant which is satisfactory for many purposes may be developed from a single experimental measurement of the compositions at equilibrium under conditions not far removed from the range of interest. The standard heat of reaction $\Delta H°$ is calculated from Equation (XVI-7) from the known entropy and equilibrium constant. Equilibrium constants at other conditions may then be calculated from Equations (XVI-6) and (XVI-11). In this manner a value of $\Delta H°$ is obtained which is consistent with the value of $\Delta S°$ in the expression for the equilibrium constant, and moderate errors in the entropies used will not seriously affect equilibrium calculations over a limited range of conditions.

EMPIRICAL CORRELATION OF ENTROPIES

Methods of correlating the entropies of inorganic compounds are reviewed by Wenner,[1] who points out that the entropies of similar compounds in the solid state vary with molecular weight in accordance with the following equation:

$$\mathrm{s} = A \log M + B \tag{2}$$

where s = molal entropy
M = molecular weight
A, B = constants characteristic of the type of compound

On this basis a group of similar compounds should yield a straight line when molecular weight is plotted on a logarithmic scale against entropy on a uniform scale. Data on two compounds serve to establish such a line, or the constants of Equation (2), from which entropies of similar materials may be estimated.

Wenner found that the points for individual compounds in each group fall fairly well on a straight line for each of the following groups of metallic compounds in the solid state: MeO_2, MeO, MeS, $MeNO_3$, and metallic halides, MeX and MeX_2. A further generalization pointed out is that the addition of 1 mole of H_2O to a solid compound causes an entropy increase of 10 units in forming a crystalline hydrate, 8.2 in forming an inorganic base from its oxide, and 7 units in forming an organic acid from its corresponding anhydride.

In the gaseous state Wenner shows that molal entropies vary with molecular weight according to the equation:

$$\log \mathrm{s(g)} = A \log M + B \tag{3}$$

[1] R. R. Wenner, "Thermochemical Calculations," McGraw-Hill Book Company (1941).

where A and B are constants determined by the number of atoms in the molecule. Thus, all monatomic gases fall reasonably on a straight line when entropy is plotted against molecular weight on logarithmic scales. Another line is formed by diatomic gases. This relationship is applicable to gases containing as many as nine carbon atoms. However, errors ranging up to six or seven units are encountered, and Wenner recommends that entropies estimated from these relationships should be used only as rough approximations.

GROUP CONTRIBUTIONS

Various methods for empirically correlating heats of formation, entropies, and heat capacities of organic compounds were reviewed by Andersen, Beyer, and Watson[2] who proposed a scheme whereby these properties for the ideal-gaseous state are resolved into contributions attributable to atomic groups. From the resulting tables the properties of complex molecules are readily estimated by summation of the contributions of their component groups. The corresponding properties in the liquid and nonideal-gaseous states may then be calculated by the methods of Chapter XII.

The molal values for the ideal-gaseous state of heats of formation at 25°C, entropies at 25°C, and 1 atm, and the constants a, b, and c of the three-term heat-capacity equation (VII-21), page 213, were resolved into group contributions on the following basis:

Each compound is considered as composed of a basic group which is modified by the substitution of other groups for atoms comprising it. For example, all paraffin hydrocarbons may be considered as derived from methane by successive substitution of CH_3 groups for hydrogen atoms. Similarly, any secondary amine can be considered as derived from the base group $NH(CH_3)_2$. The contributions of nine base groups are given in Table XXXIX.

The contributions resulting from the primary substitution of a methyl group for a hydrogen atom in any one of the base groups is given in Table XL. In the cases of benzene, naphthalene, and cyclopentane, the base group contains several carbon atoms, and successive substitutions on different carbon atoms involve different contributions depending upon the number and position of the substituted groups. For the naphthenes the terms ortho, meta, and para are taken as indicating a minimum separation of the two substituted groups by respectively 0, 1, and 2 intermediate carbon atoms in the ring. The heat-capacity coefficients, a, b,

[2] J. W. Andersen, G. H. Beyer, and K. M. Watson, *Nat'l Petroleum News, Tech. Sec.*, **36**, *R*476 (July 5, 1944). Also "Process Engineering Data," National Petroleum Publishing Company, Cleveland (1944).

and c, apply to ideal gases where temperature is expressed in degrees Kelvin.

TABLE XXXIX

BASE GROUP PROPERTIES

Group	$\Delta H^\circ_{f298.1}$ (g) kcal per g-mole	$S^\circ_{298.1}$ (g) cal/(g-mole)(°K)	a	$b(10^3)$	$c(10^6)$
			Ideal Gas at T°K		
Methane	−17.9	44.5	3.42	17.85	−4.16
Cyclopentane	−21.4	70.7	2.62	82.67	−24.72
Benzene	18.1	64.4	0.23	77.83	−27.16
Naphthalene	35.4	80.7	3.15	109.40	−34.79
Methylamine	−7.1	57.7	4.02	30.72	−8.70
Dimethylamine	−7.8	65.2	3.92	48.31	−14.09
Trimethylamine	−10.9	—	3.93	65.85	−19.48
Dimethyl ether	−46.0	63.7	6.42	39.64	−11.45
Formamide	−49.5	—	6.51	25.18	−7.47

TABLE XL

CONTRIBUTIONS OF PRIMARY CH_3 SUBSTITUTION GROUPS REPLACING HYDROGEN

Base Group	$\Delta(\Delta H^\circ_{f298.1})$ (g) kcal per g-mole	$\Delta S^\circ_{298.1}$ (g) cal/(g-mole)(°K)	Δa	$\Delta b(10^3)$	$\Delta c(10^6)$
			Ideal Gas at T°K		
1. Methane	−2.2	10.4	−2.04	24.00	−9.67
2. Cyclopentane					
(*a*) Enlargement of ring	−9.3	0.7	−1.04	19.30	−5.79
(*b*) First substitution	−5.2	11.5	−0.07	18.57	−5.77
(*c*) Second substitution:					
ortho	−12.2	—			
meta	−8.4	—	−0.24	16.56	−5.05
para	−7.1	—			
(*d*) Third substitution	−7.0	—	—	—	—
3. Benzene and naphthalene					
(*a*) First substitution	−4.5	12.0	0.36	17.65	−5.88
(*b*) Second substitution:					
ortho	−6.3	8.1	5.20	6.02	1.18
meta	−6.5	9.2	1.72	14.18	−3.76
para	−8.0	7.8	1.28	14.57	−3.98
(*c*) Third substitution (sym)	—	8.0	0.57	16.51	−5.19
4. Methylamine	−5.7	—			
5. Dimethylamine	−6.3	—	−0.10	17.52	−5.35
6. Trimethylamine	−4.1	—			
7. Formamide					
Substitution on C atom	−9.0	—	6.11	−1.75	4.75

If more than one substitution is made on a single carbon atom of a base group the additional contributions are treated as secondary substitutions and evaluated from Table XLI.

The contributions resulting from the secondary substitution of methyl groups for hydrogen atoms are classified in Table XLI according to *A*, the type of the carbon atom on which the substitution is made, and *B*, the highest type number of an adjacent carbon atom. The carbon-atom types are defined on the basis of the number of hydrogen atoms attached. Thus:

Type 1	$—CH_3$
2	$—\overset{|}{C}H_2$
3	$—\overset{|}{\underset{|}{C}}H$
4	$—\overset{|}{\underset{|}{C}}—$
5	C atom in benzene or naphthalene ring

Two special secondary substitutions are defined in Table XLI for use in calculating the properties of esters and ethers. One is the substitution of a methyl group for the hydrogen of a carboxyl group to form a methyl ester. The other is the substitution of a methyl group for one of the hydrogens of a methyl ester or ether to form an ethyl ester or ether. If additional substitutions are made on this same carbon atom, the contributions are evaluated from the corresponding values of *A* and *B* based on the type numbers of the carbon atoms involved.

In Table XLII are the contributions resulting from the substitution of multiple bonds for single bonds between two carbon atoms of types *A* and *B*, respectively. An additional contribution must be added for each pair of conjugated double bonds formed by such substitutions.

In Table XLIII are the incremental contributions resulting from the substitution of various groups for one or two methyl groups. Thus, if a methyl group is replaced by an OH group the contribution to the heat of formation is −32.7. If two methyl groups are replaced by an oxygen atom to form an aldehyde the contribution is −12.9. It may be noted that the phenyl group is included in Table XLIII in addition to being designated as a base group in Table XXXIX. This substitution contribution is used in calculating the properties of complex compounds in which several base groups are combined, as, for example, in polybasic aromatic acids.

The contributions to heats of formation resulting from substitution of chlorine for methyl groups vary with the number of substitutions made on a single carbon atom. Corresponding variations were not found for the contributions to entropy or heat capacity or for the substitution of the other halogens. As noted in the table a correction must be applied to the entropies calculated for the halogenated methanes. In general

TABLE XLI

SECONDARY METHYL SUBSTITUTIONS REPLACING HYDROGEN

A	B	$\Delta(\Delta H_f^\circ)_{298.1}$ (g) kcal per g-mole	$\Delta S^\circ_{298.1}$ (g) cal/(g-mole)(°K)	Δa	$\Delta b(10^3)$	$\Delta c(10^6)$
				Ideal Gas at T°K		
1	1	−4.5	9.8	−0.97	22.86	−8.75
1	2	−5.2	9.2	1.11	18.47	−6.85
1	3	−5.5	9.5	1.00	19.88	−8.03
1	4	−5.0	11.0	1.39	17.12	−5.88
1	5	−6.1	10.0	0.10	17.18	−5.20
2	1	−6.6	5.8	1.89	17.60	−6.21
2	2	−6.8	7.0	1.52	19.95	−8.57
2	3	−6.8	6.3	1.01	19.69	−7.83
2	4	−5.1	6.0	2.52	16.11	−5.88
2	5	−5.8	2.7	0.01	17.42	−5.33
3	1	−8.1	2.7	−0.96	27.47	−12.38
3	2	−8.0	4.8	−1.19	28.77	−12.71
3	3	−6.9	5.8	−3.27	30.96	−14.06
3	4	−5.7	1.7	−0.14	24.57	−10.27
3	5	−9.2	1.3	0.42	16.20	−4.68
1	—0—in ester or ether	−7.0	14.4	−0.01	17.58	−5.33
Substitution of H of OH group to form ester		+9.5	16.7	0.44	16.63	−4.95

TABLE XLII

MULTIPLE-BOND CONTRIBUTIONS REPLACING SINGLE BONDS

A	Type of Bond	B		$\Delta(\Delta H_f^\circ)_{298.1}$ (g) kcal per g-mole	$\Delta S^\circ_{298.1}$ (g) cal/(g-mole)(°K)	Δa	$\Delta b(10^3)$	$\Delta c(10^6)$
						Ideal Gas at T°K		
1	=	1		32.8	−2.1	1.33	−12.69	+4.77
1	=	2		30.0	0.8	1.56	−14.87	+5.57
1	=	3		28.2	2.2	0.63	−23.65	+13.10
2	=	2		28.0	−0.9	0.40	−18.87	+9.89
2	=	2	cis	28.4	−0.6	0.40	−18.87	+9.89
2	=	2	trans	27.5	−1.2	0.40	−18.87	+9.89
2	=	3		26.7	1.6	0.63	−23.65	+13.10
3	=	3		25.5	—	−4.63	−17.84	+11.88
Additional correction for each pair of conjugated double bonds				−3.8	−10.4	Approximately zero		
1	≡	1		74.4	−6.8	5.58	−31.19	+11.19
1	≡	2		69.1	−7.8	6.42	−36.41	+14.53
2	≡	2		65.1	−6.3	4.66	−36.10	+15.28
Correction for double bond adjacent to aromatic ring				−5.1	−4.3	Approximately zero		

TABLE XLIII

SUBSTITUTION GROUP CONTRIBUTIONS REPLACING CH_3 GROUP

Group	$\Delta(\Delta°H_f)_{298.1}$ (g) kcal per g-mole	$\Delta S°_{298.1}$ (g) cal/(g-mole)(°K)	Δa	$\Delta b(10^3)$	$\Delta c(10^6)$
			Ideal Gas at T°K		
–OH (aliphatic, meta, para)	–32.7	2.6	3.17	–14.86	5.59
–OH ortho	–47.7	—	—	—	—
$-NO_2$	1.2	2.0	6.3	–19.53	10.36
–CN	39.0	4.0	3.64	–13.92	4.53
–Cl	0‡ for first Cl on a carbon; 4.5 for each additional	0	2.19	–18.85	6.26
–Br	10.0	3.0‡	2.81	–19.41	6.33
–F	–35.0	–1.0‡	2.24	–23.61	11.79
–I	24.8	5.0‡	2.73	–17.37	4.09
=O aldehyde	–12.9	–12.3	3.61	–55.72	22.72
=O ketone	–13.2	–2.4	5.02	–66.08	30.21
–COOH	–87.0	15.4	8.50	–15.07	7.94
–SH	15.8	5.2	4.07	–24.96	12.37
$-C_6H_5$	32.3	21.7	–0.79	53.63	–19.21
$-NH_2$	12.3	–4.8	1.26	–7.32	2.23

‡ Add 1.0 to the calculated entropy contributions of halides for methyl derivatives; for example, methyl chloride = 44.5 (base) + 10.4 (primary CH_3) – 0.0 (Cl substitution) + 1.0.

the calculated results tend to be uncertain for single carbon-atom compounds.

The suggested sequence of operations in estimating the properties of a complex compound is as follows:

1. Select the base group and determine its properties from Table XXXIX. Where a choice of base group is possible, select the group having the largest entropy. Proceed to build up the desired compound with as few substitutions as possible.

2. Add the contributions given in Tables XL and XLI which result from all CH_3 substitutions replacing hydrogen required to establish the carbon skeleton of the compound. In this operation the longest straight chain should be built up first and then the longest side chain. Where the same compound may be arrived at by alternate substitutions an average result is used.

3. Add the contributions from Table XLI which result from additional CH_3 substitutions replacing hydrogen in the positions occupied in the compound by other groups which are listed in Table XLIII.

4. Add the contributions given in Table XLII which result from multiple bonds.

5. Add the incremental contributions given in Table XLIII which result from replacement of CH_3 groups by substitution groups.

Illustration 1. Approximate the standard heat of formation $\Delta H_{f\,298.1}$ (g) of 2,2,4 – trimethylpentane at 25°C.

Base group (methane)	Table XXXIX	−17.9
Primary CH_3	Table XL	−2.2

Secondary CH_3 substitutions by successive replacement of hydrogen (Table XLI)

	A	*B*	
Secondary methyl groups in 5 carbon chain	1	1	−4.5
	1	2	−5.2
	1	2	−5.2
Side methyl groups	2	2	−6.8
	2	2	−6.8
	3	2	−8.0
			−56.6

The value found by Rossini and co-workers[3] is −56.2 kcal per g-mole.

Illustration 2. Approximate the value of $\Delta H_{f\,298.1}$ (g) for dimethyl phthalate.

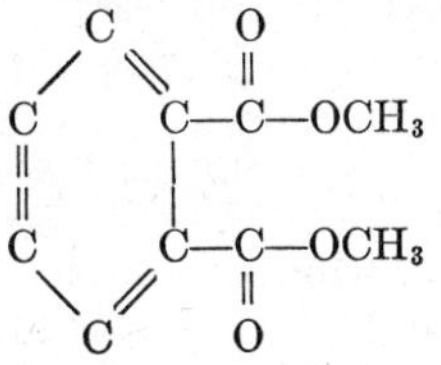

Base group (benzene)	Table XXXIX	18.1
Primary CH_3 replacing H	Table XL	−4.5
Ortho CH_3 replacing H	Table XL	−6.3
−COOH replacing CH_3	Table XLIII	−87.0
−COOH replacing CH_3	Table XLIII	−87.0
CH_3 replacing H of −COOH	Table XLI	9.5
CH_3 replacing H of −COOH	Table XLI	9.5
	$\Delta H_{f\,298.1}$ =	−147.7

The value of $\Delta H_{f\,298.1}$ calculated from the heat of combustion listed by Kharasch[4] is −147.1 kcal per g-mole.

Illustration 3. Approximate the entropy of 2,2,3,3-tetramethyl butane.

(*a*) Base group (methane)	Table XXXIX			44.5
Primary CH_3	Table XL			10.4
Secondary CH_3 substitutions replacing hydrogen	Table XLI			
	A		*B*	
Secondary CH_3 groups in 4 C chain	1		1	9.8
	1		2	9.2
Side methyl group clockwise	2		2	7.0
	2		3	6.3
	3		3	5.8
	3		4	1.7
				94.7

(*b*) Base Group (methane)	Table XXXIX	44.5
Primary CH_3	Table XL	10.4
Secondary CH_3 substitutions replacing hydrogen	Table XLI	

[3] K. S. Pitzer, *Chem. Rev.*, **27**, 39 (1940).
[4] M. S. Kharasch, *Bur. Standards J. Research*, **2**, 359 (1929).

	A	B	
Secondary CH_3 groups in 4 C chain	1	1	9.8
	1	2	9.2
Side methyl groups counterclockwise	2	2	7.0
	3	2	4.8
	2	4	6.0
	3	4	1.7
			93.4

The average result for the entropy of 2,2,3,3-tetramethyl butane is 94.05 cal/(g-mole)(°K) which is in agreement with the value found by Pitzer.[3]

Illustration 4. Approximate the entropy of 1,2-dibromoethane.

Base group (methane)		Table XXXIX	44.5
Primary CH_3 replacing hydrogen		Table XL	10.4
Secondary CH_3 substitutions replacing hydrogen:			
A	B		
1	1	Table XLI	9.8
1	2		9.2
Br substitution replacing CH_3		Table XLIII	3.0
Br substitution replacing CH_3		Table XLIII	3.0
			79.9

This result is in agreement with the value reported by Pitzer.[3]

Illustration 5. Calculate the heat required to raise 1 mole of 2,3-dimethylpentane in the ideal-gaseous state from 298.1°K to 1000°K.

	A	B	Δa	$\Delta b(10^3)$	$\Delta c(10^6)$
Base (Table XXXIX)			3.42	17.85	−4.16
Primary (Table XL)			−2.04	24.00	−9.67
5 C chain (Table XLI)	1	1	−0.97	22.86	−8.75
	1	2	1.11	18.47	−6.85
	1	2	1.11	18.47	−6.85
Side methyl groups	2	2	1.52	19.95	−8.57
	2	3	1.01	19.69	−7.83
			5.16	141.29	−52.68

$$H_{1000} - H_{298.1} = \int_{298.1}^{1000} (5.16 + 141.29(10^{-3})T - 52.68(10^{-6})T^2)\, dT$$

$$= \left[5.16T + 70.64(10^{-3})T^2 - (17.56)(10^{-6})T^3\right]_{298.1}^{1000}$$

$$= 50{,}890 \text{ cal per g-mole}$$

Pitzer[3] reports a value of 50,550 cal per g-mole.

Extensive comparisons with the data from the literature lead Andersen, Beyer, and Watson to conclude that in general molal heats of formation and entropies calculated by the group-contribution method differ from the better experimental values by less than 4.0 kcal and 2.0 entropy units, respectively. Calculated heat capacities appear to be within 5 per cent of the accepted values, although serious discrepancies exist between the experimental values of different investigators. In many cases, particularly for heats of formation, it is believed that the calculated values may be preferable to the experimental.

The contributions in Tables XXXIX to XLIII in some cases are based on data of questionable accuracy and will be subject to revision as better experimental values become available. This method of calculation may be extended to other series of compounds if data are available to establish the necessary base-group contribution. In many cases the properties of such base groups may be estimated by the theoretical methods described in the following sections. Properties of higher homologs or derivatives may then be estimated from the group contributions. It is believed that this procedure in many cases is more reliable than direct application of the simplified forms of statistical calculations to the complex molecules.

Heats of formation and heat capacities estimated from group contributions are sufficiently accurate for use in energy balances except where heats of reaction are very small or must be known with unusual accuracy. Equilibrium constants based on the estimated entropies and heats of formation are satisfactory for predicting the feasibility of a reaction but not for accurate calculation of equilibrium compositions. For accurate equilibrium constants the estimated entropy and heat-capacity data are best combined, as previously outlined, with a measured equilibrium constant to establish a consistent heat of reaction.

STATISTICAL METHODS

In recent years development of the principles of statistical mechanics has lead to methods for calculating precise heat-capacity and absolute entropy data for the ideal-gaseous state from spectroscopic measurements. Further generalization of the fundamental data has resulted in means for estimating these properties for many compounds even in the absence of direct spectroscopic measurements, from a knowledge of the structure of the molecule. A few of these developments which are of practical applicability are presented in the following pages, and their uses are illustrated. Derivation of the relations employed and complete discussions of the principles may be found in standard texts.[5–9] The

[5] "Statistical Thermodynamics," J. Mayer and M. Mayer, John Wiley & Sons (1940).

[6] "Statistical Thermodynamics," R. H. Fowler and E. A. Guggenheim, Cambridge Press (1939).

[7] "Statistical Mechanics," R. C. Tolman, Oxford Press (1938).

[8] "Introduction to Quantum Mechanics," L. Pauling and F. B. Wilson, McGraw-Hill Book Company (1935).

[9] H. Eyring, J. Walter, and G. E. Kimball, "Quantum Chemistry," John Wiley & Sons (1944).

development presented represents an amplification of the outline published by Kelley[10] with an extensive bibliography of current literature.

Molecular Energy of an Ideal Gas. The internal energy of a molecule is present in kinetic and potential forms resulting from the motion and attractive forces of the molecule and its parts. These forms of energy may be classified in the following five groups:

1. Energy of translation of the molecule as a whole designated as U_t per mole or ϵ_t per molecule. This form of energy was discussed in Chapter II and is equal to $(3/2)RT$ per mole or $(3/2)(RT/N) = (3/2)\boldsymbol{k}T$ per molecule where N is the Avogadro number and $\boldsymbol{k}$ is termed the Boltzmann constant.

2. External rotational energy resulting from rotation of the molecule as a whole, designated as U_r or ϵ_r. Just as in the rotation of large bodies, such rotational energy may be resolved into components about each of three perpendicular axes passing through the center of gravity of the molecule and is a function of the frequency of rotation and the moments of inertia about these three axes.

3. Internal rotational energy resulting from the rotation of groups of atoms in the molecule with respect to other groups in the same molecule and designated as $\text{U}_{r'}$ or $\epsilon_{r'}$. A simple example is the ethane molecule in which one CH_3 group rotates with respect to the other about the C–C bond. Significant internal rotational energy is believed to result only from the rotation of groups of atoms, not single atoms, about single-valence bonds. It is a function of the moments of inertia of the rotating groups about the axis of rotation, the frequency of rotation, and any attractive forces between the rotating groups which hinder rotation. Rotation which occurs without such restrictive forces is termed *free rotation*, and, if significant attractive forces are encountered, it is termed *hindered rotation.* For example, in the ethane molecule as a CH_3 group rotates about the C–C bond, at three points in each revolution its hydrogen atoms attain minimum distances from the hydrogen atoms of the other CH_3 group. At these three positions the restrictive force hindering rotation is a maximum. The energy required to rotate such a group from a position of minimum restrictive force to a position of maximum force is termed the *potential barrier* restricting rotation and is expressed in molal energy units. The energy associated with hindered rotation is less than the energy of free rotation of the same group by an amount depending on the magnitude of the potential barrier and the number of times it is encountered per revolution. In complicated molecules ro-

[10] K. K. Kelley, "Contributions to the Data of Theoretical Metallurgy; IX The Entropies of Inorganic Substances," Revision (1940) of Data and Methods of Calculation, *U.S. Bur. Mines Bull.*, 434.

tation of all possible combinations of groups and subgroups must be considered. It is assumed that rotation does not occur about multiple-valence bonds and that the energy associated with movements of groups about such bonds can be classified as vibrational in character.

4. Vibrational energy designated as U_ω or ϵ_ω results from deformation of valence bonds. In order to simplify consideration of the many possible types of vibration about an elastic bond, all vibrational contributions are resolved into those which result from vibrations along the axis of the bond, termed valence or stretching vibrations ν, and those which occur perpendicular to the axis of the bond, termed deformation δ or bending vibrations. Deformation vibrations change the angle between two adjacent bonds. The corresponding energy contributions are designated respectively as U_ν and U_δ per mole or ϵ_ν and ϵ_δ per molecule.

5. Electronic energy designated as U_e or ϵ_e results from the motion and forces of attraction of the electrons and nuclei comprising the molecule. These energy contributions are relatively very large, and changes in electronic arrangements are accompanied by large energy changes. However, except at high temperatures or in the presence of electric forces or radiation, it may be assumed that the electronic arrangements are stable and remain in a state of constant electronic energy. On this basis electronic energy frequently can be neglected in considering the changes of thermodynamic functions resulting from changes of state. This assumption may lead to serious errors at high temperatures, 600°C or above. In this range corrections based on spectroscopic measurements are required.

Molecular Degrees of Freedom. Based upon the foregoing classification, the number of independent component forms of internal energy other than electronic which can be possessed by a molecule in the ideal-gaseous state is termed its number of *degrees of freedom.* It may be shown that a molecule comprising n atoms has $3n$ degrees of freedom. For example, a monatomic gas molecule can possess energy only as a result of translation which can be resolved into three directions; hence, it has three degrees of freedom. A diatomic molecule such as nitrogen similarly has three degrees of translational freedom and in addition two degrees of external rotational freedom and one degree of vibrational bond-stretching freedom, a total of six. In such a linear molecule the moment of inertia is negligible about the axis joining the atoms, and external rotational energy results only from rotation about the other two perpendicular axes. There is no energy of internal rotation, because there is no group having a moment of inertia, and there is no bending or deformation energy, since there is only one bond. In a nonlinear triatomic molecule such as water there are six degrees of translational and rotational freedom, two degrees of vibrational bond-stretching freedom, and one degree of bond-bending freedom.

The degrees of freedom of a molecule exclusive of its translational freedom are termed its *internal degrees of freedom*, which include all freedoms for external and internal rotation, bond stretching, and bond bending. From the foregoing discussion it is evident that any molecule possesses $3n - 3$ internal degrees of freedom.

If only the degrees of freedom for internal rotation and vibration are considered, for a linear molecule,

$$n_{r'} + n_v + n_\delta = 3n - 5 \tag{4}$$

For a nonlinear molecule,

$$n_{r'} + n_v + n_\delta = 3n - 6 \tag{5}$$

where

$n_{r'}$ = degrees of freedom for internal rotation
n_v = degrees of freedom for stretching vibration
n_δ = degrees of freedom for bending vibration

The Quantum Theory. According to the quantum theory variation of the amount of energy possessed by a molecule in any degree of freedom does not occur as a continuous function but takes place in steps corresponding to the absorption or release of definite increments of energy termed quanta. This principle of discontinuity corresponds to the establishment of definite *energy levels* in each degree of freedom, which differ from each other by finite increments. The lowest possible level in each degree of freedom is termed the *ground level.* In a change of state from one energy level to another one quantum of energy is absorbed or released. The quantum does not correspond to a fixed quantity of energy, however, but varies widely in energy content depending upon the difference of the energy levels involved.

The levels of translational energy are so closely spaced that it may be considered to vary continuously. However, all rotational, vibrational, and electronic energies can exist only at definite levels differing from each other by finite increments. The different energy levels in a particular degree of freedom are identified by *quantum numbers* which are integers starting with 1 for the lowest energy level. When a change in energy level occurs in any degree of freedom, one quantum of energy is emitted or absorbed as one *photon* of radiation. The frequency or wave length of this radiation is related to the energy content of the quantum by the following equation:

$$\Delta\epsilon = \mathbf{h}\nu = \mathbf{h}c/\lambda = \mathbf{h}c\omega \tag{6}$$

or

$$\Delta E = 2.857\omega \text{ cal per g-mole} \tag{7}$$

where

ϵ = energy per quantum, ergs per molecule
$\mathbf{h}$ = Planck's constant = $(6.624)(10^{-27})$ erg sec
ν = frequency, reciprocal seconds
c = velocity of light = $(2.99776)(10^{10})$ cm per sec
λ = wave length, cm
ω = wave number, $1/\lambda$ = reciprocal cm

Because of their convenient numerical magnitude, wave numbers ω are commonly used to designate the frequency or wave length of radiation. Wave numbers are often referred to as frequencies, whereas actually the wave number has the dimension of reciprocal centimeters. The distinction between wave numbers ω and true frequencies ν should be carefully noted.

From Equation (6) it is evident that changes between widely separated energy levels emit or absorb radiation of high frequency and wave number or short wave length. By spectroscopic measurement of the frequency of emitted or absorbed radiation it is possible to evaluate the various levels of molecular and atomic energy. Thus, the large energy changes resulting from shifts of electrons correspond to radiation of short wave length in the ultraviolet or visible regions of the spectrum. The radiation of medium wave length corresponding to changes in vibrational energy is in the visible or infrared region, while the frequencies of rotational energy changes are detected by Raman spectra.

Translational Contributions to Thermodynamic Properties. In the development of relationships between molecular energies and thermodynamic properties it is convenient to consider separately the contributions resulting from the various forms of internal energy. It has been pointed out that the total internal energy of a gas is the sum of the individual contributions resulting from the five forms of internal energy. Since entropy is an extensive property, determined by energy content and its availability, the total entropy can also be expressed as the sum of the entropy contributions resulting from the five forms of internal energy. Similar summations represent the total enthalpy, free energy, and total work function. Derived properties such as the heat capacities can also be expressed as summations of the contributions due to each form of energy.

It was pointed out in Chapter II that the pressure and volume relations of an ideal gas are completely determined by the translational energy of its molecules. It was also shown in Chapter XI that, although the internal energy of an ideal gas is dependent only upon temperature, its entropy is a function of pressure or volume as well as temperature.

It follows that for an ideal gas the variation of entropy and other thermodynamic properties with pressure and volume result solely from translational contributions and that contributions from the other four forms of energy are independent of pressure or volume. For this reason it is convenient to develop separately all translational contributions to the thermodynamic properties.

The translational contribution to internal energy is given by Equation (II-7), page 31, as

$$\text{U}_t^\circ = \tfrac{3}{2}RT \tag{8}$$

Combination of Equations (XI–134, 137, 139), page 475, with Equation (8) gives

$$c_{vt}^\circ = \tfrac{3}{2}R \tag{9}$$

$$c_{pt}^\circ = \tfrac{5}{2}R \tag{10}$$

$$\text{H}_t^\circ = \tfrac{5}{2}RT \tag{11}$$

These are the properties of a perfect monatomic gas which possesses only translational energy. As was pointed out in Chapter XI, the absolute entropy of such a gas cannot be arrived at by simple integration of the thermodynamic relations because of the indeterminant lower limits encountered at the absolute zero. However, the translational contributions to entropy have been derived by statistical methods and are expressed by the Sackur–Tetrode equation which has been verified by experimental measurements on monatomic gases.

$$\text{S}_t^\circ = \tfrac{3}{2}R \ln M + \tfrac{5}{2}R \ln T - R \ln p - 2.298 \tag{12}$$

or at 298.1°K and 1 atm,

$$\text{S}_t^\circ = 6.861 \log M + 25.98 \tag{13}$$

where M is molecular weight, and p is expressed in atmospheres and T in °K. From Equations (8), (11), (12), and (XI-4 and 8),

$$\frac{\text{G}_t^\circ}{T} = R\left[-\frac{3}{2}\ln M - \frac{5}{2}\ln T + \ln p + 3.664\right] \tag{14}$$

$$\frac{\text{A}_t^\circ}{T} = R\left[-\frac{3}{2}\ln M - \frac{5}{2}\ln T + \ln p + 2.664\right] \tag{15}$$

or at 1 atm and 25°C, in calories per gram-mole,

$$\text{G}_{t298.16}^\circ = -2045.7 \log M - 6266.5 \tag{16}$$

$$\text{A}_{t298.16}^\circ = -2045.7 \log M - 6858.8 \tag{17}$$

Illustration 6. Calculate the translational contributions to the heat capacities, internal energy, enthalpy, entropy, free energy, and total work function of methyl alcohol in the ideal-gaseous state at temperatures of 298.1°K, 500°K, and 1,000°K and a pressure of 1 atm.

1. *Heat Capacities:* From Equations (9) and (10)

$$c^\circ_{vt} = \tfrac{3}{2}R = 2.9798 \text{ cal/(g-mole)(°K)}$$
$$c^\circ_{pt} = \tfrac{5}{2}R = 4.9663 \text{ cal/(g-mole)(°K)}$$

For a perfect gas the heat capacities c_v and c_p are independent of the temperature.

2. *Internal Energy:* From Equation (8), $U^\circ_t = 2.980T$
3. *Enthalpy:* From Equation (11), $H^\circ_t = 4.966T$
4. *Entropy:* Since $M = 32.043$ and $p = 1$, from Equation (12),

$$S^\circ_t = \tfrac{3}{2}R \ln 32.043 + \tfrac{5}{2}R \ln T - R \ln 1 - 2.298 = 8.035 + 11.437 \log T$$

5. *Free Energy:* From Equation (14),

$$\frac{G^\circ_t}{T} = R[-\tfrac{3}{2} \ln (32.043) - \tfrac{5}{2} \ln T + \ln 1 + 3.664] = -3.055 - 11.437 \log T$$

6. *Total Work Function:* From Equation (15),

$$\frac{A^\circ_t}{T} = R[-\tfrac{3}{2} \ln 32.043 - \tfrac{5}{2} \ln T + \ln 1 + 2.664] = -5.041 - 11.437 \log T$$

T	U°_t	H°_t	S°_t	$\frac{G^\circ_t}{T}$	$\frac{A^\circ_t}{T}$
298.16°K	888.4	1480.7	36.315	−31.349	−33.336
500°K	1489.9	2483.1	38.882	−33.916	−35.903
1000°K	2979.8	4966.3	42.325	−37.359	−39.346

All the preceding values are expressed in cal/(g-mole)(°K).

The Partition Function. One mole of an ideal gas comprising N molecules contains individual molecules possessing many varying energy levels, which result from various combinations of different levels in the several degrees of freedom other than translation. Of the N molecules, NN_a may have an energy level a, NN_b an energy level b, and so on, where $N_a, N_b, \cdots$ represent the fractions of the total number of molecules having energy levels $a, b, \cdots$. Then

$$N = NN_a + NN_b + NN_c + \cdots = \Sigma_i NN_i \tag{17a}$$

where $\Sigma_i NN_i$ represents the summation of the numbers of molecules of all energy levels.

Since the total internal-energy content of matter is unknown, it is possible to express an energy level only relative to a selected reference or *ground* state. It is customary to choose the ground state for a particular molecule as its lowest possible energy level. The energy of any other level resulting from any mode of motion or degree of freedom is then designated as ϵ ergs per molecule or E cal per g-mole, referred to this ground state. The internal energy of the molecule as an ideal gas in its ground state at 0°K is designated as U°_0 cal per g-mole. On this basis the molal internal energy is represented by

$$U^\circ - U^\circ_0 = U^\circ_t + N_a E_a + N_b E_b + N_c E_c + \cdots = U^\circ_t + \Sigma_i N_i E_i \tag{18}$$

U_t° = the translational contribution to internal energy from Equation (8)

N_a = the fraction of the molecules existing in state a with an energy level E_a

Useful application of summations such as Equation (18) is restricted to the ideal-gaseous state in order to avoid the complications which are introduced if the energy of 1 molecule is influenced by the proximity of others and cannot be considered individually.

It frequently occurs that two or more different molecular states will result in substantially the same energy level. In such cases this multiple contribution to a given energy level b is designated by a factor g_b representing the number of different molecular states which result in the single energy level b. This factor is termed the *statistical weight* of the energy level or sometimes the *quantum weight* or *degeneracy* and is an integer.

According to the Maxwell–Boltzmann distribution law the fraction of molecules N_b having any specified energy level E_b is given by the following expression,

$$N_b = \frac{g_b e^{-\frac{\text{E}_b}{RT}}}{\Sigma_i g_i e^{-\frac{\text{E}_i}{RT}}} = \frac{g_b e^{-\frac{\text{E}_b}{RT}}}{Q} \tag{19}$$

where $$Q = g_0 + g_a e^{-\frac{a}{RT}} + g_b e^{-\frac{\text{E}_b}{RT}} + g_c e^{-\frac{\text{E}_c}{RT}} + \cdots = \Sigma_i g_i e^{-\frac{\text{E}_i}{RT}} \tag{20}$$

The derivation of Equation (19) is based on the laws of probability and may be found in the standard texts on statistical mechanics. The group $e^{-\text{E}_b/RT}$ is termed the Boltzmann factor for the particular b energy level.

The summation in the denominator of Equation (19) includes the Boltzmann factors for all possible energy levels, including the ground level, in all degrees of freedom and is termed the *partition function Q*. Since by definition the relative energy E of the ground level is zero, its Boltzmann factor is unity, and the corresponding contribution to the partition function is g_0, the statistical weight of the ground level. It may be noted that the form of the Boltzmann factor is such that its magnitude is greatest for low energy levels. Consequently Equation (19) denotes that the states of lowest energy level are most probable, but that the probability of occurrence of high energy levels increases as the temperature is increased. Similarly, the greatest contributions to the summation of the partition function are from low energy levels. Molecular states corresponding to very high energy levels are negligible in the determination of the partition function except at very high temperatures.

An expression for molal internal energy is obtained by combining Equations (18) and (19):

$$\text{U}^\circ - \text{U}_0^\circ = \text{U}_t^\circ + \frac{1}{Q}\left(E_a g_a e^{-\frac{E_a}{RT}} + E_b g_b e^{-\frac{E_b}{RT}} + E_c g_c e^{-\frac{E_c}{RT}} + \cdots\right)$$

$$= \text{U}_t^\circ + \frac{\Sigma_i E_i g_i e^{-\frac{E_i}{RT}}}{Q} \tag{21}$$

Also, differentiation of Equation (20) with respect to temperature gives

$$\frac{dQ}{dT} = \frac{\Sigma_i E_i g_i e^{-\frac{E_i}{RT}}}{RT^2} \tag{22}$$

Combining (21) and (22) gives

$$\text{U}^\circ - \text{U}_0^\circ = \text{U}_t^\circ + \frac{RT^2}{Q}\frac{dQ}{dT} = \text{U}_t^\circ + RT\frac{d \ln Q}{d \ln T} \tag{23}$$

Equation (23) permits calculation of internal energy from the variation of the partition function with temperature.

Thermodynamic Properties from Partition Functions. The contributions of the thermodynamic properties which result from energy forms other than translation are readily expressed in terms of their individual partition functions by combination of Equation (23) with the thermodynamic relations of Chapter XI. Since rotational, vibrational, and electronic contributions are independent of pressure or volume, these variables need not be considered in the development. The subscript m is used to designate contributions resulting from both external and internal rotations, vibrations, and electronic energy. The relationships are developed in terms of the combined group of contributions but are equally applicable to any one type of contribution considered individually. The total thermodynamic properties are obtained by adding the m group of contributions to the translational values of the preceding section.

Equation (23) is directly applicable to any type of contribution to internal energy. Thus, since the m contributions to internal energy are independent of pressure, $\text{U}_m^\circ = \text{H}_m^\circ$, and

$$(\text{U}^\circ - \text{H}_0^\circ)_m = (\text{H}^\circ - \text{H}_0^\circ)_m = RT\left(\frac{d \ln Q}{d \ln T}\right) \tag{24}$$

Because the m contributions are independent of pressure or volume, the heat-capacity contributions at constant pressure and volume are equal. Then, since $c_v = d\text{U}/dT = (1/T)(d\text{U}/d \ln T)$,

$$c_{vm}^\circ = c_{pm}^\circ = R\frac{d \ln Q}{d \ln T} + R\frac{d^2 \ln Q}{d(\ln T)^2} \tag{25}$$

Since entropy is independent of pressure for the contributions under consideration, from Equation (o) of Table XXIV, page 472

$$d\mathrm{s}_m = c_{vm}\,dT/T = c_{vm}\,d\ln T \tag{26}$$

Substituting (25) in (26) and integrating,

$$(\mathrm{s}^\circ - \mathrm{s}_0^\circ)_m = R\ln Q\Big|_0^T + R\frac{d\ln Q}{d\ln T}\Big|_0^T \tag{27}$$

From Equation (20) defining the partition function it is evident that at the absolute zero $Q_0 = g_0$, which is the statistical weight of the ground energy level or the number of different molecular states possessing substantially the ground energy level. Since Q_0 is finite it follows that $d\ln Q/d\ln T$ is equal to zero where $T = 0$, and Equation (27) becomes:

$$\mathrm{s}_m^\circ = R\ln Q + R\frac{d\ln Q}{d\ln T} + (\mathrm{s}_{0m}^\circ - R\ln g_0) \tag{28}$$

However, the most general statement of the third law of thermodynamics derived by statistical mechanics states that *the entropy of a substance at the absolute zero is equal to R times the logarithm of the statistical weight of the ground energy level, or* $\mathrm{s}_0^\circ = R\ln g_0$. For substances in the crystalline state at 0°K the statistical weight of the ground level is generally unity, leading to the more common statement of the third law. Equation (28) then becomes

$$\mathrm{s}_m^\circ = R\ln Q + R\frac{d\ln Q}{d\ln T} \tag{29}$$

Combining Equations (24) and (29) gives expressions for the m contributions to G° and A°, thus,

$$\frac{(\mathrm{G}^\circ - \mathrm{H}_0^\circ)_m}{T} = -R\ln Q = \frac{(\mathrm{A}^\circ - \mathrm{H}_0^\circ)_m}{T} \tag{30}$$

By adding the m contributions calculated from Equations (24), (25), (29), and (30) to the translational contributions from Equations (8) to (17) the thermodynamic functions c_v°, c_p°, s°, $\mathrm{U}^\circ - \mathrm{H}_0^\circ$, $\mathrm{H}^\circ - \mathrm{H}_0^\circ$, $\mathrm{G}^\circ - \mathrm{H}_0^\circ$, and $\mathrm{A}^\circ - \mathrm{H}_0^\circ$ can all be completely evaluated if the partition function is known as a function of temperature. If the contributing molecular energy levels and their statistical weights are known, Q may be evaluated by the summation of Equation (20). Similarly, from Equation (22),

$$\frac{d\ln Q}{d\ln T} = \frac{\Sigma_i E_i g_i e^{-\frac{E_i}{RT}}}{RTQ} \tag{31}$$

Also, since $d^2 \ln Q/d(\ln T)^2 = Td(d \ln Q/d \ln T)/dT$

$$\frac{d^2 \ln Q}{d(\ln T)^2} = \frac{\Sigma_i E_i^2 g_i e^{-\frac{E_i}{RT}}}{R^2T^2Q} - \frac{\Sigma_i E_i g_i e^{-\frac{E_i}{RT}}}{RTQ} - \left(\frac{\Sigma_i E_i g_i e^{-\frac{E_i}{RT}}}{RQ^2}\right)\left(\frac{dQ}{dT}\right)$$

$$= \frac{\Sigma_i E_i^2 g_i e^{-\frac{E_i}{RT}}}{R^2T^2Q} - \frac{d \ln Q}{d \ln T} - \left(\frac{d \ln Q}{d \ln T}\right)^2 \tag{32}$$

From Equations (31) and (32) it is seen that all the derivatives of the partition function which are required for the evaluation of thermodynamic properties can be calculated from two summations of all significant energy levels, one $\Sigma_i E_i g_i e^{-\frac{E_i}{RT}}$ and the other $\Sigma_i E_i^2 g_i e^{-\frac{E_i}{RT}}$. For simple molecules the individual energy levels can be evaluated by application of Equation (6) to spectroscopically determined frequencies of emission or absorption.

Illustration 7. The development and application of these methods for calculating thermodynamic properties is outlined by Kelley[10], who developed the following calculation of the thermodynamic properties of nickel vapor in the state of an ideal gas at 298°K and 1 atm. This is an entirely hypothetical state in which nickel vapor cannot exist but is legitimately used as a reference from which values for actual states may be derived by the temperature and pressure relations developed in Chapters XI and XII.

Nickel vapor, Ni (g), is a monatomic gas possessing only translational and electronic-energy contributions. In Table A the spectroscopic notations for the various levels of electronic energy are given in column 1 and the corresponding frequencies in terms of wave numbers and statistical weights, derived from the spectrum, in columns 2 and 3, respectively. The molal energy levels in column 4 are calculated from Equation (6) in calories per gram-mole. The three summations required for calculation of thermodynamic properties are given in columns 5, 6, and 7. The electronic contributions to the thermodynamic functions are calculated directly from these summations. Thus, from Equations (20), (31), and (32), and Table A,

$$Q = 11.690$$

$$\frac{d \ln Q}{d \ln T} = \frac{1753}{(1.987)(298.1)(11.69)} = 0.2532$$

$$\frac{d^2 \ln Q}{d(\ln T)^2} = \frac{1,533,000}{(1.987)^2(298.1)^2(11.69)} - 0.2532 - (0.2532)^2 = 0.0565$$

TABLE A

1	2	3	4	5	6	7
State	ω	g_i	E_i	$g_i e^{-\frac{E_i}{RT}}$	$g_i E_i e^{-\frac{E_i}{RT}}$	$g_i E_i^2 e^{-\frac{E_i}{RT}}$
$^3F_4$	0.0	9	0	9.0000	0	0
$^3D_3$	204.82	7	585.27	2.6058	1525.1	892,610
$^3D_2$	879.82	5	2513.98	0.0717	180.3	453,200
$^3F_3$	1332.15	7	3806.62	0.0113	43.0	164,060
$^3D_1$	1713.11	3	4895.21	0.0008	3.9	18,500
$^3F_2$	2216.55	5	6333.79	0.0001	0.6	4,551
				$Q = 11.690$	1753	1,533,000

The complete thermodynamic functions are obtained as follows:

Eq. (8 and 21),

$$U^\circ - H_0^\circ = \frac{(3)(1.987)(298.16)}{2} + \frac{1753}{11.69} = 1038 \text{ cal per g-mole}$$

Eq. (11 and 21),

$$H^\circ - H_0^\circ = \frac{(5)(1.987)(298.16)}{2} + 150 = 1630 \text{ cal per g-mole}$$

Eq. (9 and 25),

$$c_v^\circ = \frac{(3)(1.987)}{2} + 1.987(0.2532 + 0.0565) = 3.595 \text{ cal/(g-mole)(°K)}$$

Eq. (10 and 25),

$$c_p^\circ = \frac{(5)(1.987)}{2} + 0.6152 = 5.581 \text{ cal/(g-mole)(°K)}$$

Eq. (13 and 29),

$$S^\circ = 6.861 \log 58.68 + 25.98 + 1.987(\ln 11.69 + 0.2532)$$
$$= 43.49 \text{ cal/g-mole (°K)}$$

Eq. (17 and 30),

$$A^\circ - H_0^\circ = -2045.7 \log 58.68 - 6858.8 - 1456 = -11{,}933 \text{ cal per g-mole}$$

Eq. (16 and 30),

$$G^\circ - H_0^\circ = -2045.7 \log 58.68 - 6266.5 - 1456 = -11{,}340 \text{ cal per g-mole}$$

THERMODYNAMIC DATA FROM MOLECULAR CONSTANTS

The method described in the preceding section requires extensive spectroscopic data and becomes tedious and complicated for the more complex molecules although it is rigorous if the significant energy levels can be identified and evaluated. Considerable simplification with little loss of accuracy results from approximate methods for calculating rotational and vibrational contributions from constants of molecular structure. Electronic contributions can be evaluated only by summations based on spectroscopic data as in Illustration 7. However, for most of the complex molecules it may be assumed that all molecules are in the ground state of electronic energy at all except very high temperatures. For organic compounds the statistical weight of the ground state is generally 1, and accordingly electronic contributions to the partition function may be safely neglected in calculations dealing with complex organic molecules.

It has been demonstrated that as an approximation which is satisfactory at temperatures above about 50°K the summation represented by the partition function may be divided into groups representing the contributions of the five different types of internal energy previously discussed. Thus,

$$Q = \Sigma_i g_i e^{-\frac{E_i}{RT}} = \left(\Sigma_i g_{ri} e^{-\frac{E_{ri}}{RT}}\right)\left(\Sigma_i g_{r'i} e^{-\frac{E_{r'i}}{RT}}\right)$$
$$\left(\Sigma_i g_{vi} e^{-\frac{E_{vi}}{RT}}\right)\left(\Sigma_i g_{\delta i} e^{-\frac{E_{\delta i}}{RT}}\right)\left(\Sigma_i g_{ei} e^{-\frac{E_{ei}}{RT}}\right) \quad (33)$$

or
$$Q = (Q_r)(Q_{r'})(Q_v)(Q_\delta)(Q_e) \tag{34}$$

where Q_r, g_{ri}, ϵ_{ri} represent, respectively, the partition function, the statistical weight, and the molecular energy contributions to the ith level which are due to external rotation, and the other symbols have similar significance for the contributions of internal rotation, stretching and bending vibrations, and electronic energy. Thus, the total partition function is the product of the individual partition functions of the contributing forms of energy. As previously mentioned, electronic energy levels are relatively high, and at ordinary temperatures the Boltzmann factors for this form of energy approach zero and the partition function Q_e approaches unity.

From Equations (23) to (30) it is seen that the thermodynamic properties are all expressed in terms of the logarithm of the partition function. It follows from Equation (34) that each of these thermodynamic properties may be evaluated as the sum of the individual contributions of the various forms of energy.

$$S_m^\circ = S_r^\circ + S_{r'}^\circ + S_v^\circ + S_\delta^\circ + S_e^\circ \tag{35}$$

Each of the individual contributions with the exception of that resulting from electronic energy may be evaluated from molecular constant data which permit calculation of the corresponding partition functions of the individual forms of energy.

External Rotation. The energy possessed by a molecule as a result of its rotation as a whole is a function of its moments of inertia and its speed of rotation. As a result of X-ray and electron diffraction measurements the distances between the atoms and the angles formed by interatomic bonds have been established with sufficient accuracy to permit quantitative prediction of the spatial arrangement of a molecule. Moments of inertia are calculated from such molecular dimensions and from the atomic masses by the methods of classical mechanics. Quantum mechanics has shown that molecular rotation is quantized in series of definite rotational states whose energy levels differ from each other by finite increments. For all complex molecules, however, it is found that at temperatures above 50°K the energy levels become so closely spaced that they can be treated as a continuous distribution. On this basis it has been shown that the partition function for the external rotation of a linear molecule at temperatures above 50°K is represented approximately by the following expression:

$$Q_r = \frac{8\pi^2 IkT}{\sigma h^2} = 2.484(10^{+38})\frac{IT}{\sigma} \tag{36}$$

or

$$\ln Q_r = \ln IT - \ln \sigma + 88.39 \tag{37}$$

$$\frac{d \ln Q_r}{d \ln T} = 1; \quad \text{and} \quad \frac{d^2 \ln Q_r}{d(\ln T)^2} = 0 \tag{38}$$

The corresponding rotational contributions to thermodynamic functions of linear molecules are obtained as follows:

From Equation (25),

$$(c_p^\circ)_r = (c_v^\circ)_r = R \tag{39}$$

From Equation (24),

$$(\text{U}^\circ - \text{H}_0^\circ)_r = (\text{H}^\circ - \text{H}_0^\circ)_r = RT \tag{40}$$

From Equation (29),

$$\text{S}_r^\circ = R(\ln T + \ln I - \ln \sigma + 89.39) \tag{41}$$

From Equation (30),

$$\left(\frac{\text{G}^\circ - \text{H}_0^\circ}{T}\right)_r = \left(\frac{\text{A}^\circ - \text{H}_0^\circ}{T}\right)_r = -R(\ln IT - \ln \sigma + 88.39) \tag{42}$$

where

I = moment of inertia for rotation about the center of gravity in the plane of the line through the atoms (gr)(cm²)

k = Boltzmann constant = $(1.3805)(10^{-16})$ erg/(molecule)(°K)

h = Planck's constant = $(6.624)(10^{-27})$ (erg)(sec)

T = degrees Kelvin

σ = the symmetry number

R = 1.987 cal/(g-mole)(°K)

The symmetry number is defined as the number of different positions into which the molecule can be rotated with identical appearance from every point of view. Thus, a diatomic molecule composed of like atoms can be rotated into two positions of identical appearance and therefore has a symmetry number $\sigma = 2$. However, if the atoms are dissimilar, $\sigma = 1$.

The symmetry number σ of complex nonlinear molecules may vary widely with the configuration of the molecule. Many such molecules are unsymmetrical and in such cases $\sigma = 1.0$. However, for H_2O, $\sigma = 2$; for NH_3, $\sigma = 3$; for C_2H_4, $\sigma = 4$; for CH_4, $\sigma = 12$; for C_2H_6, $\sigma = 6$; and for C_6H_6, $\sigma = 12$. Symmetry numbers are most easily established by inspection of three-dimensional molecular models.

By a similar analysis it has been shown that at temperatures above

50°K the partition function of a nonlinear molecule, having moments of inertia about three axes, is represented by the following:

$$Q_r = \frac{\sqrt{\pi}}{\sigma}\left(\frac{8\pi^2\mathbf{k}T}{\mathrm{h}^2}\right)^{\frac{3}{2}}(I_xI_yI_z)^{\frac{1}{2}} = \frac{6.940(10^{57})T^{\frac{3}{2}}(I_xI_yI_z)^{\frac{1}{2}}}{\sigma} \tag{43}$$

or

$$\ln Q_r = \tfrac{3}{2}\ln T + \tfrac{1}{2}\ln (I_xI_yI_z) - \ln \sigma + 133.18 \tag{44}$$

$$\frac{d \ln Q_r}{d \ln T} = \frac{3}{2}; \quad \frac{d^2 \ln Q_r}{d(\ln T)^2} = 0 \tag{45}$$

From these relations the contributions of external rotation to the thermodynamic properties of nonlinear molecules are obtained, thus:

Equations (24) and (45):

$$(\mathrm{U}^\circ - \mathrm{H}_0^\circ)_r = \tfrac{3}{2}RT = (\mathrm{H}^\circ - \mathrm{H}_0^\circ)_r \tag{46}$$

Equations (25) and (45):

$$c_{p_r}^\circ = \tfrac{3}{2}R = c_{v_r}^\circ \tag{47}$$

Equations (29) and (44):

$$\mathrm{S}_r^\circ = \tfrac{3}{2}R\ln T + \tfrac{1}{2}R\ln (I_xI_yI_z) - R\ln\sigma + 267.54 \tag{48}$$

Equations (30) and (44):

$$\frac{(\mathrm{G}^\circ - \mathrm{H}_0^\circ)_r}{T} = \frac{(\mathrm{A}^\circ - \mathrm{H}_0^\circ)_r}{T} = R[-\tfrac{3}{2}\ln T - \tfrac{1}{2}(\ln I_xI_yI_z) + \ln\sigma - 133.18] \tag{49}$$

It is evident that the rotational contributions to internal energy, enthalpy, and heat capacity are easily calculated for any compound, whereas determination of the entropy and free-energy contributions requires evaluation of the product of the principal moments of inertia.

Calculation of the moments of inertia I_x, I_y, and I_z from data on atomic masses, angles, and distances is relatively simple for symmetrical molecules whose center of gravity may be located by inspection. A diagram or model of the molecule is prepared with the center of gravity located at the intersection of the three perpendicular axes, x, y, and z. The moment of inertia about each axis is then the summation of md_i^2 for all the atoms, where m is the weight of the atom and d_i its distance from the axis.

For unsymmetrical molecules the calculation of the moments of inertia is more complicated. The following method was suggested by Hirschfelder.[11] A model or diagram is prepared and placed in a convenient orientation, generally with one atom at the origin and as many other atoms as possible lying on the co-ordinate axes. The position of each atom is then expressed in terms of its Cartesian co-ordinates,

[11] J. O. Hirschfelder, *J. Chem. Phys.*, **8**, 431 (1940).

x_i, y_i, z_i, where i designates a particular atom. Then the product of the three principal moments of inertia is equal to a determinant:

$$\left.\begin{aligned}
I_xI_yI_z &= \begin{vmatrix} A & -D & -E \\ -D & B & -F \\ -E & -F & C \end{vmatrix} = ABC - AF^2 - CD^2 - 2DEF - BE^2 \\
&\text{where} \\
A &= \Sigma_i m_i(y_i^2 + z_i^2) - \frac{1}{m}(\Sigma_i m_i y_i)^2 - \frac{1}{m}(\Sigma_i m_i z_i)^2 \\
B &= \Sigma_i m_i(x_i^2 + z_i^2) - \frac{1}{m}(\Sigma_i m_i x_i)^2 - \frac{1}{m}(\Sigma_i m_i z_i)^2 \\
C &= \Sigma_i m_i(x_i^2 + y_i^2) - \frac{1}{m}(\Sigma_i m_i x_i)^2 - \frac{1}{m}(\Sigma_i m_i y_i)^2 \\
D &= \Sigma_i m_i x_i y_i - \frac{1}{m}(\Sigma_i m_i x_i)(\Sigma_i m_i y_i) \\
E &= \Sigma_i m_i x_i z_i - \frac{1}{m}(\Sigma_i m_i x_i)(\Sigma_i m_i z_i) \\
F &= \Sigma_i m_i y_i z_i - \frac{1}{m}(\Sigma_i m_i y_i)(\Sigma_i m_i z_i) \\
M &= \Sigma_i m_i = \text{mass per molecule}
\end{aligned}\right\} \tag{50}$$

For convenience in calculating it is customary to express interatomic distances in Angstrom units equal to 10^{-8} centimeter and to use grams per mole M as the unit of atomic mass instead of grams per molecule m. If the moment of inertia calculated in these units is designated as I',

$$I_x = \frac{I'_x}{6.023(10^{23})(10^8)^2} = \frac{I'_x}{(6.023)(10^{39})} \tag{51}$$

and

$$I_xI_yI_z = \frac{I'_xI'_yI'_z}{(218.5)(10^{117})} \tag{52}$$

$$R \ln I_x = R \ln I'_x - 181.96 \tag{53}$$

$$R \ln I_xI_yI_z = R \ln I'_xI'_yI'_z - 545.87 \tag{54}$$

Substituting Equation (53) into (41) and (42) yields:

For linear molecules:

$$\text{S}_r^\circ = 4.575 \log I'T - 4.575 \log \sigma - 4.39 \tag{55}$$

$$\frac{(\text{G}^\circ - \text{H}_0^\circ)_r}{T} = -4.575 \log I'T + 4.575 \log \sigma + 6.37 \tag{56}$$

Substituting Equation (54) into (48) and (49) gives:

For nonlinear molecules:

$$s_r^\circ = 2.287 \log I'_x I'_y I'_z + 6.861 \log T - 4.575 \log \sigma - 5.40 \qquad (57)$$

$$\frac{(G^\circ - H_0^\circ)_r}{T} = -6.681 \log T - 2.287 \log I'_x I'_y I'_z + 4.575 \log \sigma + 8.38 \qquad (58)$$

The data necessary for estimating the interatomic distances and bond angles required for calculating values of I' are summarized in Table XLIV for some atoms in their common configuration. In the second column are values of covalent radii recommended by Hirschfelder[12] from the electron diffraction measurements of Pauling[13] and others. It is assumed that internuclear distances are the sum of the two covalent bond radii. Thus the internuclear separation for C−H bond is always 0.30A + 0.77A = 1.07A.

TABLE XLIV

ATOMIC COVALENT RADII AND BOND ANGLES

	Covalent Radius Ångstroms	*Bond*	*Angles between Bonds*
I. Hydrogen	0.30 A		
II. Carbon			
1. Single-bond carbon	0.77 A		Regular tetrahedral angles, 108° between bonds
2. Double-bond carbon	0.67 A	(for double bonds)	Bonds all lie in one plane
	0.77 A	(for single bonds)	124° 112° >C= 124°
3. Triple-bond carbon	0.60 A	(for triple bond)	Linear
	0.77 A	(for single bond)	≡C—
4. Benzene carbon	0.695 A	(for each of the two C–C bonds)	Planar 120°
	0.77 A	(the bond extending outward)	120° >C— 120°
III. Oxygen			
1. Single-bond oxygen	0.66 A		111° —O/
2. Double-bond oxygen	0.57 A		=O
IV. Nitrogen			
1. Amino nitrogen	0.70 A	(flat pyramid with three bonds making tetrahedral angles with each other)	108° | 108° /N\\ 108°
2. Nitrate nitrogen	0.65 A	(double bond)	Planar
	0.70 A	(single bond)	120 \\N/ ‖ 120 120

[12] J. O. Hirschfelder, private communication (1942).

[13] L. Pauling, *Proc. Nat. Acad. Sci. U.S.*, **18**, 293 (1932).

TABLE XLIV — *Continued*

ATOMIC COVALENT RADII AND BOND ANGLES

Covalent Radius		*Bond*	*Angles between Bonds*
3. Isonitrile nitrogen	0.55 A	(double bond)	Linear
	0.70 A	(single bond)	—N=
4. Cyanide nitrogen 0.55 A			≡N
V. Sulfur			
1. Single-bond sulfur	1.04 A		105° —S/
2. Double-bond sulfur	0.95 A		=S
3. Sulfate sulfur	0.95 A	(double bond)	Tetrahedral angles
	1.04 A	(single bond)	108° ‖ 108° /S\ 108° | 108°
VI. Sodium	1.81 A		
VII. Chlorine	0.99 A		
VIII. Bromine	1.14 A		
IX. Iodine	1.33 A		

The angles between the bonds are given in the last column of Table XLIV. These data permit construction of molecular diagrams or models from which values of I' are calculated by application of Equation (50).

Illustration 8. Calculate the rotational contribution to the entropy of NaCl (g) at 298.1°K and 1.0 atm.

Solution: From Table XLIV the interatomic distance, $d = 1.81 + 0.99 = 2.80$ A. The moment of inertia I' of a diatomic molecule is calculated as follows:

Let x_1, x_2 = distances from center of gravity of atoms 1 and 2 of masses m_1 and m_2, respectively

d = distance between atomic centers

$$d = x_1 + x_2 = 2.80 \tag{a}$$

From the principle of moments,

$$m_1x_1 = m_2x_2 \tag{b}$$

Also
$$I' = m_1x_1^2 + m_2x_2^2 \tag{c}$$

Combination of Equations (a), (b), and (c), and elimination of x_1 and x_2 gives

$$I' = \frac{m_1m_2d^2}{(m_1 + m_2)} = \frac{(23)(35.5)(2.80)^2}{(23 + 35.5)} = 109.5$$

Substitution in Equation (55), since $\sigma = 1$, gives

$$s_r^\circ = 4.575\ (\log 109.5 + \log 298.1) - 4.39 = 16.25$$

Illustration 9. Calculate the rotational contribution to the entropy of ethylene, $H_2C = CH_2$ at 25°C and 1.0 atm pressure.

Solution: Reference to Table XLIV shows that the bonds of a double-bond carbon atom are in one plane; therefore the molecular model of ethylene may be represented by a two-dimensional diagram with interatomic spacings, as shown in Fig. 164, and the center of gravity located midway between the carbon atoms. The co-ordinates of the atoms are, then,

$$
\begin{array}{llll}
H^1: & x_1 = -(1.07 \cos 56° + 0.67) = -1.27 \text{ A}; & & \\
 & y_1 = 1.07 \sin 56° = +0.89 \text{ A}; & & z_1 = 0 \\
H^2: & x_2 = -1.27 \text{ A}; & y_2 = -0.89 \text{ A}; & z_2 = 0 \\
H^3: & x_3 = +1.27 \text{ A}; & y_3 = +0.89 \text{ A}; & z_3 = 0 \\
H^4: & x_4 = +1.27 \text{ A}; & y_4 = -0.89 \text{ A}; & z_4 = 0 \\
C^1: & x_5 = -0.67 \text{ A}; & y_5 = 0; & z_5 = 0 \\
C^2: & x_6 = +0.67 \text{ A}; & y_6 = 0; & z_6 = 0
\end{array}
$$

By using these co-ordinates the product of the moments of inertia could be obtained from Equation (50). Since the origin is located at the center of gravity, however, it is simpler merely to sum up the moments of inertia of each of the atoms about each axis. Thus:

$$I'_x = 4(1.008)(0.89)^2 = 3.19$$

$$I'_y = 4(1.008)(1.27)^2 + 2(12.00)(0.67)^2 = 6.51 + 10.78 = 17.29$$

$$I'_z = 4(1.008)[(0.89)^2 + (1.27)^2] + 2(12.00)(0.67)^2 = 9.70 + 10.78 = 20.48$$

It may be noted as a general rule that when all atoms lie in the plane of the x, y axes

$$I_z = I_x + I_y$$

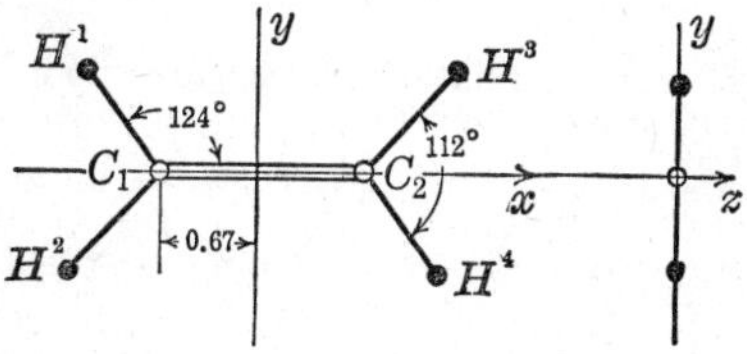

FIG. 164. The Ethylene Molecule.

The symmetry number of the molecule is arrived at by study of Fig. 164. The molecule as drawn can be rotated about the x axis into two positions of identical appearance. Two more positions of identical appearance can be obtained by the further rotation about the y axis. Therefore, the symmetry number σ is 4. Substituting in Equation (57) gives

$$
\begin{aligned}
s_r^\circ &= 2.287 \log [(3.19)(17.29)(20.48)] + 6.861 \log 298.1 - 4.575 \log 4 - 5.40 \\
&= (2.287)(3.0529) + (6.861)(2.4744) - (4.575)(0.6021) - 5.40 \\
&= 15.80 \text{ cal/(g-mole)(°K)}
\end{aligned}
$$

From a study of the band spectrum of ethylene Badger[14] has been able to determine the three moments of inertia directly. These results are as follows, compared with the calculated values of Illustration (9) which are converted to (g)(cm²) by Equation (51).

	Calculated	*Experimental*
I_x (g)(cm²)	5.3 (10^{-40})	5.7 (10^{-40})
I_y (g)(cm²)	28.7 (10^{-40})	27.5 (10^{-40})
I_z (g)(cm²)	34.0 (10^{-40})	33.2 (10^{-40})
$(I_xI_yI_z)$	5.17(10^{-117})	5.20(10^{-117})

The agreement in the product of the moments of inertia is satisfactory and indicates that the methods used may be expected to yield reliable results for more complex molecules.

Illustration 10. Calculate the external rotational contribution to the heat capacity, enthalpy, entropy, and free energy of methyl alcohol at 1.0 atm and 298.1, 500, and 1000°K, for the ideal gaseous state.

[14] R. M. Badger, *Phys. Rev.*, **45**, 648 (1934).

Solution: The first step in dealing with an unsymmetrical molecule is the construction of a diagram or model to serve as a guide in determining the co-ordinates of the atoms. A diagram for methyl alcohol is shown in Fig. 165 in three projected views using values from Table XLIV. For convenience the carbon atom is placed at the origin and the molecule and its parts so oriented that as many atoms as possible lie in the xy plane. The molecule could equally well be drawn with other relative positions of orientation of the oxygen with re spect to the carbon atom, but the product of the three moments of inertia is negligibly affected by such changes. The interatomic distances from Table XLIV are C–H = 1.07; C–O = 1.43, and O–H = 0.96. From the geometry of the diagram the co-ordinates of the atoms are calculated as follows:

H^1:	$x = -1.07 \sin 18°$ $= -0.331;$	$y = -1.07 \sin 72°$ $= -1.018;$	$z = 0$
H^2:	$x = -0.331;$	$y = 1.018 \sin 30°$ $= 0.509;$	$z = -1.018 \sin 60°$ $= -0.88$
H^3:	$x = -0.331;$	$y = 0.509;$	$z = 0.88$
H^4:	$x = 1.43 + 0.96 \sin 21°$ $= 1.77;$	$y = 0.96 \sin 69°$ $= 0.894;$	$z = 0$
O:	$x = 1.43;$	$y = 0;$	$z = 0$

These co-ordinates and the corresponding atomic weights M are substituted in Equation (50):

Summations	$M(y^2 + z^2)$	My	Mz	Mx
H^1	1.0446	−1.0261	0	−0.3336
H^2	1.0453	+0.5131	−0.8890	−0.3336
H^3	1.0453	+0.5131	+0.8890	−0.3336
H^4	0.8092	+0.9032	0	+1.7882
O	0.0	0.0	0	+22.8800
Sum	3.9444	0.9033	0	23.6674

	$M(xy)$	$M(x^2 + y^2)$	Mxz	$M(x^2 + z^2)$	Myz
H^1	+0.3397	1.1551	0	0.1105	0
H^2	−0.1698	0.3716	+0.2943	0.8946	−0.4515
H^3	−0.1698	0.3716	−0.2943	0.8946	+0.4515
H^4	+1.6022	3.9815	0	3.1723	0
O	0	32.7184	0	32.7184	0
Sum	1.6023	38.5982	0	37.7904	0

From Equation (50):

$$A = 3.9444 - \frac{1}{32.042}(0.9033)^2 = 3.9189$$

$$B = 37.7904 - \frac{1}{32.042}(23.6674)^2 = 20.3088$$

$$C = 38.5982 - \frac{1}{32.042}(23.6674)^2 - \frac{1}{32.042}(0.9033)^2 = 21.0911$$

$$D = 1.6023 - \frac{1}{32.042}(23.6674)(0.9033) = 0.9351$$

$$E = 0 - 0 = 0$$

$$F = 0 - 0 = 0$$

$$I'_x I'_y I'_z = ABC - CD^2 = 1660$$

Substitution in Equation (57), since $\sigma = 1.0$, gives

$$T = 298.16°\text{K} \qquad s_r^\circ = 1.96 + 16.98 = 18.94$$
$$T = 500°\text{K} \qquad s_r^\circ = 1.96 + 18.52 = 20.48$$
$$T = 1000°\text{K} \qquad s_r^\circ = 1.96 + 20.59 = 22.55$$

The other thermodynamic properties are similarly calculated from Equations (46), (47), and (58), and the results are summarized in Table LI, page 802.

Complex molecules in general are most conveniently treated by preparing an accurate large-scale diagram similar to Fig. 165, using pro-

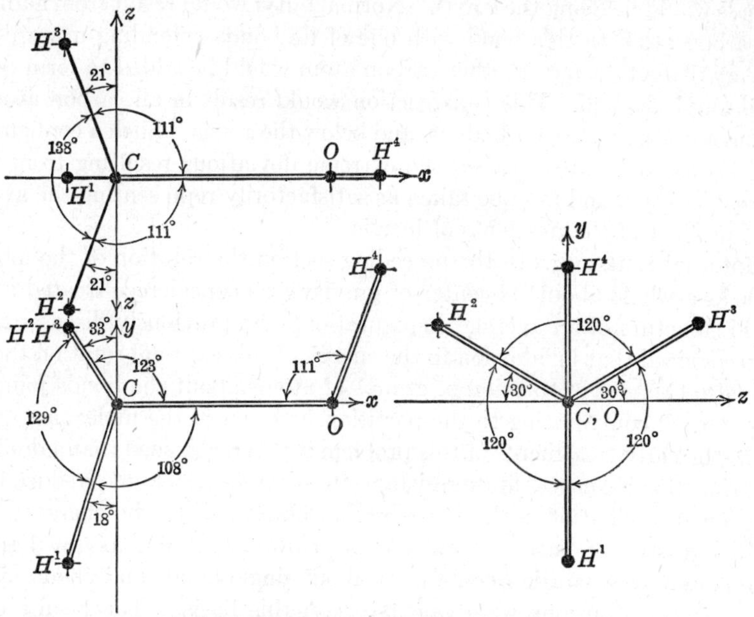

FIG. 165. The Methyl Alcohol Molecule.

jected angles and dimensions. The co-ordinates of the individual atoms are then determined by direct measurement. The projected angles necessary for such graphical construction are shown for the methyl alcohol molecule in Fig. 165. It may be noted that the projected angles of the carbon atom may also be used for sulfate sulfur while the angles of the group $CH^1H^2H^3$ of Fig. 165 are the same as those of the amino nitrogen atom.

Many different diagrams are possible for a complex molecule, depending on the positions into which the various atoms are rotated. As is discussed in the next section, these different possibilities are represented in the actual molecules as a result of either continuous or periodic rotation of the atoms and groups of atoms about the bonds. However, the resulting changes in configuration have only small effects on the product of the three principal moments of inertia of the molecule as a whole. Satisfactory results are obtained by assuming positions of rotation corresponding as closely as possible to a linear structure where straight chains are involved and approaching a single plane for chain structures. Thus, if Fig. 165 were extended to represent ethyl alcohol, hydrogen atom H^1 would be replaced with a carbon rotated with one bond pointing upward in the xy plane. Normal propyl alcohol would be represented by attaching another carbon to this bond, so oriented that one of its bonds would lie along the x axis. Normal butyl would result from adding a carbon atom to this bond with one of its bonds extending upward in the xy plane to which another carbon atom would be added to form normal amyl alcohol. This construction would result in all carbon atoms lying in the xy plane, both above and below the x axis. Such a configuration is intermediate between the extreme deviations resulting from internal rotation and may be taken as satisfactorily representing the average product of the moments of inertia.

Internal Rotation. In the preceding section the rotation of the molecule as a whole about its center of gravity was considered, treated as a rigid structure of an average configuration. As previously discussed, it is recognized that in addition to the energy of this external rotation there are also energies of rotation of groups of atoms about the bonds joining them, each contributing to the partition function of the molecule.

In the early treatments of this problem it was recognized that a double or triple bond exerts definite resistance to rotation which may reduce the motion to oscillation with only periodic rotation when the energy level of the restrictive barrier is exceeded. However, it was assumed that completely free rotation could exist about single bonds and energy and entropy contributions were calculated on this basis. The results obtained were too high to agree with experimental data and Pitzer and coworkers[15, 16] extended these methods to allow for a potential barrier restricting every internal rotation.

The partition function Q_f, resulting from the free rotation of a group with respect to another group or with respect to the molecule as a whole, is given by the following equation:

$$Q_f = \frac{(8\pi^3 I_{\text{red}} \mathbf{k} T)^{\frac{1}{2}}}{\mathbf{h} N_h} \tag{59}$$

or, if I_{red} is expressed in (gram)(cm)^2 and T in degrees Kelvin,

$$Q_f = \frac{2.7934[(I_{\text{red}} T)(10^{38})]^{\frac{1}{2}}}{N_h} \tag{60}$$

[15] K. S. Pitzer, *J. Chem. Phys.*, **5**, 469, 473, 752 (1937).
[16] K. S. Pitzer and W. D. Gwinn, *J. Chem. Phys.*, **10**, 428 (1942).

where

I_{red} = the reduced moment of inertia of the internal rotation

N_h = the symmetry number of the internal rotation, the number of points of maximum attraction existing per revolution between the atoms of the rotating groups.

For example, in the ethane molecule at three positions in each revolution the hydrogen atoms of one group reach points of minimum separation from those of the other. Hence, for this rotation $N_h = 3$.

The reduced moment of inertia of one group about another is given by the following expression if the axis of the internal rotation coincides with one of the principal axes of the molecule as a whole.

$$\frac{1}{I_{\text{red}}} = \frac{1}{I_A} + \frac{1}{I_B} \tag{61}$$

where I_A, I_B are the moments of inertia of the A and B groups, respectively, about the bond joining them. Equation (61) may be applied as an approximation where two or more internal rotations are involved if the axis of rotation approximately coincides with a principal axis of the molecule. A more exact relationship which is applicable to the rotation of several groups with respect to a rigid molecular framework is proposed by Pitzer and Gwinn.[16]

The contributions of free internal rotation to the thermodynamic properties are evaluated by combining Equation (60) with Equations (24), (25), (29), and (30), thus:

$$(c_p^\circ)_f = (c_v^\circ)_f = \tfrac{1}{2}R \tag{62}$$

$$(\text{U}^\circ - \text{H}_0^\circ)_f = (\text{H}^\circ - \text{H}_0^\circ)_f = \tfrac{1}{2}RT \tag{63}$$

$$\text{s}_f^\circ = \tfrac{1}{2}R + R \ln Q_f \tag{64}$$

$$\frac{(\text{G}^\circ - \text{H}_0^\circ)_f}{T} = -R \ln Q_f \tag{65}$$

The contributions of hindered internal rotation to the thermodynamic properties have been evaluated by Pitzer and Gwinn[16] and are presented in Tables XLV to XLVII as functions of the partition function of free rotation Q_f and the potential barrier V_h cal per g-mole, which hinders the rotation V_h times per revolution. In Table XLVI the entropy contributions of hindered rotation $\text{s}_{r'}^\circ$ are tabulated directly for values of $1/Q_f$ greater than 0.30. For lower values of $1/Q_f$ the table gives the value of $(\text{s}_f^\circ - \text{s}_{r'}^\circ)$. The corresponding values of $\text{s}_{r'}^\circ$ are calculated from Equation (64). Thus:

$$\text{s}_{r'}^\circ = \text{s}_f^\circ - (\text{s}_f^\circ - \text{s}_{r'}^\circ) = \tfrac{1}{2}R + R \ln Q_f - (\text{s}_f^\circ - \text{s}_{r'}^\circ) \tag{66}$$

TABLE XLV

INTERNAL ROTATIONAL CONTRIBUTIONS TO $\left(\frac{H_T^\circ - H_0^\circ}{T}\right)_{r'}$, cal/(°K)(g-mole)

$1/Q_f$ / V_h/RT	0.0	0.10	0.20	0.30	0.40	0.50	0.60	0.70	0.80
0.0	0.9934	0.993	0.993	0.993	0.993	0.993			
0.2	1.1822	1.106	1.050	1.022	1.008	1.000			
0.4	1.3513	1.249	1.151	1.073	1.036	1.015			
0.6	1.5011	1.374	1.251	1.138	1.072	1.030			
0.8	1.6324	1.482	1.340	1.211	1.114	1.048			
1.0	1.7460	1.576	1.418	1.275	1.155	1.065			
1.5	1.9607	1.753	1.561	1.385	1.230	1.103			
2.0	2.0934	1.854	1.636	1.440	1.265	1.120			
2.5	2.1657	1.900	1.662	1.448	1.260	1.104			
3.0	2.1971	1.909	1.651	1.426	1.224	1.060	0.92		
3.5	2.2030	1.893	1.621	1.382	1.176	1.006	0.88		
4.0	2.1944	1.864	1.577	1.329	1.121	0.947	0.82		
4.5	2.1788	1.829	1.529	1.273	1.061	0.884	0.75		
5.0	2.1607	1.794	1.481	1.218	1.002	0.824	0.67	0.59	
6.0	2.1261	1.727	1.392	1.115	0.893	0.714	0.56	0.47	
7.0	2.0984	1.670	1.315	1.029	0.802	0.624	0.482	0.38	0.31
8.0	2.0781	1.623	1.251	0.955	0.725	0.549	0.418	0.31	0.25
9.0	2.0634	1.583	1.196	0.892	0.661	0.488	0.363	0.269	0.20
10.0	2.0526	1.548	1.147	0.838	0.608	0.437	0.319	0.231	0.17
12.0	2.0382	1.492	1.067	0.745	0.519	0.356	0.244	0.169	0.116
14.0	2.0292	1.441	0.997	0.672	0.450	0.295	0.195	0.128	0.084
16.0	2.0229	1.401	0.937	0.613	0.394	0.249	0.157	0.099	0.063
18.0	2.0182	1.363	0.886	0.561	0.347	0.211	0.128	0.077	0.047
20.0	2.0147	1.329	0.841	0.515	0.307	0.181	0.105	0.061	0.036

Values of $(G^\circ - H_0^\circ)_{r'}/T$ are readily calculated from the entropy and enthalpy contributions taken from the tables. Thus:

$$\frac{(G^\circ - H_0^\circ)_{r'}}{T} = \frac{(H^\circ - H_0^\circ)_{r'}}{T} - S_{r'}^\circ \tag{67}$$

The potential barriers V_h which hinder internal rotation are evaluated by comparison of calculated with experimentally determined entropies or free energies from equilibirum or third-law measurements. For this reason all errors in any of the calculated contributions and in the experimental values result in corresponding errors in the apparent potential barrier, and the reported values for given combinations of groups vary widely. No satisfactory method of generalizing potential barriers has been developed, but approximations can be arrived at from the values of Table XLVIII taken from a summary prepared by Aston.[17] The potential barrier is largely determined by the atoms directly attached to each of the atoms joined by the bond under consideration.

[17] J. G. Aston, *Ind. Eng. Chem.*, **34**, 514 (1942).

TABLE XLVI

INTERNAL ROTATIONAL CONTRIBUTIONS TO ENTROPY, cal/(°K)(g-mole)

	←——($s_f^\circ - s_{r'}^\circ$)——→				←——————$s_{r'}^\circ$——————→					
$1/Q_f$	0.0	0.10	0.20	0.30	0.30	0.40	0.50	0.60	0.70	0.80
V_h/RT										
0.0	0.0000	0.000	0.000	0.000	3.386	2.814	2.371			
0.2	0.0049	0.004	0.004	0.004	3.382	2.811	2.369			
0.4	0.0198	0.018	0.018	0.016	3.370	2.801	2.359			
0.6	0.0440	0.043	0.040	0.039	3.347	2.780	2.340			
0.8	0.0771	0.077	0.072	0.068	3.318	2.750	2.315			
1.0	0.1185	0.117	0.112	0.107	3.279	2.714	2.279			
1.5	0.2527	0.250	0.242	0.230	3.156	2.600	2.173			
2.0	0.4182	0.415	0.402	0.382	3.004	2.458	2.048			
2.5	0.6001	0.594	0.577	0.550	2.836	2.303	1.907			
3.0	0.7856	0.777	0.757	0.719	2.667	2.138	1.756	1.45		
3.5	0.9660	0.957	0.929	0.886	2.500	1.978	1.610	1.34		
4.0	1.1356	1.126	1.094	1.043	2.343	1.834	1.475	1.22		
4.5	1.2918	1.280	1.244	1.187	2.199	1.698	1.348	1.10		
5.0	1.4339	1.421	1.380	1.318	2.068	1.579	1.233	0.97	0.81	
6.0	1.6781	1.662	1.616	1.542	1.844	1.370	1.040	0.79	0.63	
7.0	1.8783	1.860	1.807	1.721	1.665	1.204	0.891	0.665	0.50	0.39
8.0	2.0447	2.024	1.962	1.867	1.519	1.071	0.770	0.564	0.41	0.31
9.0	2.1864	2.163	2.095	1.989	1.397	0.962	0.674	0.482	0.348	0.25
10.0	2.3095	2.284	2.208	2.091	1.295	0.872	0.596	0.418	0.295	0.21
12.0	2.5155	2.485	2.394	2.261	1.125	0.728	0.476	0.315	0.212	0.143
14.0	2.6847	2.650	2.547	2.392	0.994	0.620	0.388	0.247	0.158	0.102
16.0	2.8289	2.788	2.674	2.496	0.890	0.533	0.322	0.196	0.120	0.075
18.0	2.9545	2.910	2.781	2.585	0.801	0.464	0.270	0.158	0.093	0.056
20.0	3.0659	3.017	2.872	2.659	0.727	0.405	0.228	0.129	0.073	0.042

In general the barriers are greater for large atoms. Approximation for specific groupings can be estimated from analogy to those of Table XLVIII.

Where long chains are involved, as in the higher normal paraffins and their derivatives, the rotational contributions of the intermediate groups are difficult to evaluate, and the entire problem is complicated by the effect of internal rotation on the external-rotational contributions. Improved methods for handling this problem have been proposed by Crawford[18] and Pitzer,[19] but in the present state of development these schemes require a high degree of judgment or are too tedious for general application. For the present the most practical procedure for developing data for a new compound is to work out in detail the lowest, or better the next to the lowest, homolog with the best possible assumptions as to potential

[18] B. L. Crawford, *J. Chem. Phys.*, **8**, 273 (1940).
[19] K. S. Pitzer, *Chem. Rev.*, **27**, 39 (1940).

TABLE XLVII

INTERNAL ROTATIONAL CONTRIBUTIONS TO HEAT CAPACITY, cal/(°K)(g-mole)

$1/Q_f$ / V_h/RT	0.00	0.10	0.20	0.30	0.40	0.50	0.60	0.70	0.80
0.0	0.9934	0.993	0.993	0.993	0.993	0.99			
0.2	1.0033	1.003	1.001	0.999	0.998	1.00			
0.4	1.0326	1.032	1.028	1.024	1.019	1.02			
0.6	1.0799	1.079	1.073	1.065	1.056	1.05			
0.8	1.1433	1.141	1.133	1.121	1.106	1.09			
1.0	1.2201	1.217	1.206	1.190	1.169	1.14			
1.5	1.4506	1.444	1.423	1.391	1.348	1.30			
2.0	1.6975	1.687	1.655	1.606	1.541	1.469			
2.5	1.9211	1.908	1.866	1.801	1.717	1.623			
3.0	2.0986	2.082	2.033	1.952	1.846	1.738	1.7		
3.5	2.2223	2.204	2.146	2.054	1.934	1.803	1.7		
4.0	2.2986	2.276	2.213	2.110	1.980	1.834	1.69		
4.5	2.3354	2.312	2.238	2.129	1.990	1.832	1.67		
5.0	2.3443	2.318	2.241	2.120	1.972	1.808	1.62	1.4	
6.0	2.3155	2.283	2.192	2.059	1.893	1.711	1.51	1.33	
7.0	2.2647	2.228	2.126	1.973	1.787	1.588	1.394	1.21	1.0
8.0	2.2157	2.174	2.058	1.888	1.684	1.468	1.260	1.07	0.91
9.0	2.1759	2.130	1.999	1.808	1.587	1.362	1.149	0.955	0.79
10.0	2.1454	2.094	1.951	1.745	1.507	1.262	1.047	0.853	0.68
12.0	2.1050	2.043	1.877	1.636	1.365	1.107	0.877	0.683	0.519
14.0	2.0810	2.009	1.814	1.546	1.254	0.978	0.744	0.555	0.408
16.0	2.0654	1.983	1.764	1.468	1.156	0.873	0.639	0.457	0.321
18.0	2.0544	1.961	1.717	1.397	1.070	0.780	0.549	0.378	0.256
20.0	2.0462	1.944	1.678	1.333	0.991	0.701	0.477	0.316	0.207

TABLE XLVIII

POTENTIAL BARRIERS HINDERING INTERNAL ROTATION

Group A	*Group B*	V_h cal per g-mole
CH_3	$—CH_3$	3150
CH_3	$—CH_2CH_3$	3300
CH_3	$—CH(CH_3)_2$	3870
CH_3	$—C(CH_3)_3$	4540
CH_3	$—C\equiv C\ CH_3$	0
CH_3	$—CH_2OH$	3000
CH_3	$—CO(CH_3)$	1000
CH_3	$—NH(CH_3)$	3460
CH_3	$—O—CH_3$	3100
OH	$—CH_3$	2700[20]
OH	$—CH_2CH_3$	10,000
OH	$—CH(CH_3)_2$	5000

[20] Revised in accordance with spectroscopic data of Borden and Barker, *J. Chem. Phys.*, **6**, 553 (1938), and the third law value of $s^\circ_{298} = 56.63$ calculated by L. S. Kassel, *J. Chem. Phys.*, **4**, 493 (1936), from the low-temperature measurements of K. K. Kelley, *J. Am. Chem. Soc.*, **51**, 181 (1929).

barriers and then extend these results to higher homologs by means of the group contributions of Tables XXXIX to XLIII.

Illustration 11. Calculate the contributions of internal rotation to the heat capacity, enthalpy, entropy, and free energy of methyl alcohol in the ideal-gaseous state at 1 atm and 298.1, 500, and 1000°K.

Solution:

Equation (59), $$Q_f = \left(\frac{8\pi^3 I \mathbf{k} T}{\mathbf{h}^2 (N_h)^2}\right)^{\frac{1}{2}} = \frac{2.794(IT(10^{38}))^{\frac{1}{2}}}{N_h}$$

Equation (61), $$\frac{1}{I} = \frac{1}{I_{\text{methyl}}} + \frac{1}{I_{\text{OH}}}$$

Equation (51) and Table XLIV,

$$I_{\text{methyl}} = \frac{3(1.008)(1.07 \sin 72°)^2}{6.023(10^{39})} = (5.203)(10^{-40})$$

$$I_{\text{OH}} = \frac{1(1.008)(0.96 \sin 69°)^2}{6.023(10^{39})} = 1.344(10^{-40})$$

$$I = (1.068)(10^{-40});\ N_h = 3$$

$$Q_f = (0.09625)T^{\frac{1}{2}};\ \frac{1}{Q_f} = \frac{10.39}{T^{\frac{1}{2}}}$$

From Table XLVIII,

$$V_h = 2700 \text{ cal per mole}$$

T	$T^{\frac{1}{2}}$	$1/Q_f$	V_h/RT
298.16	17.267	0.602	4.56
500	22.361	0.465	2.72
1000	31.623	0.329	1.36

Interpolation in Tables XLV to XLVII and with Equation (67) gives

T	$\frac{(H_T^\circ - H_0^\circ)_{r'}}{T}$	$S_{r'}^\circ$	$C_{r'}^\circ$	$-\frac{(G_T^\circ - H_0^\circ)_{r'}}{T}$	$(H_T^\circ - H_0^\circ)_{r'}$
298.16	0.74	1.08	1.67	0.34	221
500	1.14	1.97	1.72	0.83	570
1000	1.31	3.02	1.32	1.71	1310

Internal rotation about a multiple bond such as exists in ethylene or acetylene is so highly restrained that it is generally treated as an oscillation or vibration by the method of the following section.

Vibrational Contributions. As previously discussed, molecular vibrational energies result from movements of the atoms which stretch individual bonds, change the angles between adjacent bonds, or result in the twisting of multiple bonds. Each such mode of motion corresponds to a single vibrational degree of freedom. The number of such vibrational degrees of freedom possessed by a molecule may be calculated from Equation (4) or (5).

If complete spectroscopic data are available the vibrational contributions to the partition function might be evaluated through summation of the Boltzmann factors of all energy levels in all vibrational degrees of freedom by the procedure demonstrated in Illustration 7. A more convenient method results from treatment of each vibrational degree of freedom as a harmonic oscillator having a fundamental frequency ν' corresponding to a wave number ω'. The partition function for such a single harmonic vibration has been derived from the principles of quantum mechanics.

$$Q_\omega = \frac{1}{1 - e^{-\frac{h\nu'}{kT}}} = \frac{1}{1 - e^{-\frac{1.4384\,\omega'}{T}}} = \frac{1}{1 - e^{-x}} \tag{68}$$

where

ν' and ω' = the fundamental frequency and wave number, respectively, of the oscillator

T = degrees Kelvin

$x = 1.4384\,\omega'/T$

Equation (68) may be differentiated to obtain $d \ln Q_\omega / d \ln T$ and $d^2 \ln Q_\omega / d\,(\ln T)^2$ for evaluation of the thermodynamic properties from Equations (24), (25), (29), and (30). Thus, since

$$d \ln Q/d \ln T = T\,dQ/Q\,dT$$

and

$$x = 1.4384\omega'/T,$$

$$\frac{d \ln Q_\omega}{d \ln T} = \frac{x}{e^x - 1} \tag{69}$$

$$\frac{d^2 \ln Q_\omega}{d(\ln T)^2} = \frac{x^2 e^x}{(e^x - 1)^2} - \frac{x}{e^x - 1} \tag{70}$$

The total vibrational contributions to the thermodynamic properties are obtained as summations of the contributions of all individual degrees of vibrational freedom, each of which corresponds to particular values of ν', ω', and x. Thus

$$(\mathrm{H}^\circ - \mathrm{H}_0^\circ)_\omega = RT\Sigma \frac{x}{e^x - 1} \tag{71}$$

$$(c_p^\circ)_\omega = (c_v^\circ)_\omega = R\Sigma \frac{x^2 e^x}{(e^x - 1)^2} \tag{72}$$

$$\mathrm{s}_\omega^\circ = R\Sigma \left[\frac{x}{e^x - 1} - \ln(1 - e^{-x})\right] \tag{73}$$

$$\left(\frac{\mathrm{G}^\circ - \mathrm{H}_0^\circ}{T}\right)_\omega = R\Sigma \ln(1 - e^{-x}) \tag{74}$$

It is evident that each of the contributions may be evaluated from a knowledge of the fundamental frequencies of all vibrational degrees of freedom. These fundamental frequencies and the corresponding statistical weights are determined from spectroscopic data or may be estimated from generalized bond assignments, as discussed in the following section. To facilitate solution of Equations (71–74) extensive tables have been prepared relating the various so-called Einstein functions to x. In Table XLIX are values of the vibrational contributions to heat capacity, enthalpy, and free energy expressed as functions of ω'/T. This table originally developed by H. L. Johnston[21] and revised by Schumann and Schwartz,[22] is more convenient to use than tables of the Einstein functions but has the disadvantage of being dependent on the values of the physical constants. Contributions to the entropy are obtained from these tables by means of the equation:

$$S^\circ_\omega = \left(\frac{H^\circ - H^\circ_0}{T}\right)_\omega - \left(\frac{G^\circ - H^\circ_0}{T}\right)_\omega \qquad (75)$$

Illustration 12. From spectroscopic data the fundamental vibrational frequencies of methyl alcohol have been evaluated by Borden and Barker[20] as follows:

	Wave No., ω'	*Statistical Wgt.*, g	$\frac{\omega'}{298.16}$	$\frac{\omega'}{500}$	$\frac{\omega'}{1000}$
1	3683	1	12.352	7.366	3.683
2	2978	2	9.988	5.956	2.978
3	2845	1	9.542	5.690	2.845
4	1477	2	4.954	2.954	1.477
5	1455	1	4.880	2.910	1.455
6	1340	2	4.494	2.680	1.340
7	1034	1	3.469	2.068	1.034
8	1030	1	3.455	2.060	1.030

Calculate the contributions of vibration to the heat capacity, enthalpy, entropy, and free energy in the ideal-gaseous state at 1 atm and 298.16, 500, and 1000°K.

The contribution corresponding to each vibrational degree of freedom is read from Table XLIX and multiplied by the corresponding statistical weights. Thus, at 298.16°K,

	$\left(\frac{H^\circ - H^\circ_0}{T}\right)_\omega$	$-\left(\frac{G^\circ - H^\circ_0}{T}\right)_\omega$	$(c^\circ)_\omega$
1	0.00000	—	—
2	0.00004	—	—
3	0.00003	—	—
4	0.02342	0.00330	0.166
5	0.01282	0.00183	0.090
6	0.04120	0.00636	0.264
7	0.06920	0.01386	0.346
8	0.07030	0.01413	0.350
Σ	0.21701	0.03948	1.216

[21] E. B. Wilson, Jr., *Chem. Rev.*, **27**, 17 (1940).

[22] S. C. Schumann and M. L. Schwartz; Taylor and Glasstone, "Treatise on Physical Chemistry," Vol. I, D. Van Nostrand Co., New York (1942) with permission.

TABLE XLIX

Vibrational Contributions to Thermodynamic Properties I

cal/(°K)(g-mole)

(ω'/T)	c°	$\frac{(H^\circ - H_0^\circ)}{T}$	$-\frac{(G^\circ - H_0^\circ)}{T}$	(ω'/T)	c°	$\frac{(H^\circ - H_0^\circ)}{T}$	$-\frac{(G^\circ - H_0^\circ)}{T}$
0.10	1.9834	1.8473	3.9942	0.45	1.9188	1.4128	1.4726
0.11	1.9827	1.8338	3.8186	0.46	1.9159	1.4016	1.4416
0.12	1.9819	1.8203	3.6596	0.47	1.9128	1.3904	1.4117
0.13	1.9810	1.8068	3.5143	0.48	1.9097	1.3793	1.3826
0.14	1.9801	1.7935	3.3807	0.49	1.9066	1.3682	1.3543
0.15	1.9791	1.7801	3.2572	0.50	1.9033	1.3573	1.3267
0.16	1.9780	1.7670	3.1428	0.51	1.9001	1.3464	1.3001
0.17	1.9768	1.7538	3.0361	0.52	1.8966	1.3354	1.2741
0.18	1.9756	1.7407	2.9362	0.53	1.8933	1.3248	1.2488
0.19	1.9744	1.7276	2.8424	0.54	1.8898	1.3140	1.2242
0.20	1.9731	1.7147	2.7542	0.55	1.8863	1.3034	1.2001
0.21	1.9717	1.7018	2.6710	0.56	1.8827	1.2929	1.1766
0.22	1.9703	1.6890	2.5920	0.57	1.8791	1.2823	1.1539
0.23	1.9687	1.6763	2.5142	0.58	1.8754	1.2719	1.1316
0.24	1.9671	1.6636	2.4662	0.59	1.8718	1.2616	1.1099
0.25	1.9655	1.6510	2.3786	0.60	1.8680	1.2513	1.0887
0.26	1.9638	1.6384	2.3140	0.61	1.8641	1.2411	1.0681
0.27	1.9620	1.6260	2.2525	0.62	1.8602	1.2308	1.0481
0.28	1.9601	1.6135	2.1935	0.63	1.8562	1.2207	1.0284
0.29	1.9583	1.6012	2.1372	0.64	1.8522	1.2107	1.0093
0.30	1.9563	1.5889	2.0831	0.65	1.8481	1.2007	0.9905
0.31	1.9542	1.5767	2.0311	0.66	1.8441	1.1908	0.9722
0.32	1.9521	1.5645	1.9813	0.67	1.8398	1.1810	0.9545
0.33	1.9500	1.5525	1.9333	0.68	1.8357	1.1711	0.9370
0.34	1.9478	1.5404	1.8872	0.69	1.8315	1.1614	0.9200
0.35	1.9454	1.5285	1.8427	0.70	1.8272	1.1517	0.9033
0.36	1.9430	1.5167	1.7998	0.71	1.8227	1.1420	0.8870
0.37	1.9406	1.5048	1.7584	0.72	1.8184	1.1325	0.8713
0.38	1.9381	1.4931	1.7185	0.73	1.8140	1.1230	0.8557
0.39	1.9356	1.4814	1.6799	0.74	1.8093	1.1135	0.8405
0.40	1.9330	1.4698	1.6425	0.75	1.8047	1.1041	0.8256
0.41	1.9303	1.4583	1.6063	0.76	1.8002	1.0948	0.8110
0.42	1.9275	1.4468	1.5713	0.77	1.7955	1.0856	0.7968
0.43	1.9247	1.4354	1.5374	0.78	1.7908	1.0764	0.7828
0.44	1.9218	1.4240	1.5045	0.79	1.7863	1.0674	0.7692

TABLE XLIX (Continued)

VIBRATIONAL CONTRIBUTIONS TO THERMODYNAMIC PROPERTIES II

(ω'/T)	$c°$	$\frac{(H°-H_0°)}{T}$	$-\frac{(G°-H_0°)}{T}$	(ω'/T)	$c°$	$\frac{(H°-H_0°)}{T}$	$-\frac{(G°-H_0°)}{T}$
0.80	1.7813	1.0582	0.7558	1.20	1.5592	0.7427	0.3894
0.81	1.7765	1.0492	0.7437	1.21	1.5531	0.7359	0.3832
0.82	1.7716	1.0403	0.7299	1.22	1.5468	0.7292	0.3773
0.83	1.7667	1.0314	0.7174	1.23	1.5406	0.7225	0.3713
0.84	1.7618	1.0226	0.7052	1.24	1.5343	0.7159	0.3656
0.85	1.7568	1.0138	0.6931	1.25	1.5281	0.7093	0.3598
0.86	1.7517	1.0050	0.6812	1.26	1.5217	0.7027	0.3542
0.87	1.7466	0.9964	0.6697	1.27	1.5156	0.6963	0.3486
0.88	1.7415	0.9878	0.6583	1.28	1.5091	0.6898	0.3432
0.89	1.7364	0.9793	0.6473	1.29	1.5030	0.6834	0.3379
0.90	1.7312	0.9708	0.6363	1.30	1.4966	0.6771	0.3326
0.91	1.7259	0.9620	0.6256	1.31	1.4902	0.6708	0.3275
0.92	1.7206	0.9541	0.6151	1.32	1.4839	0.6646	0.3225
0.93	1.7154	0.9458	0.6049	1.33	1.4774	0.6585	0.3175
0.94	1.7099	0.9377	0.5948	1.34	1.4710	0.6523	0.3126
0.95	1.7045	0.9295	0.5848	1.35	1.4646	0.6462	0.3078
0.96	1.6991	0.9212	0.5751	1.36	1.4583	0.6402	0.3030
0.97	1.6935	0.9132	0.5657	1.37	1.4520	0.6342	0.2983
0.98	1.6882	0.9051	0.5563	1.38	1.4455	0.6283	0.2937
0.99	1.6826	0.8971	0.5471	1.39	1.4390	0.6223	0.2893
1.00	1.6770	0.8893	0.5382	1.40	1.4324	0.6164	0.2848
1.01	1.6714	0.8815	0.5294	1.41	1.4259	0.6107	0.2804
1.02	1.6658	0.8737	0.5208	1.42	1.4194	0.6049	0.2761
1.03	1.6601	0.8659	0.5123	1.43	1.4129	0.5998	0.2719
1.04	1.6544	0.8582	0.5040	1.44	1.4065	0.5935	0.2677
1.05	1.6487	0.8507	0.4959	1.45	1.3999	0.5879	0.2636
1.06	1.6429	0.8430	0.4877	1.46	1.3934	0.5823	0.2595
1.07	1.6371	0.8355	0.4800	1.47	1.3868	0.5768	0.2556
1.08	1.6312	0.8281	0.4722	1.48	1.3802	0.5713	0.2518
1.09	1.6254	0.8207	0.4647	1.49	1.3736	0.5658	0.2471
1.10	1.6196	0.8133	0.4572	1.50	1.3671	0.5604	0.2441
1.11	1.6137	0.8060	0.4499	1.51	1.3605	0.5550	0.2404
1.12	1.6077	0.7988	0.4426	1.52	1.3538	0.5493	0.2367
1.13	1.6017	0.7916	0.4357	1.53	1.3473	0.5445	0.2332
1.14	1.5957	0.7845	0.4287	1.54	1.3406	0.5392	0.2297
1.15	1.5897	0.7774	0.4218	1.55	1.3340	0.5340	0.2262
1.16	1.5837	0.7703	0.4151	1.56	1.3274	0.5288	0.2227
1.17	1.5776	0.7634	0.4086	1.57	1.3208	0.5238	0.2193
1.18	1.5715	0.7564	0.4021	1.58	1.3143	0.5187	0.2161
1.19	1.5653	0.7495	0.3957	1.59	1.3077	0.5138	0.2129

TABLE XLIX (Continued)

VIBRATIONAL CONTRIBUTIONS TO THERMODYNAMIC PROPERTIES III

(ω'/T)	c°	$\frac{(H^\circ - H_0^\circ)}{T}$	$-\frac{(G^\circ - H_0^\circ)}{T}$	(ω'/T)	c°	$\frac{(H^\circ - H_0^\circ)}{T}$	$-\frac{(G^\circ - H_0^\circ)}{T}$
1.60	1.3010	0.5088	0.2097	2.00	1.0398	0.3412	0.1152
1.61	1.2943	0.5039	0.2065	2.02	1.0272	0.3342	0.1118
1.62	1.2878	0.4988	0.2034	2.04	1.0146	0.3275	0.1085
1.63	1.2812	0.4942	0.2004	2.06	1.0020	0.3208	0.1054
1.64	1.2746	0.4893	0.1974	2.08	0.9897	0.3142	0.1023
1.65	1.2680	0.4847	0.1944	2.10	0.9772	0.3077	0.0992
1.66	1.2613	0.4799	0.1914	2.12	0.9649	0.3014	0.09640
1.67	1.2547	0.4753	0.1885	2.14	0.9526	0.2952	0.09362
1.68	1.2481	0.4706	0.1858	2.16	0.9405	0.2891	0.09091
1.69	1.2415	0.4659	0.1830	2.18	0.9284	0.2832	0.08828
1.70	1.2349	0.4613	0.1802	2.20	0.9161	0.2773	0.08576
1.71	1.2283	0.4568	0.1775	2.22	0.9041	0.2716	0.08326
1.72	1.2217	0.4523	0.1749	2.24	0.8921	0.2659	0.08084
1.73	1.2151	0.4478	0.1723	2.26	0.8799	0.2605	0.07850
1.74	1.2086	0.4435	0.1698	2.28	0.8682	0.2551	0.07622
1.75	1.2018	0.4391	0.1672	2.30	0.8570	0.2497	0.07402
1.76	1.1952	0.4348	0.1647	2.32	0.8455	0.2445	0.07189
1.77	1.1887	0.4304	0.1622	2.34	0.8339	0.2394	0.06982
1.78	1.1822	0.4262	0.1599	2.36	0.8225	0.2344	0.06780
1.79	1.1756	0.4219	0.1575	2.38	0.8112	0.2295	0.06585
1.80	1.1690	0.4177	0.1551	2.40	0.8000	0.2246	0.06395
1.81	1.1625	0.4135	0.1528	2.42	0.7885	0.2198	0.06213
1.82	1.1559	0.4096	0.1505	2.44	0.7778	0.2151	0.06032
1.83	1.1493	0.4053	0.1484	2.46	0.7669	0.2106	0.05859
1.84	1.1428	0.4011	0.1461	2.48	0.7560	0.2061	0.05690
1.85	1.1363	0.3972	0.1440	2.50	0.7453	0.2017	0.05526
1.86	1.1298	0.3932	0.1418	2.52	0.7346	0.1974	0.05370
1.87	1.1233	0.3893	0.1397	2.54	0.7240	0.1932	0.05217
1.88	1.1167	0.3855	0.1377	2.56	0.7135	0.1891	0.05065
1.89	1.1102	0.3816	0.1356	2.58	0.7031	0.1850	0.04919
1.90	1.1038	0.3777	0.1336	2.60	0.6928	0.1810	0.04778
1.91	1.0973	0.3739	0.1316	2.62	0.6826	0.1771	0.04640
1.92	1.0909	0.3701	0.1297	2.64	0.6725	0.1732	0.04507
1.93	1.0845	0.3665	0.1278	2.66	0.6625	0.1695	0.04379
1.94	1.0780	0.3627	0.1259	2.68	0.6527	0.1656	0.04253
1.95	1.0716	0.3591	0.1240	2.70	0.6429	0.1622	0.04131
1.96	1.0653	0.3554	0.1222	2.72	0.6332	0.1587	0.04014
1.97	1.0589	0.3518	0.1205	2.74	0.6236	0.1553	0.03901
1.98	1.0525	0.3482	0.1187	2.76	0.6140	0.1519	0.03788
1.99	1.0461	0.3446	0.1169	2.78	0.6046	0.1483	0.03679

TABLE XLIX (Continued)

VIBRATIONAL CONTRIBUTIONS TO THERMODYNAMIC PROPERTIES IV

(ω'/T)	c°	$\frac{(H^\circ - H_0^\circ)}{T}$	$-\frac{(G^\circ - H_0^\circ)}{T}$	(ω'/T)	c°	$\frac{(H^\circ - H_0^\circ)}{T}$	$-\frac{(G^\circ - H_0^\circ)}{T}$
2.80	0.5953	0.1454	0.03572	5.00	0.0775	0.01074	0.00150
2.82	0.5862	0.1422	0.03471	5.10	0.0697	0.00948	0.001290
2.84	0.5771	0.1391	0.03371	5.20	0.0628	0.00836	0.001116
2.86	0.5680	0.1361	0.03275	5.30	0.0563	0.00740	0.000965
2.88	0.5591	0.1331	0.03181	5.40	0.0506	0.00652	0.000837
2.90	0.5502	0.1301	0.03089	5.50	0.0454	0.00575	0.000727
2.92	0.5417	0.1272	0.03003	5.60	0.0409	0.00507	0.000629
2.94	0.5331	0.1244	0.02917	5.70	0.0367	0.00447	0.000545
2.96	0.5246	0.1216	0.02834	5.80	0.0330	0.00393	0.000472
2.98	0.5163	0.1189	0.02752	5.90	0.0295	0.00347	0.000409
3.00	0.5080	0.1162	0.02673	6.00	0.0265	0.00305	0.000353
3.05	0.4876	0.1099	0.02486	6.10	0.0236	0.00269	0.000306
3.10	0.4680	0.1038	0.02311	6.20	0.0212	0.00236	0.000266
3.15	0.4489	0.0981	0.02150	6.30	0.0189	0.00208	0.000231
3.20	0.4305	0.0927	0.02000	6.40	0.0169	0.00183	0.000200
3.25	0.4127	0.0875	0.01861	6.50	0.0151	0.00161	0.000173
3.30	0.3956	0.0826	0.01731	6.60	0.0134	0.00142	0.000151
3.35	0.3787	0.0780	0.01608	6.70	0.0120	0.00125	0.000131
3.40	0.3627	0.0736	0.01498	6.80	0.0107	0.00110	0.000114
3.45	0.3471	0.0696	0.01393	6.90	0.00960	0.000962	0.0000993
3.50	0.3323	0.0656	0.01297	7.00	0.00852	0.000845	0.0000865
3.55	0.3177	0.0619	0.01207	7.10	0.00758	0.000742	0.0000750
3.60	0.3038	0.0584	0.01124	7.20	0.00674	0.000651	0.0000651
3.65	0.2903	0.0550	0.01045	7.30	0.00600	0.000571	0.0000565
3.70	0.2775	0.0519	0.00972	7.40	0.00534	0.000502	0.0000491
3.75	0.2651	0.0489	0.00904	7.50	0.00474	0.000441	0.0000427
3.80	0.2532	0.0461	0.00842	7.60	0.00423	0.0003868	0.0000369
3.85	0.2417	0.0435	0.00783	7.70	0.00376	0.0003497	0.0000320
3.90	0.2307	0.0410	0.00729	7.80	0.00332	0.0002979	0.0000278
3.95	0.2201	0.0387	0.00678	7.90	0.00298	0.0002618	0.0000241
4.00	0.2099	0.0364	0.00631	8.00	0.00265	0.0002292	0.0000209
4.10	0.1907	0.0322	0.00547	8.10	0.00234	0.0002008	0.0000181
4.20	0.1732	0.0286	0.00473	8.20	0.00208	0.0001759	0.0000156
4.30	0.1571	0.0253	0.00409	8.30	0.00185	0.0001542	0.0000135
4.40	0.1424	0.0225	0.00354	8.40	0.00164	0.0001352	0.0000117
4.50	0.1286	0.0200	0.00306	8.50	0.00145	0.0001187	0.0000101
4.60	0.1166	0.0176	0.00265	8.60	0.00129	0.0001038	0.0000088
4.70	0.1054	0.0156	0.00230	8.70	0.00113	0.0000908	0.0000077
4.80	0.0952	0.0138	0.00199	8.80	0.00101	0.0000795	0.0000067
4.90	0.0859	0.0122	0.00173	8.90	0.00089	0.0000696	0.0000058

TABLE XLIX (Continued)

VIBRATIONAL CONTRIBUTIONS TO THERMODYNAMIC PROPERTIES V

(ω'/T)	c°	$\frac{(H^\circ - H_0^\circ)}{T}$	$-\frac{(G^\circ - H_0^\circ)}{T}$	(ω'/T)	c°	$\frac{(H^\circ - H_0^\circ)}{T}$	$-\frac{(G^\circ - H_0^\circ)}{T}$
9.00	0.000792	0.0000610	0.0000050	9.50	0.000431	0.0000312	0.0000024
9.10	0.000701	0.0000535	0.0000043	9.60	0.000380	0.0000273	0.0000021
9.20	0.000620	0.0000466	0.0000037	9.70	0.000335	0.0000239	0.0000018
9.30	0.000549	0.0000408	0.0000032	9.80	0.000297	0.0000209	0.0000016
9.40	0.000486	0.0000357	0.0000028	9.90	0.000262	0.0000184	0.0000014
				10.00	0.000232	0.0000161	0.0000012

From Equation (75),

$$s^\circ_{298.16} = 0.2170 + 0.03948 = 0.2565$$

Calculations at other temperatures are carried out by the same procedure. The results are summarized in Table LI, page 802.

Generalized Bond Frequencies. Useful approximations to the vibrational contributions may be obtained by assuming that a particular type of bond possesses two characteristic fundamental frequencies, which are independent of the nature of the molecule in which the bond occurs. Generalized average values of the stretching and bending frequencies may be arrived at on this basis from compounds for which spectroscopic data are available. This general procedure was developed by Bennewitz and Rossner[23] and modified by Dobratz.[24] A complication arises from the fact that there are generally more degrees of freedom than are accounted for by assuming two vibrational degrees per bond. Dobratz proposed the following equation for handling this problem for calculating heat capacities of nonlinear molecules,

$$c_\omega = \Sigma_i q_i c_{\nu_i} + \left(\frac{3n - 6 - n_{r'} - \Sigma q_i}{\Sigma q_i}\right)\Sigma(q_i c_{\delta i}) \qquad (76)$$

where

c_ω = total vibrational contribution to the heat capacity
Σq_i = total number of bonds in the molecule
c_{ν_i} = contribution to the heat capacity from the stretching vibrations of bond i
$c_{\delta i}$ = contribution to the heat capacity from the bending of bond i
n = number of atoms in the molecule
$n_{r'}$ = number of single bonds about which internal rotation of groups can take place

[23] Bennewitz and Rossner, *Z. physik. Chem.*, **39B**, 126 (1938).
[24] C. J. Dobratz, *Ind. Eng. Chem.*, **33**, 759 (1941).

Equation (76) assumes that all degrees of freedom not accounted for by internal rotation, stretching or bending, make the same average contribution to the heat capacity as do the recognized bending contributions. This same assumption can be made for calculating the contributions to the other thermodynamic properties and equations developed similar to (76).

In Table L are the fundamental frequencies of various bonds expressed in wave numbers, as recommended by Bennewitz and Rossner[23] and modified by Stull and Mayfield[25] for hydrocarbons.

TABLE L

FUNDAMENTAL BOND FREQUENCY ASSIGNMENTS

Bond	*Stretching* ω'_ν	*Deformation* ω'_δ	*Bond*	*Stretching* ω'_ν	*Deformation* w'_δ
C—H (aliphatic)	2914	1247	C=S	1050	530
C—C (aliphatic)	989	390	S—S	500	260
C=C (aliphatic sym.)	1618	599	S—H	2570	1050
C=C (aliphatic unsym.)	1664	421	C—N	990	390
C≡C (aliphatic)	2215	333	C=N	1620	845
C—H (aromatic)	3045	1318	N—N	990	390
C—C (aromatic)	989	390	N—H	2920	1320
C=C (aromatic)	1618	844	N—O	1030	205
C—I	500	260	N=O	1700	390
C—Br	560	280	C—O	1030	205
C—Cl	650	330	C=O	1700	390
C—F	1050	530	O—H	3420	1150
C—S	650	330			

For the calculation of heat capacities above 300°K Dobratz assumed that each degree of internal rotation contributed its full value of $R/2$. The complete expression for heat capacity at constant pressure then becomes

$$c_p^\circ = 4R + n_{r'}(R/2) + \Sigma_i q_i c_{\nu_i} + \left(\frac{3n - 6 - n_{r'} - \Sigma q_i}{\Sigma q_i}\right)\Sigma q_i c_{\delta i} \qquad (77)$$

Equation (77) may be evaluated by summation of the contributions from Table XLIX corresponding to the various frequencies at a selected temperature. The procedure is simplified by use of Fig. 166, in which the contributions corresponding to various wave numbers are plotted as a function of temperature.

Illustration 13. From Fig. 166 and the frequencies of Table L calculate the heat capacity at constant pressure of methyl alcohol as an ideal vapor at 298.1, 500, and 1000°K.

[25] D. R. Stull and F. D. Mayfield, *Ind. Eng. Chem.*, **35**, 639 and 1303 (1943).

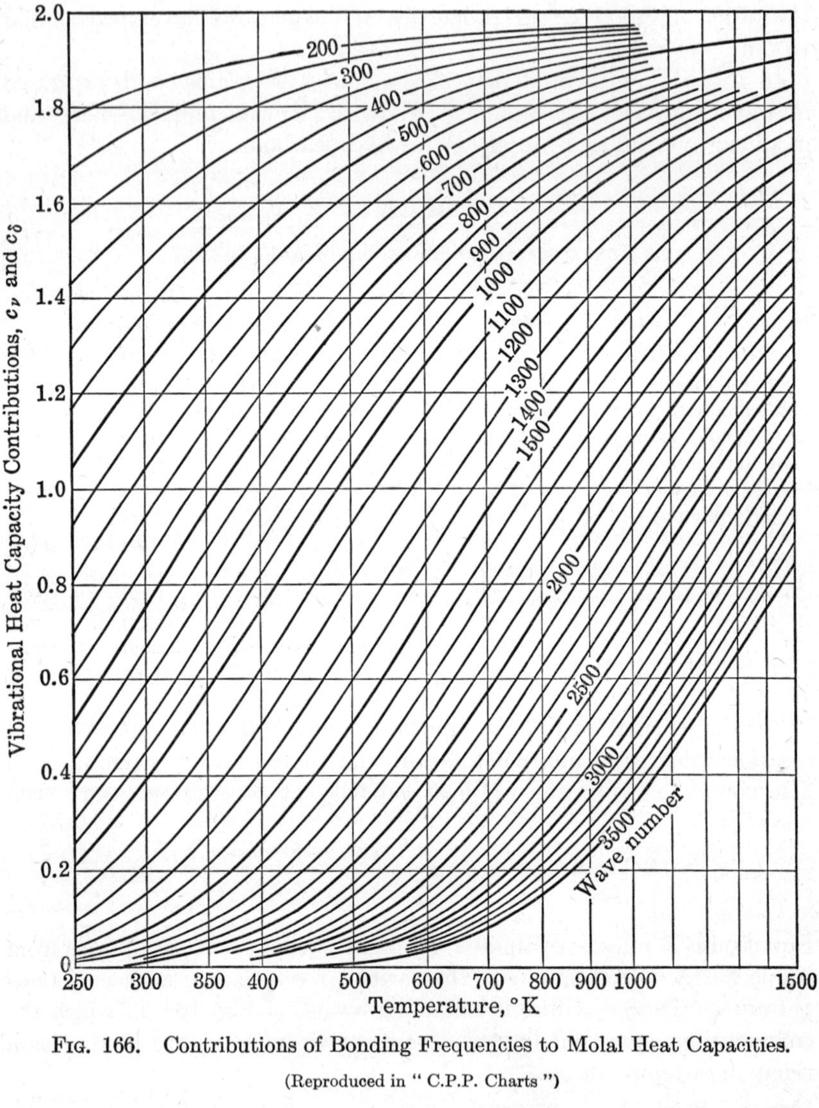

FIG. 166. Contributions of Bonding Frequencies to Molal Heat Capacities.

(Reproduced in "C.P.P. Charts")

Solution: The frequencies of the bonds are obtained from Table L, and the corresponding contributions to heat capacity at 298.1°K are read from Fig. 166:

Bond	q_i	ω'_ν	ω'_δ	$q_i c_{\nu i}(298.1)$	$q_i c_{\delta_i}(298.1)$
C—H	3	2914	1247	0.000	0.529
C—O	1	1030	205	0.346	1.833
O—H	1	3420	1150	0.000	0.240
Σ	5			0.346	2.602

Substitution in Equation (77), since $n_{r'} = 1$, gives

$$c^\circ_{p(298.1)} = 7.95 + 0.99 + 0.346 + \left(\frac{(3)(6) - 6 - 1 - 5}{5}\right)(2.602) = 12.41$$

At other temperatures the same procedure is followed. Thus:

T	c°_p
298.16	12.4
500	16.1
1000	22.4

A three-term equation is readily derived to express heat capacities calculated in this manner as a function of temperature. This method may be used for estimating the heat capacities of a variety of compounds. The results become uncertain for complex molecules, and, as previously mentioned, it is believed that this method is best applied only to small molecules. The results obtained may then be extended to larger homologs by means of the group contributions of Tables XXXIX–XLIII.

Total Thermodynamic Properties. Total values of $(H^\circ - H^\circ_0)$, c°_p, S°, and $(G^\circ - H^\circ_0)/T$ are obtained by summation of the contributions resulting from translation, external and internal rotation, and vibration each of which is calculated by the methods demonstrated in the preceding sections. Electronic contributions may be neglected when one is dealing with organic compounds at moderate temperatures.

Illustration 14. Using the results of Illustrations 6, 10, 11, 12, calculate the values of $(H^\circ - H^\circ_0)$, c°_p, S°, and $(G^\circ - H^\circ_0)/T$ for methyl alcohol in the ideal-vapor state at a pressure of 1 atm and temperatures of 298.1, 500, and 1000°K.

The results of the individual calculations from the preceding illustrations are summarized in Table LI.

Where reliable values for potential barriers and vibration frequencies can be obtained, it is desirable to calculate tables of thermodynamic properties as functions of temperature, as is done in Illustration 14. However, where potential barriers are uncertain and generalized bond-frequency assignments are used, this rather tedious procedure is not justified, and it is more convenient to derive an empirical equation for heat capacity from values at three temperatures and calculate the entropy at 298.1°K. Free-energy changes at any temperature may be calculated from these data and standard heats of reaction by the methods demonstrated in Chapter XV.

TABLE LI

	298.1°K	500°K	1000°K
	$(H^\circ - H_0^\circ)$		
Translation	1481	2483	4966
External rotation	888	1490	2980
Internal rotation	221	570	1310
Vibration	65	685	5220
Total	2655	5228	14476
	C_p°		
Translation	4.97	4.97	4.97
External rotation	2.98	2.98	2.98
Internal rotation	1.67	1.72	1.32
Vibration	1.22	5.04	12.38
Total	10.84	14.71	21.65
	S°		
Translation	36.316	38.883	42.326
External rotation	18.941	20.482	22.547
Internal rotation	1.080	1.97	3.02
Vibration	0.257	1.76	7.75
Total	56.594	63.095	75.643
	$(G^\circ - H_0^\circ)/T$		
Translation	−31.349	−33.916	−37.359
External rotation	−15.961	−17.502	−19.567
Internal rotation	−0.339	−0.820	−1.700
Vibration	−0.040	−0.390	−2.530
Total	−47.689	−52.628	−61.156

Illustration 15. Calculate the entropy of methyl alcohol as an ideal vapor at 298.1°K and 1 atm using the moment of inertia calculated in Illustration 10, the internal rotational contributions of Illustration 11, and the generalized bond frequencies of Table L for the vibrational contributions. Compare this result with that of Illustration 14.

Solution: All contributions are the same as those in Table LI except those of vibration. These are read from Table XLIX with Equation (75) to correspond to the values of (ω_ν'/T) and (ω_δ'/T) calculated in Illustration 13.

					←Stretching		→	←Bending		→
Bond	ω_ν'	ω_δ'	$\frac{\omega_\nu'}{T}$	$\frac{\omega_\delta'}{T}$	$\frac{H^\circ - H_0^\circ}{T}$	$\frac{G^\circ - H_0^\circ}{T}$	S°	$\frac{H^\circ - H_0^\circ}{T}$	$\frac{G^\circ - H_0^\circ}{T}$	S°
C—H	2914	1247	9.78	4.18	0.0000	0.0000	0.0000	0.0295	0.0048	0.0343
C—H	2914	1247	9.78	4.18	0.0000	0.0000	0.0000	0.0295	0.0048	0.0343
C—H	2914	1247	9.78	4.18	0.0000	0.0000	0.0000	0.0295	0.0048	0.0343
C—O	1030	205	3.45	0.688	0.0696	0.0139	0.0835	1.1634	0.9234	2.0868
O—H	3420	1150	11.48	3.86	0.0000	0.0000	0.0000	0.0430	0.0077	0.0507
							$\Sigma = 0.0835$			2.2404

The vibrational contribution, by an equation similar to (76), is then

$$S^\circ_{\omega(298.1)} = 0.0835 + (6/5)(2.2404) = 2.772$$

Comparing this value with that of 0.257 in Table LI indicates an error of 2.515 in the total entropy as a result of the use of the generalized bond-frequency assignments.

The large error in the vibrational contribution to the entropy of methyl alcohol calculated from the generalized bond frequencies results from the low frequency of 205 for the deformation of the C–O bond. The actual spectrum of methyl alcohol summarized in Illustration 12 indicates that no frequency actually exists in this range. The effect of this erroneous frequency on the calculated heat capacity is shown by comparison of the values of Illustration 13 with those of Table LI. The heat capacity at 298°K is too high by 14 per cent. At higher temperature the agreement becomes better in this case.

It is thus evident that the bond-frequency assignments are compromise values which may lead to serious errors particularly when one is working with the first member of a series. This method appears to be best suited for handling second or third homologs. As further data are accumulated it is hoped that improved generalizations of fundamental frequencies may be developed.

Thermodynamic Properties of Liquids and Solids. The statistical methods of the preceding sections yield results which are directly applicable only to the ideal-gaseous state. However, by the methods developed in Chapters VII and XII these data are readily converted to apply to the liquid state at any conditions and also to the solid state if data on the heat of fusion and heat capacity of the solid are available. To obtain the properties of the saturated liquid state, at a specified temperature T, the properties of the ideal-gaseous state at 1 atm are first corrected to the ideal state at the vapor pressure of the liquid at temperature T. Corrections for deviations from ideality are then applied, and the changes accompanying vaporization are subtracted. This procedure is described in detail for enthalpy and entropy calculations in Chapter XII. It should be remembered that the free-energy change in vaporization at saturation is zero. The effect of pressure on the free energy of an ideal gas is obtained by integrating Equation (k) of Table XXIV.

$$\mathrm{G}^* = \mathrm{G}^\circ + RT \ln p \tag{78}$$

where

G^* = free energy of ideal gas at pressure p atm

G° = free energy of ideal gas in the standard state of 1 atm

Corrections to free energy for deviation from ideal behavior may be obtained by the combined use of Figs. 106 and 107. Thus,

$$\frac{(\mathrm{G}^* - \mathrm{G})}{T} = \frac{(\mathrm{H}^* - \mathrm{H})}{T} - (\mathrm{S}^* - \mathrm{S}) \tag{79}$$

PROBLEMS

1. Estimate from the group contributions of Tables XXXIX–XLIII the heat of formation, entropies, and heat-capacity equations for the following compounds in the ideal-gaseous state:

(*a*) Isoprene (2-methyl-1,3-butadiene)
(*b*) Meta di-isopropyl benzene
(*c*) Iso butyronitrile
(*d*) Diphenyl
(*e*) *o*-Nitrobenzoic acid

2. The boiling point of isobutyro nitrile is 108°C, and its liquid density at 20°C is 0.773 g per cc. From the result of problem 1*c* and the generalizations of Chapters III, VII, and XII calculate the entropy and heat of formation of liquid isobutyro nitrile at 25°C under its vapor pressure.

3. From the covalent radii and bond angles of Table XLIV and the generalized bond frequencies of Table L, calculate values of s°, $(H^\circ - H_0^\circ)$, $(G^\circ - H_0^\circ)/T$, and c_p^* at temperatures of 298.16, 500, and 1000°K for dimethyl sulfide.

4. From the bond-frequency assignments of Table L derive three-term equations for the heat capacity at constant pressure in the range 298–1500°K for the following compounds in the ideal-gaseous state:

(*a*) CCl_3F
(*b*) N_2O_4
(*c*) Dimethyl amine

APPENDIX

ATOMIC WEIGHTS OF THE MORE COMMON ELEMENTS

Aluminum	Al	26.97	Manganese	Mn	54.93
Antimony	Sb	121.76	Mercury	Hg	200.61
Argon	A	39.944	Molybdenum	Mo	95.95
Arsenic	As	74.91	Neon	Ne	20.183
Barium	Ba	137.36	Nickel	Ni	58.69
Bismuth	Bi	209.00	Nitrogen	N	14.008
Boron	B	10.82	Oxygen	O	16.000
Bromine	Br	79.916	Phosphorus	P	30.98
Cadmium	Cd	112.41	Platinum	Pt	195.23
Calcium	Ca	40.08	Potassium	K	39.096
Carbon	C	12.010	Selenium	Se	78.96
Chlorine	Cl	35.457	Silicon	Si	28.06
Chromium	Cr	52.01	Silver	Ag	107.880
Cobalt	Co	58.94	Sodium	Na	22.997
Copper	Cu	63.57	Strontium	Sr	87.63
Fluorine	F	19.00	Sulfur	S	32.06
Gold	Au	197.2	Tellurium	Te	127.61
Helium	He	4.003	Tin	Sn	118.70
Hydrogen	H	1.008	Titanium	Ti	47.90
Iodine	I	126.92	Tungsten	W	183.92
Iron	Fe	55.85	Uranium	U	238.07
Lead	Pb	207.21	Vanadium	V	50.95
Lithium	Li	6.940	Zinc	Zn	65.38
Magnesium	Mg	24.32	Zirconium	Zr	91.22

Source: *J. Am. Chem. Soc.*, **63**, 850 (1941).

CONVERSION FACTORS AND CONSTANTS

Analysis of Air

By weight: oxygen, 23.2%; nitrogen, 76.8%
By volume: oxygen, 21.0%; nitrogen, 79.0%
Average molecular weight of air.................................. 29
Average molecular weight of atmospheric nitrogen................ 28.2

Physical Constants

The Gas Law Constant R

R = 1.987 (calories)/(g-mole)(°K)
R = 82.06 (cu cm)(atm)/(g-mole)(°K)
R = 10.71 (lb/sq in.)(cu ft)/(lb-mole)(°R)
R = 0.729 (atm)(cu ft)/(lb-mole)(°R)

1 faraday = 96,500 coulombs
Avogadro constant = 6.023×10^{23} per gram-atom

Density

1 gram-mole of an ideal gas at 0°C, 760 mm of mercury = 22.414 liters
1 pound-mole of an ideal gas at 0°C, 760 mm of mercury = 359.0 cubic feet
Density of dry air at 0°C, 760 mm of mercury = 1.293 grams per liter or.. 0.0807 pound per cubic foot
1 gram per cc......................... 62.4 pounds per cubic foot
1 gram per cc......................... 8.337 pounds per U. S. gallon

Energy*

	Calories	Btu	Joules	Foot-pounds	Kilogram-meters
Calorie	1	3.968×10^{-3}	4.185	3.087	0.4267
Btu	252	1	1055	777.9	107.5
Joule	0.2389	9.482×10^{-4}	1	0.73756	0.1019
Foot-pound	0.3240	1.286×10^{-3}	1.356	1	0.13826
Kilogram-meter	2.343	9.298×10^{-3}	9.806	7.2327	1
Liter-atmos	24.21	9.607×10^{-2}	101.32	74.733	10.333
1 Chu	453.6	1.8	1899	1400	193.5

	Liter-atm	Cu ft-atm	Foot-poundals	Horsepower hours
Calorie	4.130×10^{-2}	1.459×10^{-3}	99.31	1.5591×10^{-6}
Btu	10.41	0.3676	25030	3.929×10^{-4}
Joule	9.869×10^{-3}	3.485×10^{-4}	23.73	3.725×10^{-7}
Foot-pound	1.3381×10^{-2}	4.7253×10^{-4}	32.174	5.0505×10^{-7}
Kilogram-meter	9.678×10^{-2}	3.4177×10^{-3}	232.7	3.6529×10^{-6}
Liter-atmos	1	3.5319×10^{-2}	2403.8	3.7734×10^{-5}
1 Chu	13.74	0.6617	45054	7.072×10^{-4}

* From Perkins, *Introduction to General Thermodynamics*, John Wiley & Sons, Inc., Publishers, with permission.

Length

1 inch	2.540 centimeters
1 micron	10^{-6} meter
1 Ångström unit	10^{-10} meter

Mass

1 pound*	16 ounces*
1 pound*	7000 grains
1 pound*	453.6 grams
1 ton (short)	2000 pounds*
1 gram	15.43 grains
1 kilogram	2.205 pounds*

* Avoirdupois.

Mathematical Constants

e	2.7183
π	3.1416
ln N	2.303 log N

Power

1 kilowatt	56.92 British thermal units per minute
1 kilowatt	1.341 horsepower
1 horsepower	550 foot-pounds per second
1 watt	44.24 foot-pounds per minute
1 watt	14.34 calories per minute

Pressure

1 pound per square inch	2.04 inches of mercury
1 pound per square inch	2.31 feet of water
1 atmosphere	14.7 pounds per square inch
1 atmosphere	760 mm of mercury
1 atmosphere	29.92 inches of mercury

Temperature Scales

Degrees F	1.8 (degrees C) + 32
Degrees K	degrees C + 273.15
Degrees R	degrees F + 459.7

Volume

1 cubic inch	16.39 cubic centimeters
1 liter	61.02 cubic inches
1 cubic foot	28.32 liters
1 cubic meter	1.308 cubic yards
1 cubic meter	1000 liters
1 U. S. gallon	4 quarts
1 U. S. gallon	3.785 liters
1 U. S. gallon	231 cubic inches
1 British gallon	1.20094 U. S. gallons
1 cubic foot	7.481 gallons
1 liter	1.057 quarts
1 U. S. fluid ounce	29.57 cubic centimeters

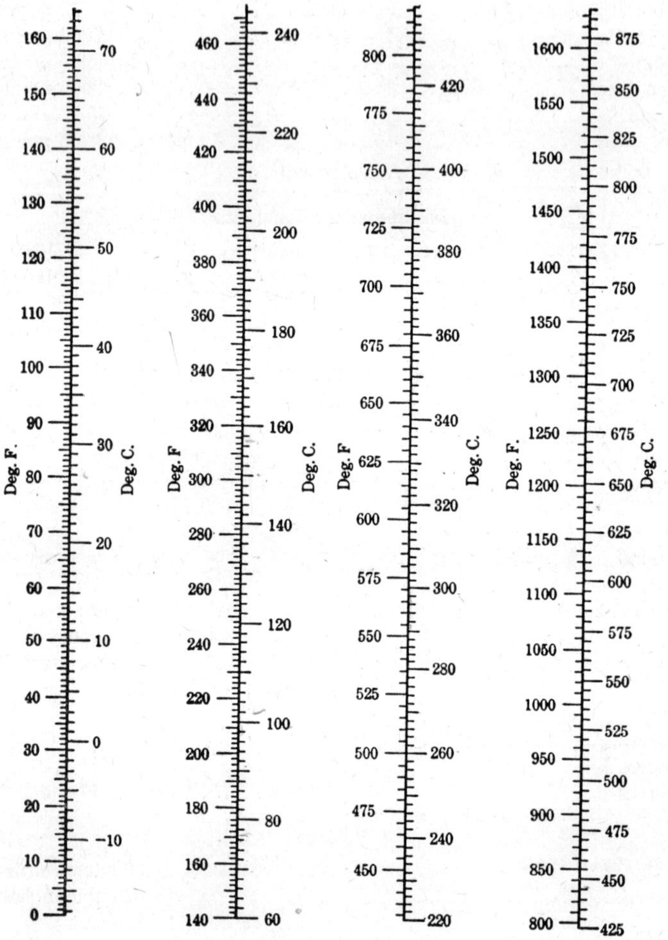

Fig. A. Temperature conversions.

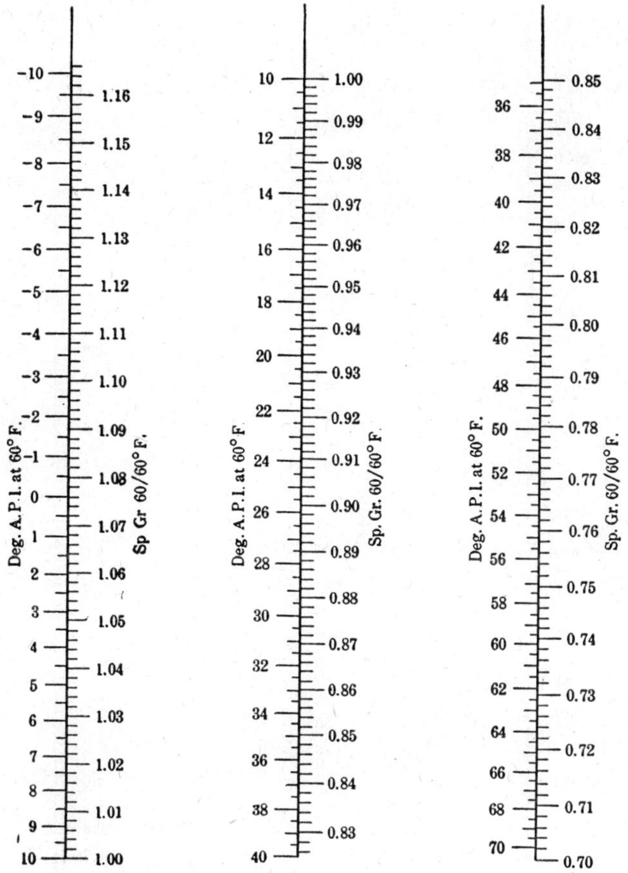

FIG. B. Gravity conversions.

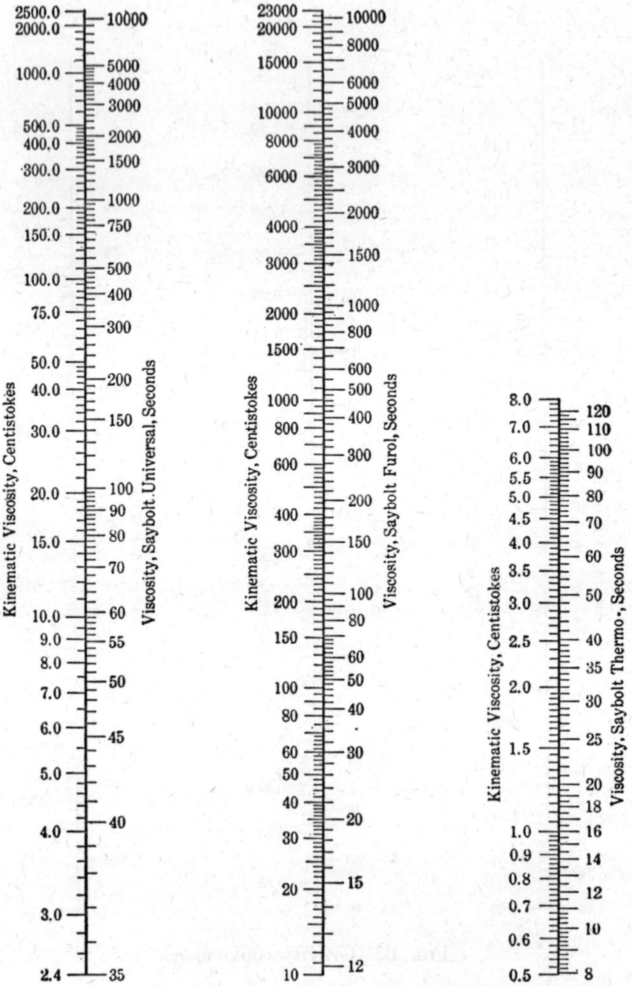

FIG. C. Viscosity conversions.

AUTHOR INDEX

Part I, pages **1–436** Part II, **437–804** Part III, *805–1107*

SUBJECT INDEX

Part I, pages 1–436 Part II, pages **437–804** Part III, pages *805–1107*